Superplasticity and Grain Boundaries in Ultrafine-Grained Materials

Superplasticity and Grain Boundaries in Ultrafine-Grained Materials

A. P. Zhilyaev
A.I. Pshenichnyuk

Viva Books

New Delhi | Mumbai | Chennai | Kolkata | Bengaluru | Hyderabad | Kochi | Guwahati

First Indian Edition 2012
Reprinted 2015, 2017

VIVA BOOKS PRIVATE LIMITED

- 4737/23 Ansari Road, Daryaganj, New Delhi 110 002
  E-mail: vivadelhi@vivagroupindia.net, Tel. 42242200
- 76, Service Industries, Shirvane, Sector 1, Nerul, Navi Mumbai 400 706
  E-mail: vivamumbai@vivagroupindia.net, Tel. 27721273, 27721274
- Megh Towers, Old No. 307, New No. 165, Poonamallee High Road, Maduravoyal
  Chennai - 600 095 E-mail: vivachennai@vivagroupindia.net, Tel. 23780991, 23780992
- B-103, Jindal Towers, 21/1A/3 Darga Road, Kolkata 700 017
  E-mail: vivakolkata@vivagroupindia.net, Tel. 22836381, 22816713
- 7, Sovereign Park Apartments, 56-58, K. R. Road, Basavanagudi, Bengaluru 560 004
  E-mail: vivabangalore@vivagroupindia.net, Tel. 26607409, 26607410
- 101-102 Mughal Marc Apartments, 3-4-637 to 641, Narayanguda, Hyderabad 500 029
  E-mail: vivahyderabad@vivagroupindia.net, Tel. 27564481, 27564482
- Beevi Towers, First Floor, SRM Road, Kaloor, Kochi 682 018, Kerala
  E-mail: vivakochi@vivagroupindia.net, Tel: 0484-2403055, 2403056
- 232, GNB Road, Beside UCO Bank, Silpukhuri, Guwahati 781 003
  E-mail: vivaguwahati@vivagroupindia.net, Tel: 0361-2666386

www.vivagroupindia.com

Published by arrangement with:
Cambridge International Science Publishing
7, Meadow Walk, Great Abington
Cambridge CB1 6AZ, UK

ISBN : 978-81-309-1991-1

Published by Vinod Vasishtha for Viva Books Private Limited, 4737/23, Ansari Road, Daryaganj, New Delhi - 110 002.

Printed and bound by Global Printing Services, New Delhi.

Contents

10

Duration of the stable flow stage in superplastic deformation **276**

11

Derivation of constitutive equations in multicomponent loading conditions **293**

Introduction

One of the methods of producing materials with special properties is, in addition to alloying, the formation in these materials of the ultrafine-grained (UFG) and nanocrystalline (NC) microstructure. A special feature of these structures is the small size of the crystals (0.1–1 μm for the UFG materials and less than 100 nm for NC) and, consequently, the higher fraction of intercrystalline (intergranular and interface) boundaries. The grain boundaries in these materials begin to play an important role and they significantly change the properties of the materials in comparison with the coarse-grained state.

There are at least ten processes [1–6] taking place in polycrystalline materials in which the grain boundaries (GB) play a controlling role. This role is especially evident and varied in deformation processes. In the case of small deformation, the grain boundaries act as dislocations sinks, and the kinetics of dissociation of the dislocations at the grain boundaries has the controlling role as it is the process with the lowest rate. The role of the grain boundaries is more active in high-temperature deformation, especially in the superplastic conditions. In this case another deformation mechanism (grain boundary sliding, GBS) begins to operate and this results in elongation of the material to hundreds or even thousands of percent, and the accommodation processes in which the grain boundaries also take part, actually determinate the kinetics of the superplasticity process (SP). In this case, in the context of the investigated problems [7] it has been established the grain boundaries can interact resulting in the merger of the boundaries into stable formation (clusters). Specifically these clusters determine the appearance of the new deformation mode – cooperative grain boundary sliding (CGBS). The diffusion processes in polycrystalline materials [5], especially at low temperatures and in materials with ultrafine-grained and nanocrystalline structures, also take place mostly in the network of the grain boundaries whose properties determine the kinetics of the diffusion process [8]. The increase of the effective diffusion coefficient in the ultrafine-grained and nanocrystalline materials may reach values several orders of magnitude higher in comparison with the coarse-crystalline state [5]. The variation of the thermal and magnetic properties (experimentally recorded displacement of Debye temperature and Curie point in the UFG and NC materials), which are regarded as structure-insensitive parameters, is evidently also determined by the increase of the concentration of intercrystalline boundaries. In all likelihood, the changes in the phonon spectrum are caused by the nature of interaction of the phonons with the atoms distributed directly in the boundary region, and the changes in the magnetic characteristics may be associated with the disruption of exchange interaction at the grain boundaries.

These facts which have been known now for a long period of time represent a strong stimulus for the development of the methods of formation of the nanocrystalline structures in bulk materials with the hope of developing new properties in the materials with customary composition and of finding qualitatively new processes which can take place in these materials. Undoubtedly, this hope is supported by the fact that the transition from the coarse-grained materials with the grain size of tens of micrometres to the materials with the grain size represented by units of micron is accompanied by a qualitatively new phenomenon – superplastic deformation (SPD). Are these expectations justified if we are able to reduce the grain size by an order of magnitude or more? At present, there is no answer to this question. We also have no answer. However, it is hoped that the experience, obtained in the investigations of superplasticity and the grain boundaries, can be useful. We would like to present these results to readers.

The book discusses a number of issues associated with the problems of description and experimental verification of the individual boundaries and grain boundary ensembles in polycrystals, and also investigations of processes such as grain boundary diffusion, relaxation and grain growth. This process is also an inseparable attribute of superplastic deformation. Our task is to explain what is specific in that grain size reduced from course grain one, at which superplastic regime becomes possible. Is this size unique or a new one, even smaller size can be achieved and it would result in a new phenomenon? The answer to this question can be provided only if a sufficiently general superplasticity model is available. Therefore, the investigations of the grain boundaries are accompanied by the construction of such a model based on the experimentally determined relationships and concerned with the bands of cooperative grain boundary sliding. In addition, attention will be given to the processes of formation and evolution of the microstructure, texture and grain boundary ensembles in the materials produced by the methods of severe plastic deformation (SPD) which include equal channel angular pressing (ECAP) and high-pressure torsion (HPT).

Since the main task is the problem of the existence (or absence) of the hierarchy of the scales and the associated fractal dimension, we begin with explanation of the most general considerations regarding the relationship of the structural levels of deformation with the characteristic spatial scales. These questions are discussed in chapter 1. The ambiguous nature of the classification of declaratively defined spatial scales is noted and it is proposed to use the concept of the 'level of description' (determined by the set of collective variables) used for formulating the given task. The spatial scales should be treated as the characteristic lengths of the produced structures. In this context, the history of the formation of considerations regarding the mechanism of superplastic deformation is briefly discussed and the main problems in describing the superplastic deformation process are formulated.

Chapter 2 discusses the methods of geometrical description of the crystallographic structure of individual boundaries and divides the methods into the main classes: low-angle, high-angle, arbitrary and special. The algorithm of constructing the basis of the coincident-site lattice is described. Possible special boundaries in the monoclinic lattice are calculated using ZrO_2 as an example.

Chapter 3 presents the mathematical description of the orientation characteristics of the polycrystal, such as the grain boundary misorientation distribution (GBMD) and the orientation distribution function (ODF). The essential and sufficient conditions for the unambiguous restoration of the spectra of the grain boundaries with respect to the crystallographic texture are formulated. The final part of the chapter deals with the main considerations regarding the spatial correlation in the orientation of the adjacent grains. The spectra of the grain boundaries for polycrystals with a modelled texture are calculated.

Chapter 4 is concerned with the main methods of experimental measurement of the misorientation of the adjacent grains. The accuracy of measurement of the crystallographic orientation of the grains and the misorientation of two grains is verified by experiments. The results of numerical simulation of the GBMD in the cubic polycrystals, susceptible and not susceptible to twinning in annealing, are presented. The calculated spectra of the grain boundaries are compared with the experimentally measured values. It is shown that the proposed model describes adequately the grain boundary misorientation distribution in the FCC polycrystals.

Determining the parameters characterising the boundary (and the set of the boundaries) and methods of direct measurement of these quantities, it is attempted to understand how these characteristics behave in the description of the process of grain boundary sliding at the grain boundaries in a bicrystal. For this purpose, chapter 5 contains experimental results confirming the dislocation nature of GBS in metallic bicrystals. On the basis of the analysis of the experimental results obtained in the measurement of grain boundary sliding in cadmium and zinc, it is concluded that the density of the grain boundary dislocations (GBD) controls the rate of stimulated grain boundary sliding. A system of equations for the density of the grain boundary dislocations is derived, solved, and the results are compared with the experimental data. A set of the processes with participation of the lattice and grain boundary dislocations for describing grain boundary sliding is proposed. The minimum expansion of the investigated system together with the model in which the grain boundaries interact is the triple junction. The experiments are described in which the reorganisation of triple junctions in the superplastic deformation conditions was observed. The mechanism of this reorganisation is proposed, the conditions in which the energy parameters of the process are favourable are determined, and the number of reorganised triple junctions in the given loading conditions and the size distribution of the grains is estimated. The factors capable of correcting the results are discussed.

Chapter 6 presents the experimental results which can be used to investigate the formation of the band of cooperative grain boundary sliding by the percolation transition mechanism when the number of the reorganised triple junctions in the material is greater than the percolation threshold. The number of the bands of cooperative grain boundary sliding is estimated, and the superplastic deformation conditions are analysed as the conditions of formation of the bands. The shear rate along a band is estimated. On the basis of the rate field formed in a macrospecimen in shear deformation along the CGBS the strain rate is expressed through the characteristics of the bands. The theoretical stress–strain rate dependence is compared with the results of experimental investigations.

Chapter 7 describes the investigations of the effect of the statistics of the grain boundaries on the processes of grain growth and diffusion in nanocrystalline films of zirconium and nickel oxides. The experiments show the self-organisation of the grain boundary ensemble in the ZrO_2 films resulting in the formation of clusters consisting only of the special type boundaries. The differences in the oxidation kinetics of the nickel single crystals are explained by the formation of different types of grain boundary misorientation distribution. The type of grain boundary misorientation distribution is determined by the crystallographic orientation of the substrate. The results show that cerium additions reduce the rate of grain boundary diffusion but do not affect the nature of formation of the grain boundary misorientation distribution.

Chapter 8 depicts in detail the effect of the parameters of severe plastic deformation on the evolution of the microstructure and the grain boundary ensemble in ultrafine-grained nickel. The possibility of formation of a homogeneous ultrafine-grained structure by the method of high-pressure torsion is indicated for the first time. The results show that the grain boundary ensemble in ultrafine-grained materials is statistically uniform over the entire volume of the specimen.

Chapter 9 presents the results of experimental investigations of the relaxation processes in ultrafine-grained nickel, produced by both the methods of severe plastic deformation and electrodeposition. The kinetics of grain growth in ultrafine-grained nickel, produced by equal channel angular pressing, is investigated. The results show that the activation energy of grain growth coincides with the activation energy of grain boundary self-diffusion. The results of experimental investigations and simulation of the grain boundary misorientation distribution in ultrafine-grained nickel, produced by torsion with annealing, are described. Differential scanning calorimetry is used to measure the activation energy and stored enthalpy for ultrafine-grained nickel. The results are presented of investigations of low-temperature superplasticity detected in nanocrystalline nickel produced by electrodeposition.

The two final chapters are concerned with the problem of transfer of the relationships from the mesolevel to the upper structural level – the level of the specimen as a whole (macrolevel). For this purpose, the model of superplastic deformation, constructed in chapter 10, is used for calculating the superplastic capacity. The limiting deformation is regarded as the formation of a difference in the thickness of the specimen in which the sections of the material characterised by the largest thickness difference in comparison with the average section of the specimen do not fit (as regards of stress) the range of superplastic deformation. Some properties of limiting deformation, in particular its dependence on characteristic lengths of the specimens, are analysed.

Chapter 11 solves the problem of transition from the deformation mechanism to the appropriate constitutive equations, which determine this mechanism. Analysis is based on the description of the movement of polycrystalline continuum, restricted by the continuity condition. The force characteristics of the deformation process and the geometrical characteristics of the appropriate flow pattern at separated. The orientation of slip systems is calculated for an arbitrary stress state. The variants of the coaxiality of the stresses deviator and the strain rate tensor are investigated.

In the conclusion, the results presented in the monograph are generalised and the main problems of advanced materials science are listed taking into account new achievements in the area of formation of ultrafine-grained structures for producing the improved (and completely new) mechanical and physical properties.

We are grateful to our colleagues – co-authors of joint publications which form the basis of this book. We are also grateful to the scientific community which has been around for the last 20 years and exerted a positive effect on our work. We are especially grateful to our families (wives and children), whose support and understanding helped us not only in the work on this book but also in all our scientific activities. We devote this book to them.

References

1. Sutton A.P. and Ballufi R.F., Interfaces in crystalline materials, Clarendon Press, Oxford, 1995.
2. Kaibyshev O.A. and Valiev R.Z., Grain boundaries and properties of metals, Metallurgiya, Moscow, 1987.
3. Orlov A.N., et al., Grain boundaries in metals, Metallurgiya, Moscow, 1975.
4. Gleiter G. and Chalmers B., High-angle grain boundaries, Mir, Moscow, 1975,.
5. Kolobov Yu.R., et al., Grain boundary diffusion and properties of nanocrystalline materials, Nauka, Novosibirsk, 2001.
6. McFadden S.X., et al., Nature, 1999, vol. 398 (6729), 684–686.
7. Pshenichnyuk A.I., Structural levels of superplastic deformation, Dissertation, Institute of Problems of Supeplasticity, Russian Academy of Sciences, Ufa, 2001.
8. Bokshtein B.S., et al., Thermodynamics and kinetics of grain boundaries in metals, Metallurgiya, Moscow, 1986.

1

STRUCTURAL SUPERPLASTICITY OF POLYCRYSTALLINE MATERIALS

1.1. Structural levels, spatial scales and description levels

The physical processes accompanying plastic deformation of polycrystalline materials are reduced to the nucleation, propagation and interaction of point (vacancies, impurity atoms), linear (dislocations, disclinations), planar (grain boundaries, interfacial boundaries) and volume (cracks, pores) defects of the crystal structure [1]. The attempts for the classification of different physical processes taking place with the participation of the above defects lead unavoidably to the considerations of the structural levels of deformation [2] and the characteristic spatial scales [3]. In the classification proposed by V.I. Vladimirov [4] there are spatial scales, defining micro- ($\ell_2 < 10^{-7}$–10^{-6} cm), meso- ($\ell_1 \sim 10^{-6}$–10^{-3} cm) and macrolevels ($\ell_0 > 10^{-3}$–10^{-1} cm). Accepting that this classification is conditional and incomplete, Vladimirov proposes to define for polycrystals an additional scale given by the mean grain size. In a later study [5], Vladimirov introduces a five-level model whose characteristic scales are determined by the quantities divisible by the previously introduced scale ℓ_2 and the mean grain size. However, even this classification is not universal: for the given conditions, some scale levels are absorbed by others. As mentioned justifiably in a study by Khannanov [6], in a general case there is an entire set of scales λ_i ($i = 1, 2, \ldots, n$).

Better results were obtained using the classification based on the concept of the representative volume which determines the size of the spatial averaging region in which the structural special features of the objects of the lower scale level are characterised by a specific set of integral characteristics and the objects themselves are structureless formations [7]. This approach automatically assumes that every description level has its own set of collective variables which are defined unambiguously by the resultant structural formations. The spatial scales are not specified *a priori* and they form the characteristic lengths of the resultant structures. Thus, the main priority is the concept of the description level and the scales are determined by the content of a specific investigation.

The investigation of the physical processes taking place with participation of defects are carried out on three description levels, depending on the given task. The microlevel is characterised by the determination of the individual properties and special features of the behaviour of every defect: appropriate collective variables – the coordinates and atomic momenta. The interaction of the atoms is defined by the interatomic potential. Undoubtedly, this description is unnecessary in the investigation of the mechanical properties of solids because its response to the external mechanical effect is determined by the collective behaviour of the previously mentioned crystal structure defects, i.e., by interactions in ensembles of defects and their individual response to the external effect. However, the laws of interaction of defects (roughly speaking, the paired potential) can be determined only on the basis of analysis of their atomic configuration. For example, the molecular static methods can be used to determine the perturbation of the lattice in the vicinity of a vacancy, an impurity atom, a dislocation core or a grain boundary. The calculated displacements of the atom from the equilibrium positions corresponding to the nodes of a regular lattice can be used to determine the excess energy and the characteristic length of the field of elastic perturbations in the lattice for any type of defect. The molecular dynamic methods can be used to calculate the vibrational spectra of new lattice configurations and, consequently, determine the special features of their behaviour at different temperatures. After fulfilling this program, we transfer to the next description level.

The evolution of an ensemble of defects in the field of external stresses is investigated on the mesolevel using the characteristics of the defects determined on the microlevel. Instead of indicating the position of every vacancy and the appropriate dilation field, the spatial distribution of vacancies and the appropriate elastic fields are introduced. Their kinetic properties are determined by the diffusion coefficient derived on the lower structural level. The impurity atoms are also described in a similar manner. The dislocation on this description level can be treated as an elastic string with a specific energy of unit length and appropriate field of elastic perturbations, defining the interaction of the dislocation with other defects. The kinetic properties of the dislocations in the field of external stresses are determined by the sliding and climbing speeds. The ensembles of the dislocations can be investigated in discrete or continuous formalism in relation to the specific content of the investigation. If the previous description level requires operation with a small number of defects (the number of atoms of different type) and with the astronomically large number of variables (the coordinates and atomic momenta), we now have a slightly larger number of objects (defects of all types) and an incomparably smaller number of collective variables: spatial–time densities of vacancy distribution, dislocations (generally speaking, tensor distribution), grains (according to size), grain boundaries (with respect to disorientation). Complete information on these quantities enables us to determine the deformation response of the microvolume to the external effect. On this description level, the material is still principally structural and is characterised by steep spatial gradients of the deformation response.

The spatial scale of the macrolevel is determined as the minimum size of the averaging region starting at which the deformation response is given only by the external loading conditions. This description is principally structureless and is characterised by the classic tensors of the mechanics of deformed solids.

At present, the investigations of the characteristics of the crystal structure defects on the microlevel can be regarded as almost completed to the extent sufficient for the requirements of plasticity physics. The main difficulty is the presence of considerable non-linearities and of self-organisation processes, determined by these non-linearities, in the ensembles of the defects. For example, the intrinsic energy of a dislocation ensemble at the given dislocation density is proportional to the first degree of density, and the interaction energy with the long-range effect taken into account is proportional to the second degree of density. The latter circumstance results in a faster transition to the non-linear regime in comparison with, for example, dense gases for which the interaction energy increases with increase of the density at a considerably lower rate because of the short range of their potential. The interaction of the dislocations of different slip systems at low dislocation densities can be ignored with a high degree of accuracy and it is sufficient to examine only their interaction with other crystal structure defects (mostly with grain boundaries and with forming dislocation clusters). At high temperatures the structure of the grain boundary can change as a result of the spreading of lattice dislocations. Beginning at some level of dislocation density, the interactions in the ensemble become controlling and the system is converted to a highly non-linear system. This results in the formation of greatly differing spatial dislocation structures (dipoles, polygonisation walls, dislocation networks, clusters, substructures, etc). This field in materials science is referred to as physical mesomechanics and is being developed by the Tomsk scientific school [8]. These processes are detected in loading of the materials with a large mean grain size and determine the wide range of different physical situations (from the stages of strain hardening to recrystallisation processes). Although these most complicated processes are difficult to formalise, so that we still have no physical theory of plasticity (and dynamic recrystallisation), the transition to the macrolevel of description for materials with a large mean grain size without additional attempts is ensured using the so-called Taylor model. The efficiency of the model in the investigated situation is determined by the frequently verified experimental fact: in deformation of the material with a large mean grain size the deformation of the grains repeats the deformation of the specimen as a whole. In other words, the representative volume is given by the mean grain size and the transition to the macrolevel is ensured by averaging of the accurately determined region. In fact, the problem of developing the physical theory of plasticity of coarse-grained materials is reduced to the correct description of the processes on the mesolevel [9].

A slightly different situation is encountered when investigating the superplasticity phenomenon. It is well known that one of the conditions of realisation of this phenomenon is the application of the material with a relatively small mean grain size. The density of the lattice dislocations in the volume of the grain does not reach high values, deformation processes are not accompanied by the transition to the non-linear evolution regime and it is sufficient to examine only the interaction of lattice dislocations with the grain boundaries. In addition, the experiments show that intragranular deformation during superplastic deformation is an accommodation process which accompanies the main deformation mechanism – grain boundary sliding. Thus, in comparison with the classic plasticity of coarse-grained materials, the mesolevel of superplastic deformation is characterised by a considerably smaller set of structures in the volume of the grains and by the transfer of the main processes to the interface (interphase, intergranular). In this case, the Taylor model cannot be used for the transfer of the relationships of the mesolevel to the macrolevel: in deformation of the specimens by thousands of percent the grains remain almost completely equiaxed, i.e., the size of the averaging region becomes indeterminate. In particular, this is the main reason for having a large number of physical models of superplasticity which reproduce on average the behaviour of the well-known sigma-shaped curve (dependence of flow stress on strain rate in the stable yielding stage), and we have no examples of application of these models for describing the relationships on the macrolevel (deformation to fracture, construction of the controlling relationships, etc).

In recent years, experimental investigations have resulted in new concepts characterising the mesolevel of superplastic deformation. It has been found that grain boundary sliding during superplastic deformation does not take place independently on different interfaces of the polycrystalline material and includes in a single process the macroscopically elongated system of conjugate boundaries in which the self-consistent shear also takes place. This system of the boundaries boundaries is referred to as the bands of cooperative grain boundary sliding (CGBS); it is a macro-object, and according to general considerations should determine the representative volume essential for the transfer of the mesolevel relationships to the macrolevel.

Superplasticity is not only the interesting physical phenomenon but also the basis of a number of unique technological processes operating by not micro- but rather by macroparameters. The existing situation with the physical models of superplasticity can not be regarded as satisfactory until it is possible to carry out the self-consistent (within the framework of a single concept) description of the meso- and macrolevels.

1.2. Structural superplasticity: from the combination of mechanisms to cooperative grain boundaries sliding

Detailed reviews of the experimental and theoretical investigations in the area of superplasticity have been published in well known monographs, dissertation and reviews by Kaibyshev [10–13], Grabskii [14], Novikov and Portnoi [15], Smirnov [16], Astanin [17], Faizova [18], Perevezentsev [19], Chuvil'deev [20], Eddington [212], Mukerjee [22] and many others. Therefore, we shall discuss only briefly the history of development of considerations regarding the superplasticity mechanisms and the experimental results obtained in recent years which are used as a basis for the proposed physical model of superplasticity.

At the start of purposeful investigation of superplasticity the researchers quickly found that: 1) superplasticity is detected in the strain rate range 10^{-4}–10^{-2} s^{-1} (in classic eutectoid alloys investigated in the initial stage) at temperatures of $T > 0.4\ T_m$, where T_m is the melting point of the material; 2) for the superplasticity regime to form, the material should have relatively small grains (<10 μm); 3) during deformation, the grains do not repeat the deformation of the specimen as a whole and the extent of elongation of the grains along the tensile loading axis is negligible in comparison with the elongation of the specimen. The latter fact indicates unambiguously that the grain is no longer the representative volume; in contrast to the plasticity of coarse-grained materials, in the superplasticity condition the main processes are transferred from the grain interior (the extent of these processes in the grains appears to be very small) to the periphery of the grain – the boundary. The system of the internal interfaces (the grain boundaries or interfacial boundaries) should create suitable conditions for these processes. This is also assured by the small grain size which increases the specific area of the grain boundaries. It was the beginning of intensive investigations of the process of grain boundary sliding (GBS).

The GBS mechanism can be comprehended by explaining first of all what the grain boundary is. Therefore, the investigations of the grain boundaries have been accompanied by permanent discussions of the nature of grain boundaries in polycrystals. Variants ranging from the amorphous (which is often incorrectly regarded as identical with the liquid structure and is referred to by the poorly defined term 'liquid-like') to the crystal structure have been discussed. The history of development of these considerations, starting with the island model proposed by Mott and ending with the current views, based on the concept of the lattice of coincident sites, developed by Ballman, which specifies the zero approximation for calculating the structure of the grain boundaries by the molecular static mechanics, has been described in a large number of reviews and monographs [23, 24]. A brief summary of this stage of investigations may be formulated as follows. The grain boundary is a two-dimensional crystal structure whose characteristics are determined by the disorientation of the adjacent crystals and by the definition of the plane of positioning of the boundary. On the background of the regular structure we can specify some minimum disorder of the regularity, or a defect which in complete analogy with the defects of the volume crystal structure is referred to as the grain boundary dislocation. It should be stressed that this holds for the grain boundaries in metallic materials. At a high temperature the grain boundaries of ceramic materials can show the formation of a liquid phase which is determined by the deviation from stoichiometry in the vicinity of the grain boundary [25, 26]. The crystal nature of the grain boundary has been confirmed in the experiments concerned with the investigations of grain boundary sliding.

The grain boundaries sliding process is studied on specially prepared simulation objects – bicrystals. The experimentally detected shear along the grain boundary plane is accompanied in a general case by volume deformation of the crystals and this complicates interpretation of the results. In [27] the authors reported on a method which can be used both to separate the effects of grain boundary and intergranular sliding and also investigate their interaction. This is achieved by selecting zinc bicrystals with the 90° symmetric inclination boundary (Fig. 1.1) as the experimental object,

The hexagonal zinc lattice has a distinctive easy slip system given by the distribution of the most closely packed (basal) atomic plane. Such a bicrystal can always be oriented in relation to the tensile loading axis in such a manner that the shear stresses on the basal planes of both crystals are zero and intragranular deformation is almost completely suppressed. In this case, the plane of the boundary coincides with the plane of the maximum shear stresses. The type of grain boundary sliding, referred to as pure, takes place in a similar situation. With any other orientation of the bicrystal in relation to the tensile loading axis the shear stresses on the basal planes of the crystals differ from zero and loading is accompanied by intragranular sliding. It has been shown that this accelerates grain boundary sliding. This grain boundary sliding variant is referred to as stimulated. Later, detailed experiments using this scheme were carried out on cadmium bicrystals (the crystal lattice with the same symmetry) [28] and zinc [29]. Measurements were taken of the time dependence of shear along the boundary plane at different temperatures and tensile forces. The results show that the rate of pure grain boundary sliding during remains almost constant for a long period of time, and its dependence on temperature T and shear stress σ in the grain boundary plane has the form $\dot{S}_0 \propto \sigma \cdot \exp(-E_{GB}/RT)$ where E_{GB} is the activation energy of grain boundary diffusion. In the case of stimulated grain boundary sliding the rate depends strongly on

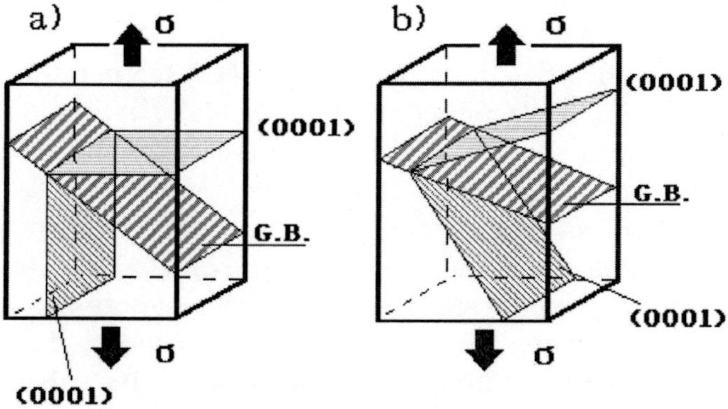

Fig. 1.1. Zinc bicrystals with the 90° inclination boundary [11$\bar{2}$0], used for investigating pure (a) and stimulated (b) grain boundary sliding [27].

time. At the initial moment of time, the sliding rate is considerably higher than the rate of pure grain boundary sliding and has the form $\dot{S}_0 \propto \sigma^2 \cdot \exp(-E_V / kT)$ where E_V is the activation energy of lattice self-diffusion. With time, the effect of stimulation monotonically decreases and at some moment the rate of stimulated grain boundary sliding decreases to the pure sliding rate. The dislocation mechanism of grain boundary sliding has been confirmed by the following experimental facts: 1) spatial inhomogeneity of shear along the sliding direction; 2) the linear relationship between sliding and migration; 3) anisotropy of the sliding rate in the grain boundary plane.

For polycrystals, the development of grain boundary sliding is complicated by the inherent structural element – triple junctions. At the triple junctions, the planes of the adjacent grain boundaries form a diheral angle close to 2π/3. The solution of the problem of the possible methods of transfer of shear deformation through a triple junction was the subject of theoretical analysis for a long period of time. At least 20 mechanisms of shear accommodation at the triple junctions, which often differed in small details, were proposed. Every mechanism pretended to play the role of the mechanism of superplastic deformation. A relatively complete collection of these methods is presented in a review in [21]. It is surprising that irrespective of the method assumed to be the method controlling the deformation rate, all the models led to the same equation for the strain rate in the stable yielding stage (when the dependence on strain disappears)

$$\dot{\varepsilon} = C\left(\frac{b}{d}\right)^{1/n}\left(\frac{\sigma - \sigma_{th}}{\mu}\right)^{1/m}\frac{\mu b D_0}{kT}\exp(-E / kT). \quad (1.1)$$

Here b is the modulus of the Burgers vector; d is the mean grain size; σ is the external stress; σ_{th} is the threshold stress; μ is the shear modulus; k is the Boltzmann constant; T is test temperature; D_0 is the pre-exponent of the diffusion coefficient; E is the activation energy of superplasticity (in a general case it is the mean weighted value of the activation energies of grain boundary and lattice diffusion) and n and m are the parameters inherited from the phenomenological form and they determine the dependence on the grain size and stress; C is a constant and it is the only value by which the available superplasticity models differ. Some differences were detected only in the evaluation of the numerical value of the dimensionless constant C. Its value depending on the model and changes from 0.6 to 200. It is well known that these models are incomplete.

1. Equation (1.1) does not describe the *S*-shaped curve. In the best case, it is a tangent to the curve at the inflection point (in the appropriate coordinates), and only under the condition that the maximum value of the parameter of rate sensitivity (i.e., the value of the logarithmic derivative of σ with respect to $\dot{\varepsilon}$ at the inflection point) is equal to 1/2 (all the models lead to the value $m = 1/2$). The introduction of the threshold stresses does not save the situation because the value of σ in the superplasticity range is considerably higher than σ_{th}.

2. The experimentators have known for a long time that superplasticity is detected only in some strain rate range (or a corresponding stress range because the strain rate and stress in the stable yielding stage are mutually linked unambiguously by the *S*-shaped curve). The models of type of (1.1) do not make it possible to determine the required interval: threshold stresses in equation (1.1) are not linked with the lower boundary of the stress range and are considerably lower than this boundary.

3. The set of the parameters characterising the material (the mean grain size made dimensionless with respect to the modulus of the Burgers vector, the shear modulus and the activation energy of grain boundary diffusion, or some effective diffusion in the Gittus model [30]) is not completely determined because many essentially different materials are defined by this set as identical. Adding to the set the parameter m which differs from 1/2 (i.e., from the value determined by the models) not only transforms the model to a phenomenological one, and this is done without arguments, but also directly contradicts the characteristics of the micromechanisms used as the basis of the model.

It should be mentioned that this incomplete success was associated with the following circumstance. The models implicitly describe some 'typical' triple junction. The erroneous nature of these considerations became evident after carrying out experiments to investigate structural changes during deformation. As established, the grains change their neighbours during deformation; in addition, larger grain groups are displaced in the specimen as an integral unit with the variation of orientation (Fig. 1.2) [31]. The first factor was used as a basis for formulating a large number of geometrical models of superplastic deformation systematised by Zelin (Table 1.1) [32]. The best known is the Ashby–Verrall model [33] in which the mechanism of rearrangement of the grains was speculatively formulated whilst retaining the integrity of the material. However, regardless of the attractive original nature of this mechanism, subsequent analysis showed that this model does not apply to three-dimensional polycrystals [34]. The models based on the

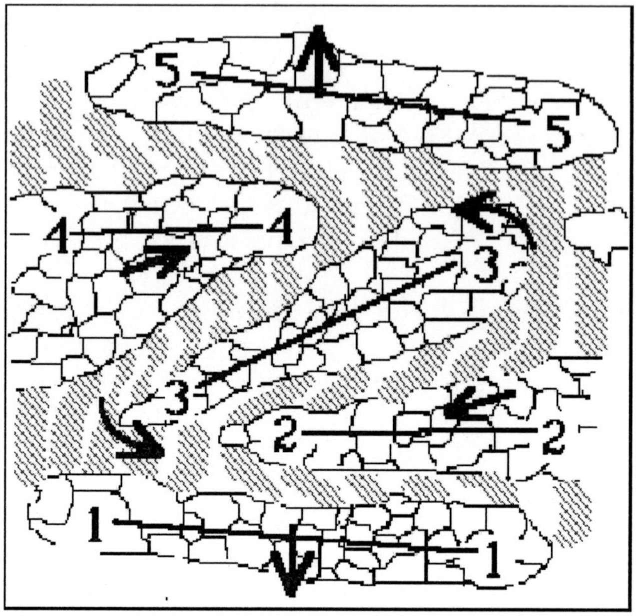

Fig. 1.2. Diagram of displacement of a group of grains during superplastic deformation.

mechanism of switching of the grains were discussed for a long time in the literature [35].

The second fact appears to be simply very unusual: on the one side, the grain size should be smaller and, on the other side, large groups of grains move as an integral unit. A large number of problems arise immediately. Firstly, why should the small grains be required if large conglomerates move as an integral unit? Why not consider the grain size equal to the size of the conglomerate? Secondly, what type of triple junction is discussed in the superplastic deformation models? Is it the junction belonging to this type of conglomerate so nothing takes place in it. How does one junction differ from another and what type of junction was considered by the authors of the models? The only answer is as follows: the triple junction is not the representative element of the structure. It became clear only later that this was the first confirmation of the action of the co-operative deformation mechanism.

Analysis of the physical mechanisms of superplasticity was accompanied by a more successful process of expansion of the range of the materials and the conditions in which superplastic deformation takes place. In ceramics and in composites superplasticity was detected at rates (in inverse seconds) approximately 3–4 orders of magnitude higher than the rates typical of the classic eutectoids. In the case

Table 1.1. Geometrical models of superplastic deformation [32]

Deformation mechanism. Authors	Schematic representation	Comment
	1. Individual GBS	
Accommodation by grain boundary migration Lee (1970) [48]	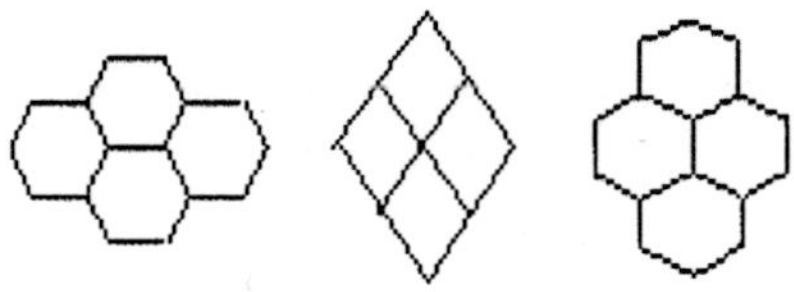	It is assumed that GB migration is interrupted after deformation. Further deformation not explained
Diffusion accommodation Ashby, Verral (1973) [33]	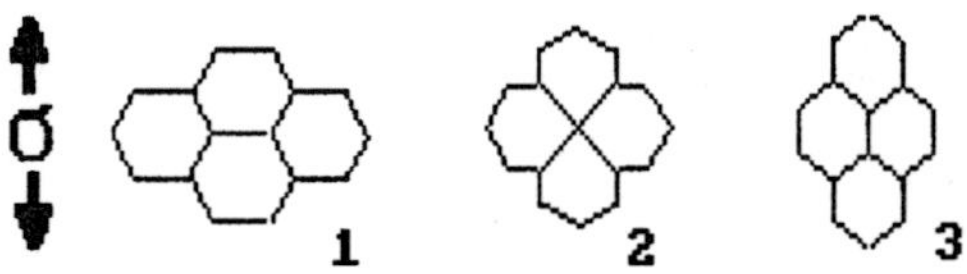	Deformation rearrangement of surrounding grains not explained
Dislocation accommodation Matsuki et al. (1977) [49]	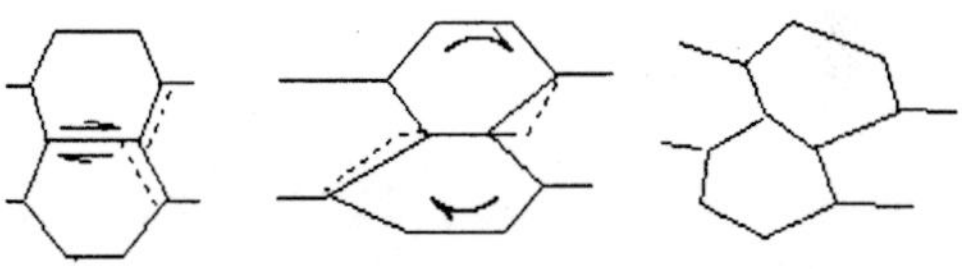	Grain rotation takes place in addition to sliding. Continuation of deformation not explained.
Grain entry Gifkins (1978) [50,51], modification Langdon (1981) [52]	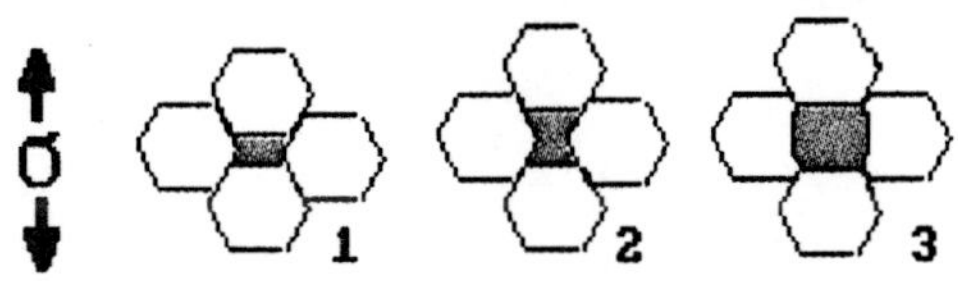	Model can explain only 110% strain
Geckinly (1983) [53]	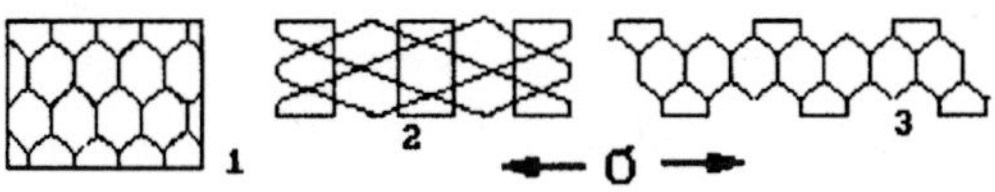	Further deformation explained by a change of trajectory of grain movement

of superfine-grained crystalline materials it was possible to reduce greatly the superplasticity temperature (low-temperature superplasticity). Investigations of nanocrystalline materials started. The hopes for the existence of a general micromechanism applicable to all materials and the conditions and responsible for the realisation of the superplasticity conditions have not been fulfilled. Studies appeared in which it was shown that the mechanisms controlling the strain rate can differ at low and high temperatures even for the same material [36].

The problem of the nature of superplasticity was transferred to another aspect: what is the common feature of greatly differing

Table 1.1. (continued)

Deformation mechanism. Authors	Schematic representation	Comment
	2. Cooperative GBS. Rigid shear	
Cannon, Nix (1973) [54]	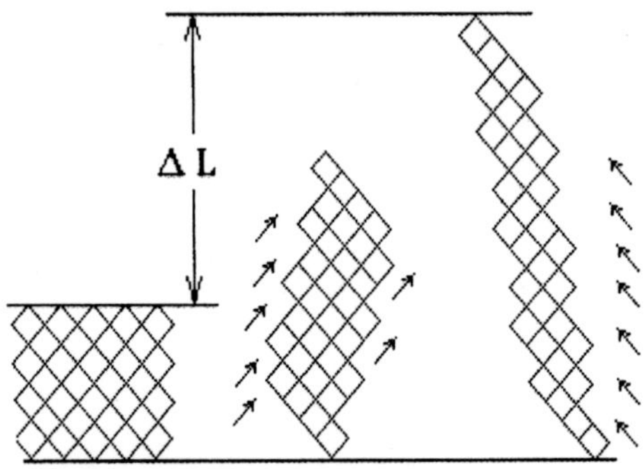	Macroscopic deformation caused by movement of a group of grains. Expected simultaneous sliding of all grains not realistic
Patridge et al. (1985) [56]	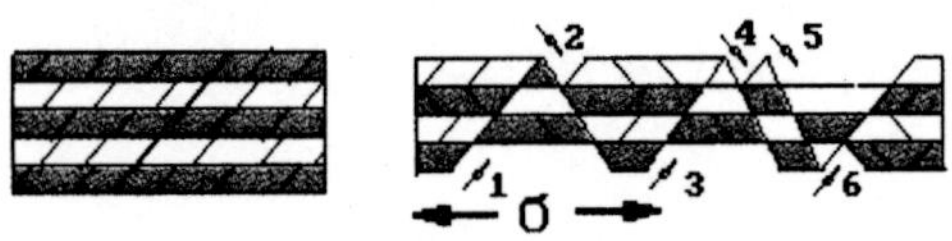	Rigid shear in two-phase maerial with different shear values along interphase boundaries of different type taken into account
Newbury, Hazzledine (1973) [57]	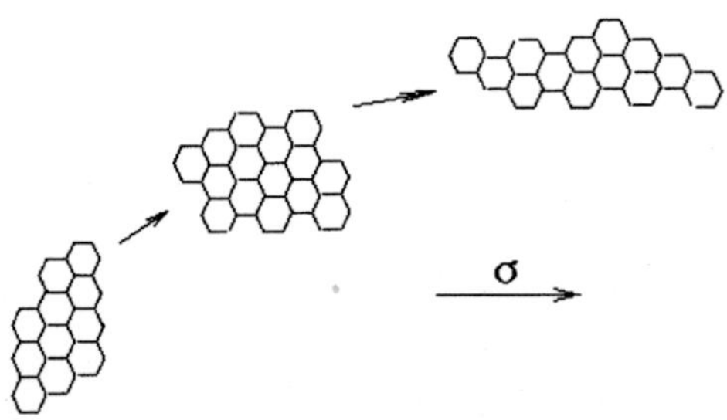	Model assumes specific trajectories of movement of grains. Entry of grain from parallel sections is taken into account

materials, tested at greatly different conditions, in such a manner that they show the same behaviour. Generally speaking, the answer is clear: they are united by the absence of conditions for fracture. This would appear to be dull tautology. However, it leads to new experiments and the formation of new views on superplasticity as a whole.

The absence of the conditions for fracture is reduced to the absence of the conditions for the localisation of deformation (leading to ductile fracture) and the stress concentration (leading, at the appropriate loading rates, to the propagation of cracks and brittle fracture surface), or, in other words, to the requirements for the homogeneity of the flow within the limits of the specimen (it should be mentioned that here we are not discussing the homogeneity of deformation on the mesolevel) of a relatively long period of time. The required homogeneity can be achieved only by specific0 organisation of the large-scale flow pattern: it should not look as the superposition of independent processes, taking place in differcnt arcas of the deformed specimen, but rather should be represented by a single process coherent on the macrolevel. In particular, investigation of the organisation of the large-scale flow pattern have also resulted in new views regarding superplasticity.

The investigations of the special features of the deformation mechanisms on the meso- and macrolevel are carried out using specially prepared specimens. A thin carbon film is sprayed on the surface of the specimen. The carbon layer fractures in tensile loading in the area of the highest intensity of deformation. The low-voltage regime of scanning electron microscopy (SEM) results in a contrast determined by the thickness of the oxide film. Consequently, it is possible to distinguish the surface, formed during deformation, from the initial surface. Precise snap-shots of the same region of the surface prior to and after deformation can be used to investigate the movement of the individual structural elements. As a result of the investigation of the deformation of relief, formed on the surface of the specimen during superplastic deformation, it was concluded that the flow on the mesolevel is not uniform: deformation is localised along the surfaces passing through the entire cross-section of the specimen and separated by the sections of the material not included in deformation. Examination of the fine structure of the strain bands shows that the position of these bands can be linked with the structural elements of the polycrystal. It was also shown that the shear surface unites the grain boundaries oriented closest to the planes with the maximum shear stresses. Under unfavourable conditions the shear surface is forced to intersect with the body of the grain.

The relationship of the structural changes in the process of deformation with the deformation stages determined by the special features of the loading curves has been used to draw conclusions regarding the dynamics of formation of the shear bands. It was established [37–39] that the stage of transfer of the loading curves to the stable flow regime is characterised by the formation in the material of individual independent flow sections represented either by nonuniform shear along the favourably oriented grain boundaries or the manifestation of the dislocation activity in relatively large grains. On reaching the stable flow stage which is already the feature of transition to the superplasticity regime [1], the individual flow sections merge and deformation takes place by means of matched shear along the surfaces passing through the entire cross–section of the specimen and situated in the immediate vicinity of the planes with the maximum shear stresses [37–43]. These surfaces are referred to as the band of cooperative grain boundary sliding (CGBS) [40]. The flow pattern, formed by the effect of the CGBS bands, ensures the maximum homogeneity of deformation within the limits of the specimen.

What is the mechanism of merger of the independent flow sections into a coherently acting shear band? Experiments were carried out to determine the distribution of the dihedral angles at triple junctions belonging to the shear band and situated outside the band [39]. The results show that the dihedral angles outside the band are equal to approximately $2\pi/3$, i.e., they are close to equilibrium configurations. The triple junctions, belonging to the band, are 'straightened', i.e. their dihedral angles show a distinctive tendency to increasing (equal to π). In addition to this, in investigating grain boundary sliding in tricrystals [44] it was found that the 'straightening' of the triple junctions takes place as a result of local migration (buckling) of one of the boundaries (with the maximum shear stresses acting in the plane of the boundary) to the rearrangement of the configuration of the triple junctions which ensures compatibility of shear at two out of three boundaries forming the given junction. Finally, in some especially unfavourable configurations (relatively rare) the band is forced to intersect the body of the grain. The flow pattern, formed by the effect of the CGBS bands, ensures the maximum uniformity of deformation within the entire specimen.

Thus, the problem of historical development of superplasticity is regarded as the problem of recurring detailization of the representative volume reproducing the main special feature of the phenomenon – high strains to fracture. The change of the investigated object in the grain–grain boundary–triple junction–the band of cooperative grain boundary sliding is quite evident. At the same time, this sequence defines the relationship of the processes on the mesolevel with the formation of the macroobject and determines the experimental tasks.

1.3. Structural superplasticity: from meso-description to macrocharacteristics

Experimental investigations have formulated the essential basis for theoretical considerations regarding the nature of superplastic deformation. The previously proposed theoretical models of superplasticity can be regarded only as the models of accommodation of shear in the isolated triple junctions. In particular, it should be mentioned that there are at least three studies in which it was attempted to investigate the self-consistent shear along the system of conjugate grain boundaries. These are the studies by Spingarn and Nix [45], Khannanov [46] and Baudelet [47]. The first two of these studies preceded the investigations by the authors of this book, the third study was published later. However, since the authors of these studies did not know the experimental results obtained in the investigation of the special features of triple junctions belonging to the CGBS band and situated outside the band, the investigations were reduced to studying a chain of equilibrium and isolated triple junctions. Nevertheless, this was still the theoretical precursor of the band.

Thus, the construction of the theoretical model of superplasticity is reduced to constructing a method of formal description of the structural level given successively by the grain boundaries, triple junctions, and the band of cooperative grain boundary sliding. This determines the formulation of specific experimental tasks.

Grain boundary sliding on the flat boundary of a bicrystal is regarded as the basis of the deformation mechanism of superplastic deformation: although the dislocation nature of the phenomenon is confirmed quite convincingly by experimental investigations, there is still no verbal description of the scenario with the participation of lattice and grain by the dislocations, not mentioning the mathematical formalism, which would make it possible to reproduce in the calculations the experimentally recorded dependence of the sliding on the loading time. This leads to the first task: construction of the model of stimulated grain boundary sliding in the bicrystal.

According to the experimental data, the natural obstacle to the development of grain boundary sliding in a polycrystal – a triple junction – undergoes configuration rearrangement ensuring the compatibility of shear on two out of three boundaries forming the given junction (Fig. 1.3). The driving force for this rearrangement is not known. This determines the second task of the investigations: construction of a model of the reorganisation of isolated triple junction in the stress state conditions. The absence of uniformity of the deformation on the mesolevel indicates convincingly that rearrangement takes place only at some of the junctions. This formulates the problem of the distributed parameters with calculation of the number of reorganised junctions in the polycrystal in the given loading conditions.

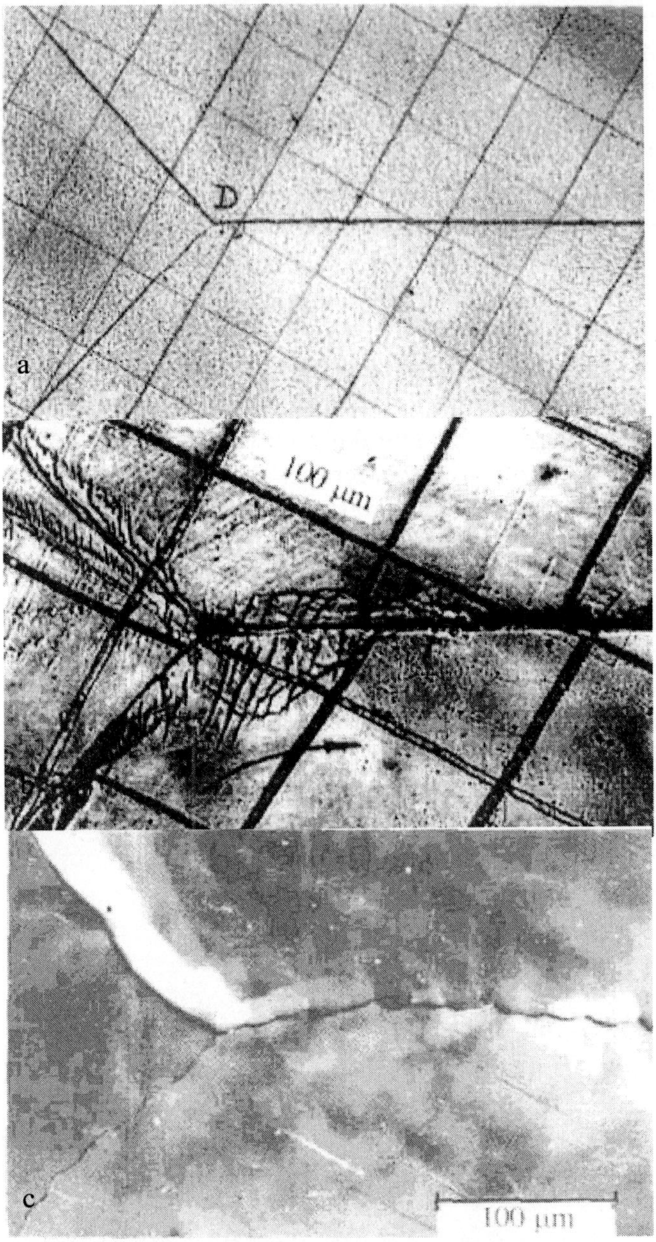

Fig. 1.3. Reorganisation of a triple junction in an aluminium tricrystal: a) initial position; b) after deformation; c) after deformation and repolishing the surface [44].

The experiments also show that in the stage of the stable flow (the indication of transition to the superplastic condition) the independent shear sections merge into shear surfaces – CGBS bands. The conditions of formation of these bands and calculation of their number in the polycrystal in the given welding conditions are the content of the third task. The final characteristic, determined by the dislocation activity of the mesolevel, is the shear rate in the CGBS band. This completes the verification of the band as an independent object on the macrolevel.

The next three tasks relate to the macrolevel onto which the characteristics of the bands as independent object, determined in the solution of the first three tasks, are transferred from the mesolevel.

The section of the dependence of the strain rate on stress in the stable flow stage – the standard test of any physical model of superplastic deformation – is the content of the fourth task. Its solution includes the development of formalism which would make it possible to link the macroscopic strain rate with the characteristics of the CGBS bands.

The presence of a relationship between meso- and macrodescriptions enables us to formulate the task to calculate the limiting strains or the duration of the stable flow stage. This calculation and investigation of its dependence on the strain rate and the dimensions of the specimen is the fifth task.

Knowing the special features of the large scale orientation of the flow it is possible to propose a reliable approach to constructing the constitutive relationships and carrying out preliminary analysis of the condition of coaxiality of the stress deviator and the strain rate tensor in superplastic deformation. This is our last, sixth task.

All these tasks will be investigated in this book.

References

1. Orlov A.N., Some problems of the kinetics of defects in crystals, in: Problems of theory of crystal defects, Nauka, Leningrad, 1986.
2. Panin V.E., et al., Structural levels of deformation of solids, Nauka, Novosibirsk, 1985.
3. Priemskii N.D. and Romanov A.E., Characteristics scales of plastic deformation, in: Disclinations: Experimental investigations and theoretical description, Nauka, Leningrad, 1982.
4. Vladimirov V.I., et al., Mescoscopic level of plastic deformation, in: Physics of strength and plasticity, Nauka, Leningrad, 1986, 69–80.
5. Vladimirov V.I., Collective effects in defect ensembles, in: Problems of theory of crystal defects, Nauka, Leningrad, 1987, 43–57.
6. Khannanov M.Kh., FMM, 1988, No. 2, 213–218.
7. Grinyaev YuV., Physical mesomechanics and computer design of materials (Panin V.E., editor), vol. 1, Nauka, Novosibirsk, 102–112.
8. Panin V.E. (editor), Physical mesomechanics and computer design of materials, in two volumes, Nauka, Novosibirsk, 1995.
9. Ivanov M.A., et al., FMM, 2000, No. 6, 5–15.
10. Kaibyshev O.A., Plasticity and superplasticity of metals, Metallurgiya, Moscow, 1975.
11. Kaibyshev O.A., Superplasticity of industrial alloys, Metallurgiya, Moscow, 1984.
12. Kuibyshev O.A., Superplasticity of alloys, intermetallides, and ceramics. Berlin; New-York: Springer-Verlag, 1992.
13. Kaibyshev O.A., Scientific fundamentals, advances and prospects for superplastic deformation, Gidem, Ufa, 2000.
14. Grabskii M.V., Structural superplasticity of metals, Metallurgyia, Moscow, 1975.
15. Novikov I.I. and Portnoi V.K., Superplasticity of alloys with ultrafine grains, Metallurgiya, Moscow, 1975.
16. Smirnov O.M., Pressure treatment of metals in the superplastic condition, Mashinostroenie, 1979.
17. Astanin V.V., Cooperative processes and localisation of superplastic deformation, Dissertation, IPSM, Ufa, 1996.
18. Faizova S.N. Localisation of superplastic deformation of Zn–22%Al alloy on different structural levels, Dissertation, IPSM, Ufa, 1994.
19. Perevezentsev V.N., Deformation-stimulated transformation of the structure of intergranular boundaries and superplasticity of metallic alloys, Dissertation, Branch of the Institute of Engineering, USSR Academy of Sciences, Gor'kii, 1987.
20. Chuvil'deev V.N., Theoretical investigations of the micro-mechanisms of deformation and failure of superplastic materials, dissertation, Branch of the Institute of Engineering, USSR Academy of Sciences, Gor'kii, 1986.
21. Edington J. W., et al., Prog. Mater. Sci., 1976, vol. 21, 61–170.
22. Mucherjee A.K., Superplasticity in metals, ceramics and intermetallic, in: Plastic deformation and fracture of materials, ed. by Mughraby H., Materials Science and Technology, 1993, 6, VCR Verlagsgesellschaft GmbH, Germany.
23. Kaibyshev O.A. and Valiev R.S., Grain boundaries and properties of metals, Metallurgiya, Moscow, 1987.
24. Orlov A.N. (editor), Atomic structure of intergranular boundaries, in: New in solid state

physics, No. 8, Collection of translations, Mir, Moscow, 1978.
25. Imaev M.F., et al., Mater. Forum, 1994, vol. 170–172, 445–449.
26. Imaev M.F., et al., Physica C, 2000, vol. 329, 75–87.
27. Kaibyshev O.A., et al., DAN, 1979, vol. 245, No. 6, 1356–1358
28. Fukutomi H., Trans. Japan Inst. Metals, 1982, vol. 23, No. 10, 579–584.
29. Valiev R.Z., et al., Izv. VUZ, Fizika, 1991, No. 3, 93–103
30. Gittus J.W., Phil. Trans., R. Soc. London, A 288, 1978, 121–146.
31. Astanin V.V., et al., Acta Metall., 1994, vol. 42, No. 8, 2617–2622.
32. Zelin M.G., J. Appl. Phys., 1993, vol. 74, No. 8, 4972–4982.
33. Ashby M.F. and Verrall R.A., Acta Metall., 1973, vol. 21, 149–163.
34. Poire G.P., Creep of crystals. The mechanism of deformation of metals, ceramics and minerals at high temperatures, Mir, Moscow, 1987.
35. Sherwood D.J., and Hamilton C.H., Phil. Mag. A, 1994, vol. 70, 109–143.
36. Hayden H.W., Metall. Trans., 1972, 3, 833–842.
37. Kaibyshev O.A., et al., Advanced Materials '93, III / B: Composites, grain boundaries and nanophase Materials, ed. M. Sakai., Trans. Mat. Res. Soc. Jpn., 1994, 1473–1476.
38. Astanin V.V., et al., Mat. Sci. Technol., 1996, vol. 12, 489–494.
39. Astanin VV. and Kaibyshev O.A., Mater. Sci. Forum, 1994, 170–172, 23–28.
40. Astanin V.V., et al., Scripta Met. Mater., 1991, vol. 25, No. 12, 2663–2668.
41. Astanin V.V., et al., Acta Metall., 1994, vol. 42, No. 8, 2617–2622.
42. Zelin M.G. et al., Acta Metall., 1994, 42, No. 10, 119–126.
43. Zelin M.G. and Mukherjee A.K., Acta Metall. Mater., 1995, vol. 43, No. 6. 2359–2372.
44. Astanin V.V., et al., Scripta Met. Mater., 1997, vol. 36, No. 1, 117–122.
45. Spingarn J.R.and Nix W.D., Acta Metall., 1978, vol. 26, No. 9. 1389–1398
46. Khannanov Sh.Kh., Metallofizika, 1983, 5, No. 4, 113–114.
47. Blandin J.J., et al., Mater. Sci. Forum, 1997, vol. 243–245, 47–52.
48. Lee D., Metall. Trans., 1970, 1, 309–319.
49. Matsuki K., et al., J. Japan. Inst. Met., 1977, vol. 41, No. 11, 1136–1144; Matsuki K., et al., Metall. Sci., 1977, No. 11, 156–161.
50. Gifkins R.S. and Langdon T.G.. Mater. Sci. Eng., 1978, vol. 36, 27–33.
51. Gifkins R.C., J. Mater. Sci., 1978, vol. 13, No. 9, 1926–1936.
52. Langdon T.G., JOM, 1981, vol. 33, No. 9, A20–A20.
53. Geckinly A.E., Metal. Sci., 1983, vol. 17, No. 1, 12–18.
54. Cannon W.R. and Nix W.D., Phil. Mag., 1973, 27, 9–29.
55. NovikovI.L., et al.,, Acta Metall., 1981, vol. 29, 1077–1090.
56. Partridge P.O., et al., Acta Metall. Mater., 1985, vol. 33, No. 4, 571–577.
57. Hazzledine P. and Newbury D., Proc. III Conf. on Strength of metals and Alloys, IC-SMA-3, 1973, 1, 202–211.

2

CHARACTERISTICS OF GRAIN BOUNDARY ENSEMBLES

2.1. Crystal geometry and structure of intercrystalline boundaries

2.1.2. Methods for describing the structure of the grain boundaries

The polycrystalline solid as a statistical ensemble of the grains forming the solid is described by a large number of statistical characteristics. A special position in this set is occupied by the characteristics of the network of intercrystalline boundaries. They include the distribution function of the grain boundaries with respect to the crystal geometry parameters and primarily with respect to misorientation. Special attention will be given to the mathematical methods of describing the crystal geometry of intercrystalline boundaries which represent the basis of the algorithms for calculating the required parameters in processing the experimental data and modelling the spectra of misorientation of the grains in the polycrystals.

The structure and position in space of the boundary of two grains (Fig. 2.1) can be defined by five macroscopic parameters, three of which (angle and axis) describe the mutual rotation of neighbours around some axis, and two of them is the plane in which the interface is situated. The parameters of the boundary determine the crystal geometry and represent the basis for describing their structure.

The misorientation of the grains is a vector quantity and, therefore, it is necessary to solve the problem of the efficient representation of the grain boundary misorientation distribution (GBMD). This problem will be discussed in detail later, here it is accepted that the GBMD is represented by three distributions. The first of them is the distribution of the grain boundaries with respect to the misorientation angles and axis. As a result of the symmetry of the lattice there are several equivalent rotations coinciding the sites of the lattice of the adjacent

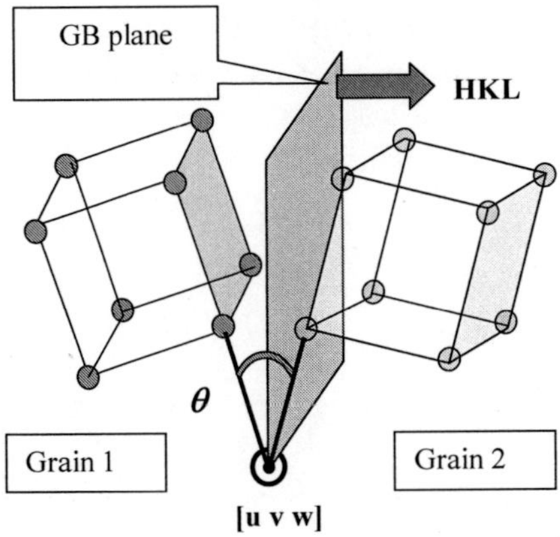

Fig. 2.1. Rotation of two adjacent grains.

grains (for example, in the case of the cubic lattice there are 24 equivalent rotations). For unambiguous selection, all the misorientations are reduced to the minimum angles θ_{min} and the directions $\langle pqr \rangle$ in the standard stereographic angle. The third distribution is based on the classification of the boundaries using the model of the lattices of the coincident sites. This is the distribution of the grain boundaries with respect to the reciprocal density of the coincident sites (Σ) characterising the fraction in the general spectrum of the boundaries with low values of Σ which usually have unique properties.

The mutual rotation of two lattices can be described using the rotating matrix [1]:

$$\widehat{R} = \begin{bmatrix} R_{11} & R_{12} & R_{13} \\ R_{21} & R_{22} & R_{23} \\ R_{31} & R_{32} & R_{33} \end{bmatrix}. \tag{2.1}$$

The elements of this matrix include the directing cosines between the coordinate axes of the first and second grain. The following equality applies to any two parallel directions into adjacent grains defined by the unit vectors **A** and **B**

$$\mathbf{A} = \widehat{R}\mathbf{B}. \tag{2.2}$$

Of the nine elements of the matrix $\widehat{R}$ only three are linearly independent quantities. The physical meaning of the matrix is that its elements represent the coordinates of the crystallographic basis of the lattice of one grain in the lattice of another grain. Knowing the rotational matrix, it is easy to determine the misorientation angle and axis. The rotation angle is determined by the trace of the matrix $\widehat{R}$:

$$\theta = \arccos\left[\frac{1}{2}\left(\sum_{i=1}^{3} R_{ii} - 1\right)\right]. \tag{2.3}$$

The components of the axis of rotation **C** for the cubic system have the form

$$\begin{aligned} C_1 &= (R_{32} - R_{23})/(2\sin\theta); \\ C_2 &= (R_{13} - R_{31})/(2\sin\theta); \\ C_3 &= (R_{21} - R_{12})/(2\sin\theta); \end{aligned} \tag{2.4}$$

The misorientation of the same boundary can have several equivalent values because as a result of crystal symmetry the selection of the system of crystallographic cooOrdinates in each grain is ambiguous. The set of the equivalent matrices, describing the same rotation, can be obtained as follows

$$\widehat{R}' = \widehat{M}_i \widehat{R} \widehat{M}_j, \tag{2.5}$$

Here the matrices $\widehat{M}_i$ describe the operations of transition from the given system of coordinates to the crystallographically equivalent one. For the cubic system there are 24 such matrices. Thus, for n equivalent transformations of the crystal lattice there are n^2 equivalent descriptions of the misorientation of two grains. In practice, it is however sufficient to find only n descriptions using the equation

$$\widehat{R}' = \widehat{M}_j \widehat{R}. \tag{2.6}$$

The description of the misorientation of the grains as the angle–axis pair and as the rotating matrix has been used most widely because of its simple form in the form of the three-dimensional Rodriguez vector (referred to usually as the Gibbs vector) [2]. The latter includes only three elements, i.e., in the explicit form the vector contains three independent crystallographic misorientation parameters:

$$\mathbf{G} = \mathrm{tg}\,(\theta/2)[C_1, C_2, C_3]. \tag{2.7}$$

In addition, the vector description of the misorientation is linked directly with quaternionic description which is more suitable for determining special misorientations [3].

In analysis of the crystal geometry of the grain boundary structure it is sometimes not rational to use the crystal lattices of the adjacent grains as the reference lattices. It is therefore important to investigate the different auxiliary lattices, common for both grains: the coincident-site lattice (CSL), the '0' lattice, the displacement shift complete (DSC) lattice [4], the grain boundary shear (GBS) lattice [5, 6].

Since the theory of the coincident-site lattice is the basis of the classification separation of the grain boundaries into special and general ones, this problem will be investigated in detail. The assumptions on the coincident-site lattice, introduced for the first time by Kronberg and Wilson [7] and developed further subsequently by many authors (the appropriate references can be found in [4, 8]; study [9]] should also be mentioned), proved to be extremely fruitful in the crystal geometry description of the grain boundaries. They are still used at the present time. The authors of [9] found that in the rotation of two identical lattices around the common crystallographic axis through a specific angle, part of the sites of the one lattice coincides with the sites of the other lattice, forming its three-dimensional lattice in space, i.e. the coincident-site lattice. This lattice is characterised by the inverse density of the coincident sites Σ, equal to the number of sites of one of the lattices in the elementary cell of the lattice of coincident sites. The concept of the coincident-site lattice enables us to construct any periodic boundary, both symmetric and non-symmetric. The division into symmetric and non-symmetric boundaries was proposed in [10] in which a general classification of the boundaries was also suggested. The boundary which contains only coincident sites is referred to as symmetric if its plane coincides with one of the planes of symmetry of the coincident-site lattice, and quasi-symmetric in the opposite case. The boundaries which contain, in addition to the coincident sites, also non-coincident sites of one or both lattices, are referred to as non-symmetric. Figure 2.1 shows an example of the projection of the (001) planes of two superposed primitive lattices disoriented by the angle of 36.7° around the [001] axis. It can be seen that every fifth node of the lattices 1 and 2 coincides, forming a spatial coincident-site lattice with the inverse density of the coincident sites $\Sigma = 5$ (the period of the coincident-site lattice in the [100] direction is equal to the period of the atomic-crystalline lattice). The graph indicates that the coincident-site lattice of the cubic lattices contains two crystallographic planes forming symmetric grain boundaries. The period of the boundary is equal to the other coincident-site lattice in the appropriate plane. Regardless of the fact that the coincident-site lattice can be constructed for the denumerable set of the angles of rotation and not for a continuous number any boundary with the given misorientation can be calculated with satisfactory accuracy as the boundary with a high but finite period. The theory of the coincident-site lattice has made it possible to explain qualitatively the existence of special grain boundaries with special properties. Evidently, there is a superposition with their minimum value of angle Σ for any misorientation axis. The appropriate periods of the symmetric grain boundaries are small in comparison with the areas of the remaining grain boundaries. As a result of the high density of the general sites in these grain boundaries, the excess energy should be minimum, and the dependence of the energy of the boundary of the misorientation angle shows dips which were also detected in experiments [11, 12].

Mathematical description of the coincident-site lattice in a simple cubic lattice was proposed in [9]. It is based on the construction of the rotational matrix generating the given coincident-site lattice. This matrix can be presented in the form

$$\hat{R}_{\mathrm{CSL}} = \frac{1}{\Sigma}\begin{bmatrix} \rho_{11} & \rho_{12} & \rho_{13} \\ \rho_{21} & \rho_{22} & \rho_{23} \\ \rho_{31} & \rho_{32} & \rho_{33} \end{bmatrix}. \tag{2.8}$$

Here Σ and ρ_{ii} are the coprime numbers (for the cubic crystals is an odd number), satisfying the relationships

$$\rho_{ik}\rho_{kj} = \Sigma^2\delta_{ij};\ \rho_{ik}\rho_{jk} = \Sigma^2\delta_{ij}, \tag{2.9}$$

where summation is carried out with respect to the same indexes. Using this method, Grimmer, Ballmann and Warrington constructed basic sectors of the coincident-site lattice and calculated a table of special misorientations with $\Sigma < 50$ for simple cubic lattices. The matrices of the basis of the coincident-site lattice and special misorientations for the FCC and BCC lattices were published in [10]. Tables of special misorientations for almost all metal lattices are also available.

The next important lattice which is used widely in analysis of the structure of the grain boundaries is the total superposition lattice, introduced for the first time by Ballmann [13]. In superposition of two lattices forming the coincident-site lattice, there is a set of the vectors $\Delta\mathbf{r} = \mathbf{r}_1 - \mathbf{r}_2$. The spatial pattern of the set does not change in mutual displacements (Fig. 2.2).

The total superposition lattice is defined by every possible translations $\Delta\mathbf{r}$, representing the linear combinations of the vectors of two crystal lattices. In displacement on the vectors CSL, DSC and '0'-lattice remain unchanged, although they can be displaced as a whole. The total superposition lattice has a special physical meaning: its basic vectors determine the Burgers vector of the total grain boundary dislocations.

The rearrangement of the structure of the grain boundaries which takes place during the mutual displacements of the boundaries and depends on the orientation of the plane in which the interface is situated is analysed using the GBS lattice (Fig. 2.2), introduced by the authors of [6]. The GBS vectors are the projections of the DSC vector on the planes parallel and perpendicular to the grain boundary plane, and the two planes normal to the boundary should have the highest reticular atomic density. In a general case, the GBS is the lattice with the finest cells and contains as sublattices CSL, DSC and the crystal lattices of the adjacent grains. In a partial case when the grain boundary plane coincides with the close-packed plane of the coincident-site lattice, the GBS coincides with DSC. The GBS vectors form all possible partial and total shears on the special boundary with the given misorientation and the orientation of the plane. It should be mentioned that the size of the elementary cells of DSC and GBS decreases with increasing Σ, and at higher values of Σ it is no longer physically justified for use in the analysis of the structure of the grain boundaries.

The authors of [14] proposed the following criterion for for the applicability of DSC and GBS. These lattices determine the periodicity of the energy relief along the grain boundary which should not be smaller than the period of thermal oscillations of the atoms. Otherwise,

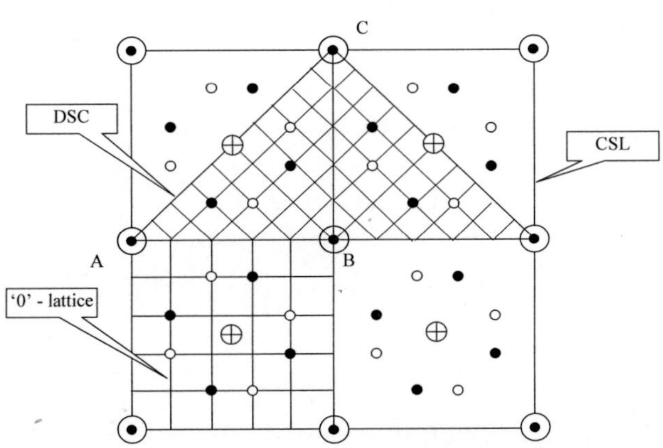

Fig. 2.2. Auxiliary lattices for the boundary $\Sigma5$; AB and AC are the planes with the highest density of the coincident sites; crosses indicate the zero lattice.

the interface would have no special properties and should be regarded as arbitrary.

In conclusion, it should be mentioned that the model of the coincident-site lattice does not generate any efficient criterion for separating the boundaries to special and arbitrary i.e., the problem of the maximum value of Σ at which the special boundary differs in the properties from the general type of the boundary has not yet been solved. The authors of [15] assumed that the special boundaries with $\Sigma \leq 25$ should exist up to the melting point, and for the majority of metals the special grain boundaries with $\Sigma \geq 65$ can be regarded as arbitrary. In this book, following this conclusion, all the boundaries with greater than $\Sigma > 65$ are general type boundaries. However, for analysis of the effect of the ensemble of the grain boundaries on the macroscopic properties of the polycrystals, we propose to use a more generalised criterion of 'speciality' for individual grain boundaries (see later), and for the ensemble of the grain boundaries it is proposed to define three types: LAB (the given ensemble contains mainly low-angle grain boundaries), HAB (the main part of the ensemble of the grain boundaries is represented by high-angle grain boundaries of arbitrary type), and CSL (with the largest fraction of the boundaries of the special type in the general spectrum).

2.1.2. Analytical representation of the basis of the coincident-site lattice for cubic lattices

The practical application of the concept of the coincident-site lattice is restricted by the absence of analytical description of its basis which must be carried out numerically for every specific misorientation using the methods discussed in [17] which can be used only in a computer. Here, we propose a method of analytical representation of the coincident-site lattice which can be used for almost any misorientation of cubic crystals.

We examine a primitive cubic lattice with a basis ($\mathbf{e}_1$, $\mathbf{e}_2$, $\mathbf{e}_3$). The set of the vectors (sites) of the coincident-site lattice is given by solutions $\mathbf{x}$ of the equation

$$(\hat{I} - \hat{A}^{-1})\mathbf{x} = \mathbf{b}, \tag{2.10}$$

whose basis ($\mathbf{e}_1$, $\mathbf{e}_2$, $\mathbf{e}_3$) contains whole components, i.e., they are lattice vectors [17]. In (2.8) $\hat{I}$ is the unit matrix, and $\mathbf{b}$ is the set of all the translational vectors of the initial lattice. The transformation of $\hat{A}$ takes place by right rotation around the axis **c** through the angle θ. The rank of the matrix $\hat{I}-\hat{A}^{-1}$ is less than three and this creates certain problems when examining equation (2.10) as the system of linear equations. For this reason, in [17] $\hat{A}^{-1}$ is represented by the transformation $\hat{U}\hat{R}^{-1}$, where $\hat{R}$ is the rotational transformation, and $\hat{U}$ is the unimodular transformation, determined by the following condition: rank $(\hat{I}-\hat{U}\hat{R}^{-1})=3$.

The formalism in which (2.10) is regarded as some vector equation is more adequate and makes it possible to avoid artificial introduction of the transformation $\hat{U}$. It is well known [1] that

$$\hat{R}\mathbf{x}=\mathbf{c}_0(\mathbf{x}\mathbf{c}_0)+\mathbf{c}_0\times(\mathbf{x}\times\mathbf{c}_0)\cos\theta+\mathbf{c}_0\times\mathbf{x}\sin\vartheta, \tag{2.11}$$

where

$$\mathbf{c}_0=\frac{\mathbf{c}}{|\mathbf{c}|}\,. \tag{2.12}$$

The equation (2.10) using (2.11) has the form

$$\mathbf{c}_0\times(\mathbf{x}\times\mathbf{c}_0)(1-\cos\theta)+\mathbf{x}\times\mathbf{c}_0\sin\theta=b \tag{2.13}$$

Since the left-hand part of the equation (2.13) contains only the vectors normal to the axis **c**, the set of the vectors **b** narrows down to the set of the translational vectors of the initial lattice, determined by the condition

$$\mathbf{cb}=0 \tag{2.14}$$

Using the well-known procedures of vector algebra, from (2.13) we obtain

$$\mathbf{c}_0\times(\mathbf{x}\times\mathbf{c}_0)=\mathbf{b}/2+\mathrm{ctg}(\theta/2)(\mathbf{c}_0\times\mathbf{b})/2.$$

Since any vector **x** is identically expanded into a sum

$$\mathbf{x}=\mathbf{c}_0(\mathbf{x}\mathbf{c}_0)+\mathbf{c}_0\times(\mathbf{x}\times\mathbf{c}_0).$$

and the projection of the lattice vector **x** to the direction of the **c** axis is divisible by the planar spacings in this direction, all the solutions of the equation (2.10) have the form

$$\mathbf{x} = n\frac{\mathbf{c}}{|\mathbf{c}|^2} + \frac{1}{2}\mathbf{b} + \mathrm{ctg}\frac{\theta}{2}\frac{\mathbf{c}\times\mathbf{b}}{2|\mathbf{c}|}, \tag{2.15}$$

where n is an arbitrary integer. The problem of recalculating the vectors of the coincident-site lattice has been reduced to separating the lattice vectors from the set of the vectors (2.15). In this case, the equation (2.15) should not contain irrational quantities, i.e., θ and $|\mathbf{c}|$ should be linked by, for example, the relationship

$$\mathrm{tg}\frac{\theta}{2} = |\mathbf{c}|\frac{p}{q},$$

p and q are coprime numbers.

We obtain sufficient conditions for the vector (2.15) to be an integer number in the basis ($\mathbf{e}_1$, $\mathbf{e}_2$, $\mathbf{e}_3$). Because of the symmetry of the crystal lattice there are several equivalent descriptions for every misorientation. It can be shown that up to the inverse density of the concident sites of Σ = 221 the group of these equivalent notations must contain at least one rotation around the axis of the type $\langle uv1\rangle$. Undoubtedly, the value Σ = 221 is sufficient for any practical purpose because high values can hardly have any physical meaning. Therefore, we restrict our considerations to the class of the axes $\mathbf{c}$= $[uv1]$. In this case, any vector $\mathbf{b}$, satisfying the condition (2.14), can be expanded on the flat basis given by the vectors

$$\mathbf{b}_1 = \mathbf{e}_1 - u\mathbf{e}_3; \quad \mathbf{b}_2 = \mathbf{e}_2 - v\mathbf{e}_3.$$

Substituting the values $\mathbf{b}$ = $K\mathbf{b}_1$+$L\mathbf{b}_2$ (where K and L are any integers), in equation (2.15) we obtain

$$x = \frac{1}{2}\left\{K - L\frac{q}{p} + \frac{u}{c^2}\left[2n + \frac{q}{p}(Lu - Kv)\right]\right\}\mathbf{e}_1 + \frac{1}{2}\left\{L + K\frac{q}{p} + \frac{v}{c^2}\left[2n + \frac{q}{p}(Lu - Kv)\right]\right\}\mathbf{e}_2 + \frac{1}{2}\left\{-Ku - Lv + \frac{1}{c^2}\left[2n + \frac{q}{p}(Lu - Kv)\right]\right\}\mathbf{e}_3.$$

Analysing the vector $\mathbf{c} \times \mathbf{x}$, it can be shown that K and L should be divisible by p, i.e., $K = kp$ and $L = lp$. Consequently

$$\mathbf{x} = \frac{1}{2}\left[(kp - lq + \frac{m}{c^2}u)\mathbf{e}_1 + (lp + kq + \frac{m}{c^2}v)\mathbf{e}_2 + (kpu - lqv + \frac{m}{c^2})\mathbf{e}_3 \right]. \tag{2.16}$$

Here $m = 2n + q\,(lu - kv)$. The arbitrary number n can always be selected to ensure that m is divisible by c^2. It is now necessary to write integers k, l and m/c^2 in such a manner as to remove the common multiplier 1/2 in equation (2.16). This selection is determined by the parity of the quantities p, q, u, v. Selecting subsequently from (2.16) three non-complanar vectors (such that the elementary cell constructed using these vectors has the minimum volume) we obtain the following variants of the basis of the coincident-site lattice in the matrix form (the vector of the basis – the matrix column):

1. At odd p, q, u and v

$$\hat{c} = \begin{pmatrix} u & (p+u)/2 & (u-q)/2 \\ v & (q+v)/2 & (v+p)/2 \\ 1 & (1-pu)/2 & (1-pv)/2 \end{pmatrix}; \tag{2.17a}$$

2. At odd p, q and even u, v

$$\hat{c} = \begin{pmatrix} u & p & (p-q)/2 \\ v & q & (p+q)/2 \\ 1 & -pu & -p(u+v)/2 \end{pmatrix}; \tag{2.17b}$$

3. At odd p, u and even q, v

$$\hat{c} = \begin{pmatrix} u & (p+u)/2 & -q \\ v & (q+v)/2 & p \\ 1 & (1-pu)/2 & -pu \end{pmatrix}; \tag{2.17c}$$

4. At odd p, v and even q, u

$$\hat{c} = \begin{pmatrix} u & p & (u-q)/2 \\ v & q & (v+p)/2 \\ 1 & -pu & (1-pu)/2 \end{pmatrix}; \tag{2.17d}$$

5. At all other q, p, u and v

$$\widehat{c} = \begin{pmatrix} u & p & -q \\ v & q & p \\ 1 & -pu & -pu \end{pmatrix}; \tag{2.17e}$$

The volume of the elementary cell of the lattice of collected and sites is equal to the determinant of the matrix and according to the definition is equal to Σ. Consequently, in case 1

$$\Sigma = (p^2c^2 + q^2)/4; \tag{2.18a}$$

in the cases 2, 3 and 4

$$\Sigma = (p^2c^2 + q^2)/2; \tag{2.18b}$$

and in the case 5

$$\Sigma = p^2c^2 + q^2. \tag{2.18c}$$

The matrix $\widehat{c}$ defines one of the infinite number of equivalent variants of the selection of the basis of the coincident-site lattice. The equivalent description is obtained from this description by constructing different linear combinations of lines and columns of the matrix $\widehat{c}$. The most suitable description can be selected only for the specific values of the misorientation parameters.

The matrices of the basis of the coincident-site lattice for the centred lattices are determined from the matrix $\widehat{c}$ for the primitive cubic lattice using the simple rules presented in [1].

We examine a specific example of construction of the basis of the coincident-site lattice. Let the misorientation of 60.77° $\langle 332 \rangle$ be given. The equivalent description with the axis of the type $\langle uv1 \rangle$ can be obtained in the quaternionic form. If the rotation is described by the quaternion [18]

$$(A,\ B,\ C,\ D), \tag{2.19a}$$

where A, B, C and D are coprime integers, the equivalent description of this quaternion in the cubic lattice is

$$(A+B, A-B, C+D, C-D); \tag{2.19b}$$

$$(A+C, A-C, B+D, B-D); \tag{2.19c}$$

$$(A+D, A-D, B+C, B-C); \tag{2.19d}$$

$$(A+B+C+D, A+B-C-D, A-B-C+D, A-B+C-D); \tag{2.19e}$$

The 'axis–angle' description is obtained from the quaternion form as follows:

$$\mathrm{tg}^2\frac{\theta}{2}=\frac{x^2+y^2+z^2}{T^2},\ \mathbf{c}\|\langle xyz\rangle,$$

where T, x, y z are the elements of the quaternion (2.19).

In this case, the misorientation is described by the quaternion (8, 3, 3, 2). From (2.19c) we obtain the equivalent description: (11, 5, 5, 1), i.e. $\mathbf{c}$ = [551), tg $\theta/2=\sqrt{51}/11$. Consequently p = 1; q = 11; u = v = 5. From (2.17a) and (2.18a) we obtain

$$\hat{c}=\begin{pmatrix}5 & 3 & -3\\ 5 & 8 & 3\\ 1 & -2 & -2\end{pmatrix},\ \Sigma=43. \tag{2.20}$$

The elementary cell is represented by the cell of the coincident-site lattice constructed on the shortest non-complanar vectors. The column vectors of the matrix (2.20) are denoted by $\mathbf{e}_1^0, \mathbf{e}_2^0, \mathbf{e}_3^0$. Its equivalent matrix with the columns is $3\mathbf{e}_1^0-2\mathbf{e}_2^0+2\mathbf{e}_3^0$, $-\mathbf{e}_1^0+\mathbf{e}_2^0-\mathbf{e}_3^0, \mathbf{e}_3^0$:

$$\hat{c}=\begin{pmatrix}3 & 1 & -3\\ 5 & 0 & 3\\ 3 & 1 & -2\end{pmatrix}, \tag{2.21}$$

and defines the basis of the coincident-site lattice with the shortest vectors.

The following matrix is presented in [17] for the misorientation Σ= 43c:

$$\hat{c}=\begin{pmatrix}3 & 1 & -5\\ 3 & 0 & 3\\ 2 & -1 & 3\end{pmatrix},$$

which is fully equivalent to the matrix (2.21) and is obtained from this matrix by the cyclic rearrangement of the columns and the rearrangement of the lines with the appropriate change of the sign of the coefficients.

2.2. Special grain boundaries in the monoclinic lattice

As already mentioned, types of special boundaries have been calculated for almost all types of lattice. However, because of the mathematical difficulties, a large class of polycrystalline materials with the monoclinic lattice does not have such classification. Therefore, analysis of the ensemble of the grain boundaries in similar materials was associated with difficulties. Since the general method of calculating the coincident-site lattice for the lattice of arbitrary type has not been developed, numerical modelling was used for producing tables of special boundaries in these materials [19]. Previously, it was assumed that the materials with a low symmetry crystal lattice cannot have special misorientations. However, as shown in [20, 21], special type boundaries can exist in these lattices. If the parameters of the crystal lattice correspond to some specific conditions, then the number of the coincident-site lattices and, consequently, of the 'special' boundaries may prove to be quite large.

Zirconia, ZrO_2, exists in three crystalline phases, monoclinic, tetragonal and cubic. The monoclinic lattice is characterised by the lowest temperature and highest stability of the three phases. If the calculations of the types of 'special' boundaries and for the tetragonal and cubic phases are not very difficult, no data of this type are available for the monoclinic phase. The experimental data, obtained by electron microscopy, indicate the existence of straight grain boundaries whose external morphology indicates that they may have special properties. The crystal geometry of the coincident-site lattices in the monoclinic lattice of zirconia will now be discussed in more detail.

The monoclinic lattice of ZrO_2 belongs to the crystallographic group $P2_1/c$ and has the following parameters of the elementary cell: a = 5.1490 Å; b = 5.2133 Å; c = 5.3161 Å; the angle β between the axes a and c is 99.228°. The elementary lattice of ZrO_2 is shown in Fig. 2.3.

Prior to investigating possible types of 'special' grain boundaries, attention will be given to the geometry of the elementary cell. In the monoclinic lattice with any parameters, two symmetric boundaries with the planes (001) and (100) should have a close-packed atomic structure. This misorientation is described by rotation through 180° around the [100] axis or the normal to the plane (001) for the first symmetric grain boundary and rotation through 180° around the axis [001] or the

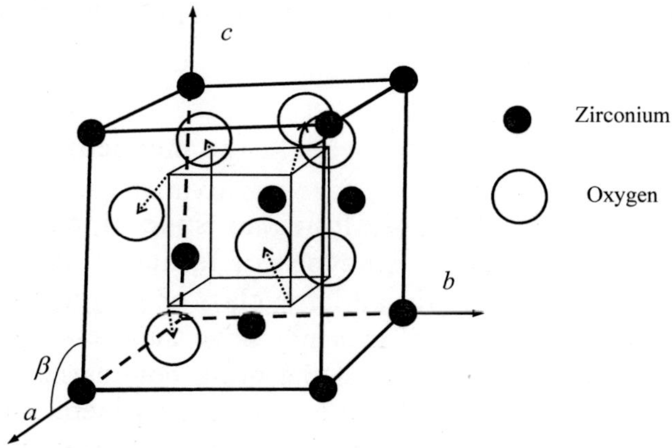

Fig. 2.3. The elementary cell of the monoclinic lattice of ZrO_2.

normal to the (100 plane) for the other symmetric grain boundary. It is clear that the three-dimensional coincident-site lattice can exist if the parameter $(c/a)\cos\beta$ is an integer. Thus, we do not calculate accurately the special boundaries but calculate accurately the boundaries close to them, using the approximation $a = b = c$ and $\cos\beta = -1/6$ (i.e., $\beta = 99.594°$) for the given lattice. We consider a prismatic cell in the monoclinic lattice ([101] [$\bar{1}$01], [010]), which is orthogonal, with the ratio of the squares of the length 5:7:3. In this case, we can use the algorithm of calculating the coincident-site lattice for orthorhombic lattices with the rational ratio of the squares of the lattice spacing values [22]. However, in this case the lattice is also base-centred, i.e., there is an additional atom inside the calculation cell. In practice, this means that the value Σ of the inverse density of the coincident sites can be calculated with the accuracy to the coefficient k (according to the conclusions made in [22]). This study shows that Σ_P (the value Σ of for the primitive orthorhombic lattice) is connected with Σ_B (Σ for the base-centred orthorhombic lattice) by the relation $\Sigma_P = k\Sigma_B$, where k has the values 1, 1/2 or 2.

The orthonormal basis $\varepsilon = a\,[\varepsilon_1, \varepsilon_2, \varepsilon_3]$, linked with the basis of the monoclinic lattice $e = [a, b, c, \beta]$ by the relationship

$$\varepsilon = \hat{S}\cdot e, \tag{2.22}$$

will now be examined. Here is the transformation matrix has the form

$$\widehat{S}=\begin{pmatrix}1 & 0 & (c/a)\cos\beta\\ 0 & b/a & 0\\ 0 & 0 & (c/a)\sin\beta\end{pmatrix}. \qquad (2.23)$$

Taking into account the given approximation ($a = b = c$ and $\cos\beta = -1/6$), equation (2.23) acquires the following form:

$$\widehat{S}=\begin{pmatrix}1 & 0 & -1/6\\ 0 & 1 & 0\\ 0 & 0 & \sqrt{35}/6\end{pmatrix}. \qquad (2.24)$$

It is assumed that using the algorithm described previously for the case of the orthorhombic lattice, we obtain the misorientation of the coincident-site lattice in the form of the 'angle–axis" pair (θ, $c = [uvw]$, $|c| = 1$). Consequently, the misorientation axes can be expressed in the orthonormal basis as $c_0 = [u_0v_0w_0] = \widehat{S}c$. Using the angle θ and the given axis, we construct the misorientation matrix [1]:

$$\widehat{R}_0\begin{pmatrix}\cos\theta+ & (1-\cos\theta)u_0v_0- & (1-\cos\theta)u_0w_0\\ +(1-\cos\theta)u_0^2 & -(\sin\theta)w_0 & -(\sin\theta)\upsilon_0\\ (1-\cos\theta)u_0v_0+ & \cos\theta+ & (1-\cos\theta)v_0w_0-\\ +(\sin\theta)w_0 & +(1-\cos\theta)v_0^2 & -(\sin\theta)u_0\\ (1-\cos\theta)u_0w_0- & (1-\cos\theta)v_0w_0+ & \cos\theta+\\ -(\sin\theta)v_0 & +(\sin\theta)u_0 & +(1-\cos\theta)w_0^2\end{pmatrix}. \qquad (2.25)$$

Consequently, the misorientation matrix in the monoclinic lattice can be determined by means of the transformation

$$\widehat{R}=\widehat{S}\widehat{R}_0\widehat{S}^{-1}. \qquad (2.26)$$

Using this equation, we can determine Σ as the so-called Σ-theorem which is valid for lattices of any symmetry [23]. The theorem shows that Σ is the positive smallest co-multiplier which is such that the matrices ΣR and ΣR^{-1} containing only rational numbers. Using the previously described algorithm, the misorientation of the coincident-site lattice was calculated to $\Sigma = 33$, inclusive. Table 2.1 shows the data on special misorientations only up to $\Sigma = 25$. The misorientations with the same value of Σ, shown in Table 2.1, are not equivalent. The columns 1 and 2 contain equivalent misorientations having however different representation. This also refers to the columns 3 and 4. At the same

Table 2.1. CSL ($\Sigma \le 25$) for the monoclinic ZrO_2 lattice

Σ	1				2				3				4			
	θ, deg	*u*	*v*	*w*	θ, deg	*u*	*v*	*w*	θ, deg	*u*	*v*	*w*	θ, deg	*u*	*v*	*w*
1	180.0	1	0	1					180.0	1	0	–1				
3	180.0	1	0	0	180.0	1	0	6	180.0	0	0	1	180.0	6	0	1
3	80.4	0	1	0	99.6	0	1	0								
6	90.0	1	0	6	180.0	1	1	0								
6	114.6	6	5	6	125.7	6	7	–6								
8	75.5	1	0	–1	180.0	3	5	3								
8	104.5	1	0	–1	180.0	1	1	1								
9	19.2	0	1	0	160.8	0	1	0								
9	180.0	3	0	4	180.0	3	0	–2	180.0	4	0	3	180.0	2	0	–3
10	66.4	1	0	1	180.0	3	7	–3								
10	87.1	2	5	2	104.5	2	7	–2								
10	113.6	1	0	1	180.0	1	1	–1								
10	143.1	1	0	6	180.0	3	1	0	143.1	6	0	1	180.0	0	1	3
11	53.8	0	1	0	126.2	0	1	0								
11	180.0	2	0	5	180.0	4	0	–1	180.0	5	0	2	180.0	1	0	–4
13	43.0	0	1	0	137.0	0	1	0								
13	101.1	4	5	4	115.0	4	7	–4								
13	112.6	1	0	6	180.0	3	2	0	112.6	6	0	1	180.0	0	2	3
13	180.0	2	0	1	180.0	4	0	11	180.0	1	0	2	180.0	11	0	–4
15	53.1	6	0	1	180.0	0	2	1								
15	93.8	3	5	3	109.5	3	7	–3								
15	126.9	1	0	6	180.0	2	1	0	126.9	6	0	1	180.0	0	1	2
15	140.1	12	5	12	146.4	12	7	–12								
16	29.0	1	0	–1	180.0	1	5	1								
16	66.0	6	7	–6	141.4	2	5	2								
16	128.7	4	0	–11	180.0	2	1	1	128.7	11	0	–4	180.0	1	1	2

Table 2.1 (Cont.)

Σ	1				2				3				4			
	θ, deg	*u*	*v*	*w*	θ, deg	*u*	*v*	*w*	θ, deg	*u*	*v*	*w*	θ, deg	*u*	*v*	*w*
16	151.0	1	0	–1	180.0	3	1	3								
17	31.5	0	1	0	148.5	0	1	0								
17	65.7	1	0	–1	180.0	1	2	1								
17	114.3	1	0	–1	180.0	6	5	6								
17	180.0	8	0	13	180.0	2	0	–1	180.0	13	0	8	180.0	1	0	–2
18	91.6	6	1	0	166.5	6	35	36	91.6	0	1	6	166.5	36	35	6
18	114.6	6	5	–4	125.7	6	7	8	114.6	4	5	–6	125.7	8	7	6
19	54.6	1	0	–1	180.0	2	5	2								
19	74.7	1	0	1	180.0	1	2	–1								
19	105.3	1	0	1	180.0	6	7	–6								
19	125.4	1	0	–1	180.0	3	2	3								
20	54.9	6	5	6	151.0	2	7	–2								
20	97.2	2	1	0	140.8	6	35	36	97.2	0	1	2	140.8	36	35	6
20	118.4	8	5	–2	128.7	4	7	10	118.4	2	5	–8	128.7	10	7	4
20	134.4	8	0	13	180.0	2	1	–1	134.4	13	0	8	189.0	1	1	–2
22	24.6	1	0	1	180.0	1	7	–1								
22	155.4	1	0	1	180.0	3	1	–3								
23	42.3	1	0	–1	180.0	3	10	3								
23	137.7	1	0	–1	180.0	2	1	2								
24	75.5	3	0	4	180.0	3	5	–2	75.5	4	0	3	180.0	2	5	–3
24	105.7	6	1	6	168.3	6	7	–6								
24	124.2	2	1	0	133.4	2	7	12	124.2	0	1	2	133,4	12	7	2
25	47.2	1	0	1	180.0	2	7	–2								
25	73.7	1	0	6	180.0	3	4	0	73.7	6	0	1	180.0	2	2	1
25	92.3	11	0	–4	180.0	1	2	2	73.7	6	0	1	180.0	0	4	3
25	95.7	12	7	–12	134.4	8	7	–8								
25	125.5	8	5	8	134.4	8	7	–8								

time, the pairs of the misorientations 1, 2 are not equivalent to the pairs 3, 4; they represent different coincident-site lattices. Two equivalent misorientations of the coincident-site lattices are connected together by rotation through 180° around the [010] angle. This direction is the only axis of symmetry of the second order in the monoclinic lattice. All other descriptions can be obtained by rearrangement of the indexes of the axes in accordance with the rule $[uvw] \Rightarrow [\bar{u}v\bar{w}] \Rightarrow [\bar{u}\,\bar{v}\,\bar{w}]$. The first line of Table 2.1 shows the misorientation with Σ = 1. Usually, in the coincident-site lattice theory Σ1 is used to denote a single crystal, i.e., rotation through 0° (360°) around any axis or rotation around the axis of symmetry through the appropriate angle (in the case of the monoclinic lattice it is the rotation through 180° around the [010] axis). In the present case, we consider the misorientations Σ1 180° [101] and Σ1 180°$[\bar{1}0\bar{1}]$ because of the following reasons. In the approximation in which $a = 2b$, the misorientation Σ1 would lead to the complete coincidence of the lattices; however, in the actual case, $a \neq b$ and the [101] and $[\bar{1}0\bar{1}]$ axes are not the second order axes of symmetry. Therefore, there is a low-angle interface between the crystals rotated through 180° around these axes. For the monoclinic lattice, there are also two symmetric boundaries (101) and $(\bar{1}0\bar{1})$. In joining of two adjacent grains, rotated through 180° around the [101] or $[\bar{1}0\bar{1}]$ axis we obtain ideal coincidence so that the appropriate boundaries are similar to the coherent grain boundaries. The atomic structure of these boundaries is almost identical with the ideal single crystal. Consequently, these grain boundaries should be of the low-energy type and detected in the experiments. This question will be discussed in a separate section.

2.3. Description of the grain boundary misorientation distribution (GBMD)

In most cases, in the graphical description of the GBMD, all the misorientations are represented by the minimum angles and the directions of the axes of misorientation, situated in the standard stereographic triangle (SST) and this is followed by the construction of the distribution of the angles and the axes of misorientation of the grain boundaries. In some cases, the standard triangle is divided into zones (Fig. 2.4) and the distribution of the misorientations with respect to the axes is presented in the form of the function of the density of probability of the axes lying in one of the zones of the standard stereographic angle. In this representation, the distributions of misorientation with respect to

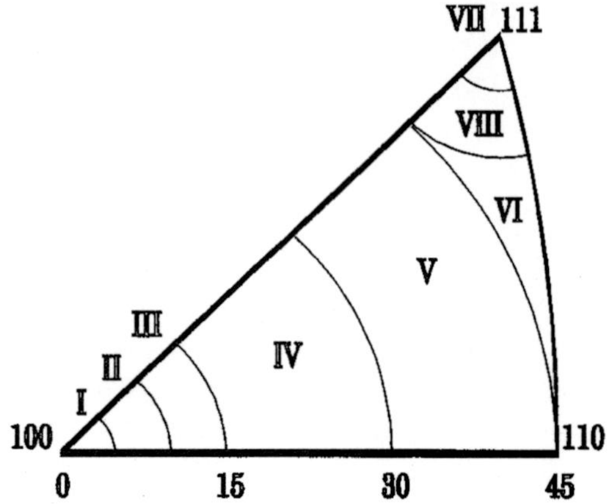

Fig. 2.4. Standard stereographic triangle divided into zones.

the angles and axes are integral characteristics, i.e., the characteristics averaged out in the first case with respect to all misorientation axes and in the second case with respect to all angles.

As a result of using the model of the coincident-site lattice for describing the structure of the grain boundaries, it was necessary to define special or close to special misorientations in the spectrum of misorientation of the grain boundaries. In this case, another distribution was introduced – the distribution of the grain boundaries with respect to the inverse density of the coincident sites (Σ). The boundary is classified in a specific class with respect to Σ if the deviation of the misorientation of the boundary satisfies the Brandon criterion [24]. Formally, the Brandon criterion was introduced on the basis of the analogy with the description of low-angle grain boundaries as some flat distribution of the dislocations. In this case, the total Burgers vector of such a wall is described by the Frank equation [25]:

$$\mathbf{B} = 2\sin\frac{\theta}{2}[\boldsymbol{\rho}\times\mathbf{u}], \tag{2.27}$$

here **B** is the total Burgers vector of the dislocations, intersected by the arbitrary vector **ρ**, situated in the grain boundary plane. For the low-angle grain boundaries, the dislocation description has not only geometrical but also physical meaning. This model has been used to calculate the dependence $\gamma(\theta)$ of specific energy on the misorientation angle. Reed and Shockley calculated for the first time the stress fields and specific energy of the grain boundaries for the cases of uniform and nonuniform distribution of the dislocations. It is assumed that the dislocation model describes efficiently the low-angle boundaries to misorientation angles equal to 15°. At high angles, the dislocations are so close to each other that the cores of the dislocations start to merge and the continual theory of dislocations is no longer valid in this case. For the high-angle arbitrary grain boundaries the initial structure is represented by the special boundary described by the coincident-site lattice with a low value of Σ. Consequently, the high-angle boundaries are presented in the form of a combination of a special boundary with a dislocation network superposed on the boundary, with the network rotating additionally through the angle $\Delta 0 = 0 - 0_0$. It is evident that, as in the case of the low-angle grain boundaries, the maximum additional misorientation cannot exceed 15°. According to Brandon [24], the high-angle grain boundary is regarded as special if its the deviation from the nearest special grain boundary satisfies the conditions

$$\Delta\theta \le \frac{15^{\circ}}{\Sigma^{1/2}}. \tag{2.28}$$

There are also other criteria for evaluating the 'special nature' of the high-angle boundaries. However, this criterion is generally recognised and will be used in our future considerations. In addition, in the case of high values of Σ it is no longer rational to talk about the 'special' nature of the grain boundaries because the period of the coincident-site lattice becomes too large. All the boundaries with $\Sigma > 65$ belong in the group of arbitrary boundaries [16]. The low-angle boundaries, denoted by Σ1, are considered separately.

In [26], the authors presented a concept of representation of the misorientation of the grain boundaries by three Euler angles (φ_1, Φ, φ_2) by analogy with the description of the orientation distribution function of the grains. The definition region of this function for cubic crystals (the point symmetry group O_h) is an asymmetric region, whose boundaries are described by the equations

$$\Phi = 0; \quad \varphi_1 = \varphi_2; \quad \varphi_1 = \pi/2 - \varphi_2; \quad \cos\Phi = \frac{\sin\varphi_1 \sin\varphi_2}{1 + \cos\varphi_1\varphi_2}. \tag{2.29}$$

In this representation, the points in the definition region, situated on the boundary surfaces, correspond to the special grain boundaries. In [26] Euler angles (φ_1, Φ, φ_2), were determined for all special boundaries for the cubic lattice with $\Sigma < 50$. This description of the spectrum of misorientation of the grain boundaries has not been used widely. It would appear that it provides more information but it is more difficult to understand.

If the misorientation is regarded as the Rodriguez vector [27], fixed at the origin of the coordinates, the misorientation distribution has the form of the distribution of vectors in a trihedral angle with the tip in the starting point restricted by two surfaces from the opposite side. The appropriate region of the variation of the misorientation of vectors for the cubic lattice is shown in Fig. 2.5 [28] which shows the limiting values of the misorientation angle for several directions of the axes. The figure indicates that by extending the faces of the trihedral angle a triangle is 'cutout' from the more distant spherical surface and this triangle is one of the 48 equivalent triangles into which the sphere is divided by cubic symmetry elements. As in the representation of the misorientation spectrum of the grain boundaries by the Euler angles, all the special misorientations (with the exception of $\Sigma = 39b : 50.31°$ [321]) have the Rodriguez vectors, situated on the faces and edges of the trihedral angle.

It should be mentioned that the previously described representations of the misorientation spectrum of the grain boundaries are in fact equivalent and their selection is dictated only by convenience in solving the specific task. The representation of the misorientation of other types of lattice (for example, monoclinic) is not suitable for perception and analysis. In this book, we used the first of the previously mentioned representations, i.e., the spectrum of the misorientation of the grain boundaries is presented in the form of three distributions: with respect to the angles and axes, and also with respect to the inverse density of the coincident angles Σ.

In order to analyse efficiently the spectra of the grain boundaries, it is desirable to have a detailed picture for the accurately defined basic

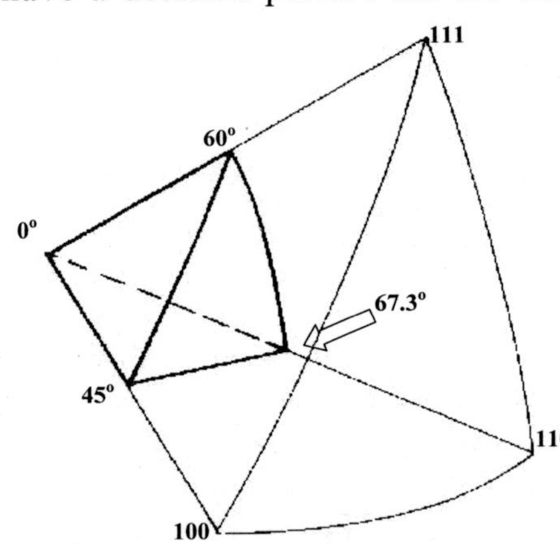

Fig. 2.5. The region of variation of the misorientation of vectors of the grain boundaries [28].

model of the polycrystals, for example, the misorientation spectrum of the grain boundaries in a polycrystal with randomly oriented grains. In the majority of investigations [29–32], the chaotically disoriented ensemble was represented by a set of chaotically disorientated grains not bonded in a polycrystal, i.e., in fact, the spectrum of the chaotically disorientated bicrystals [33, 34]. Mackenzie derived an analytical expression for the distribution function of misorientation angles [29] and axes [30] for chaotically disoriented grains. The principle of determination of the spectrum of the misorientation of the grain boundaries of the chaotically disoriented ensemble may be described as follows. Let us consider a sphere with a radius θ = 180° (where θ is the misorientation angle). The misorientation axis is the vector with the spherical coordinates v and ρ:

$$C = (\sin\varphi \sin \text{v}, \cos\varphi \sin \text{v}, \cos \text{v}), \tag{2.30}$$

where $0 \le v \le \pi$ and $0 \le \rho \le 2\pi$. In the case of the chaotic distribution, the radial density of the points, corresponding to the ends of the vectors θ = (θ, ν, φ) is determined by the formula [31]

$$\rho(\theta, v, \varphi) d\theta\, dv\, d\varphi = \frac{1}{4\pi}(1 - \cos\theta) d\theta \, \sin v\, dv\, d\varphi. \tag{2.31}$$

Consequently, the probability of the misorientation angle lying in the interval (θ, θ+Δθ) is:

$$P(\theta) = \frac{1}{n}\frac{\Delta n}{\Delta\theta} = \frac{1 - \cos\theta}{4\pi^2} \int_0^{\pi} \sin v\, dv \int_0^{2\pi} d\varphi = \frac{2}{\pi}\sin^2(\theta/2). \tag{2.32}$$

The introduction of the spectrum of the minimum misorientations, obtained taking into account the rotational symmetry of the lattice, leads to a change of the type of dependence $P(\theta)$. For a cubic crystal $0 \le \theta \le 62.7°$ (Fig. 2.6). In addition to this, the lattice symmetry results in the redistribution of the outputs of the axes to the spherical surface (Table 2.2). In a polycrystal having the form of a bonded aggregate of grains, the topologically essential conditions of the content of the grains impose additional restrictions on the misorientation spectrum of the grain boundaries [35]. However, for a chaotically oriented polycrystals these restrictions do not change the type of distribution of the grain boundaries with respect to misorientation.

Table 2.2. Distribution of the misorientation axes in a standard stereographic angle [29, 33, 34]

Zone	I	II	III	IV	V	VI	VII	VIII
Mackenzie	0.70	2.10	3.60	22.70	53.40	9.10	2.10	6.00
Garbach and Grabskii	0.75	2.27	3.78	23.32	55.38	6.47	2.14	5.89

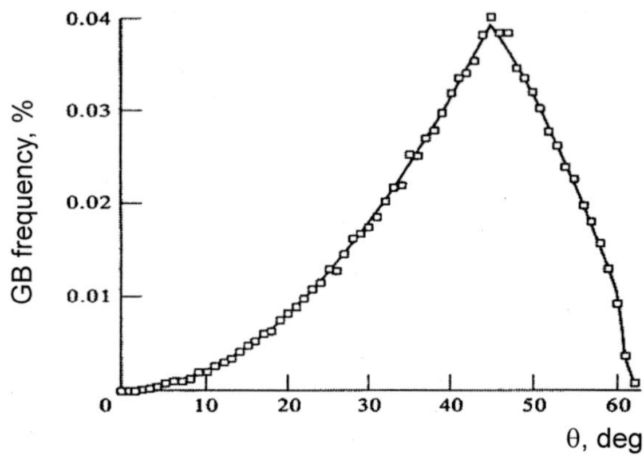

Fig. 2.6. Distribution of the misorientation angles in modelling spectra of the chaotically oriented cubic crystals [29, 33, 34]

2.4. Computer model of a polycrystal. A calculation algorithm

One of the simple and convenient models of the grain structure of the polycrystal is the volume model consisting of cubic polyhedra. The volume model of the polycrystal, developed in [36], and used in modelling of the spectrum of the misorientation of the grain boundaries [35, 37] will be described. The computer model of the polycrystal has the following form: firstly, the grain structure – aggregate of ideal polyhedra, filling the space without pores. Secondly, the crystal basis is defined for every grain and determines the orientation of the grain in relation to the laboratory coordinate system. Of the polyhedra filling the space of the polycrystal, the cubic polyhedron and the Williams tetrahedron are closest to the ideal polyhedron [36]. In this book, the polycrystal grains are represented by the cubic octahedron shown in Fig. 2.7. The polycrystal itself has the form of a cube with different numbers of the grains along the axes *x, y* and *z*. The aggregate contains 3375 grains and 21 671 boundaries between them. The basis of the crystal lattice is specified for each grain in accordance with the texture of the material. The algorithm of definition of the basis in the polycrystal with the chaotically oriented grains has the following

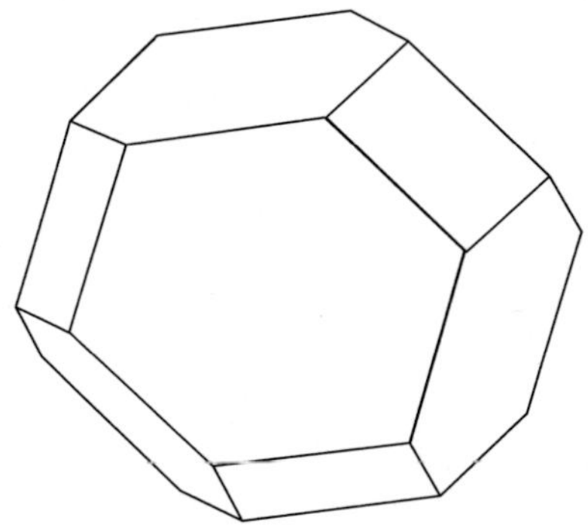

Fig. 2.7. Cubic octahedron used as the crystal forming the aggregate.

form. The basis in each grain is defined by the trio of the vectors $\mathbf{e}_1$, $\mathbf{e}_2$, $\mathbf{e}_3$, which determines the orientation of the grains in the laboratory coordinate system. For the chaotic definition of the orientation, the basis in each grain is calculated using the generator of random numbers. The pair of the random numbers (β_1, β_2), uniformly distributed in the interval [0, 1], is used to construct the unit vector $\mathbf{e}_1$, whose direction is uniformly distributed in the solid angle 4π [38]:

$$e_{1x} = 2\beta_1 - 1;\ \ e_{1y} = 2\sqrt{\beta_1 - \beta_1^2}\sin(2\pi\beta_2);\\ e_{1z} = 2\sqrt{\beta_1 - \beta_1^2}\cos(2\pi\beta_2). \tag{2.33}$$

Subsequently, in the plane normal to the vector $\mathbf{e}_1$ we construct the vector $\mathbf{e}_2$ which forms the angle $\Phi = 2\pi\beta_3$, with the vector (for example, $[-e_{1y}, e_{1x}, 0]$) situated in the same plane. Here β_3 is the third random number ($\beta_3 \in [0, 1]$). If the vector $\mathbf{e}_2$ is expanded with respect to two mutually perpendicular unit vectors:

$$\xi = \frac{\left[-e_{1y}, e_{1x}, 0\right]}{\sqrt{e_{1x}^2 + e_{1y}^2}} \quad \text{and} \quad \left[\mathbf{e}_1 \times \xi\right], \tag{2.34}$$

then the vector can be expressed in the following form:

$$\mathbf{e}_2 = \xi\cos\Phi + \left[\mathbf{e}_1 \times \xi\right]\sin\Phi. \tag{2.35}$$

The last vector, $\mathbf{e}_3$, is additionally constructed in the orthogonal direction in relation to the first two vectors so that all three vectors form a regular trio. The procedure used for defining the basis in the case of the simulated axial texture is the same as that discussed previously but here we select only the trio for which the deviation of the given crystallographic direction [*hkl*] from the texture axes in the laboratory system does not exceed the given angle φ.

The model was tested in the calculations of the spectra of misorientations in a textureless polycrystal. At that time, it had not been shown or confirmed by analytical or direct measurements that the experimentally determined distribution functions of the misorientations of the grain boundary, measured in thin foils, are representative of the entire volume. For the numerical verification of this hypothesis we calculated the misorientation distribution of the grain boundaries in the volume polycrystal with a chaotic texture and with a modelling axial texture of a different type, and also the misorientation distributions of the grains in arbitrary cross-section of the given modelling polycrystal. Three types of axial texture were selected: $\langle 100 \rangle$, $\langle 110 \rangle$, and $\langle 111 \rangle$. The misorientation distribution in the flat cross-section was simulated using four types of cross-section: the cross-section normal to the axis of the texture; the cross-section containing the given axis of the texture; two cross-sections forming the angles of 45° and 54.7° with the texture axis. (If we use the index system of notation of the axes and planes for the laboratory coordinate system, these sections will be denoted by $(100)_L$, $(010)_L$, $(110)_L$ and $(111)_L$).

The orientations of the individual grains were used to calculate the misorientation of all the grains with the common boundaries. From the group of the crystallographically equivalent descriptions of each misorientation we selected the description with a minimum angle and the axis situated in the standard stereographic angle. The boundary was related to the specific type with respect to Σ, if the deviation of its misorientation from the ideal misorientation did not exceed the maximum permissible deviation according to the Brandon criterion. All the boundaries with $\Sigma > 65$ were regarded as arbitrary boundaries.

The misorientation distributions were determined for both the polycrystal with chaotically oriented grains and for all three cases of the axial texture. In all cases of the axial texture the angle φ of deviation from the given direction did not exceed 15°. Figure 2.8 shows the distribution with respect to the misorientation angles for the chaotic and texturised polycrystals. The appropriate distributions with respect to the axes and Σ are presented in Tables 2.3 and 2.4. The resultant data coincide (within the limits of the statistical error) with the analytical results obtained by Mackenzie and computer calculations by Garbach and Grabskii. Misorientation distributions in the four types of flat cross-section were constructed for the textures and for the textureless case. The results obtained for them were identical. As an example, we

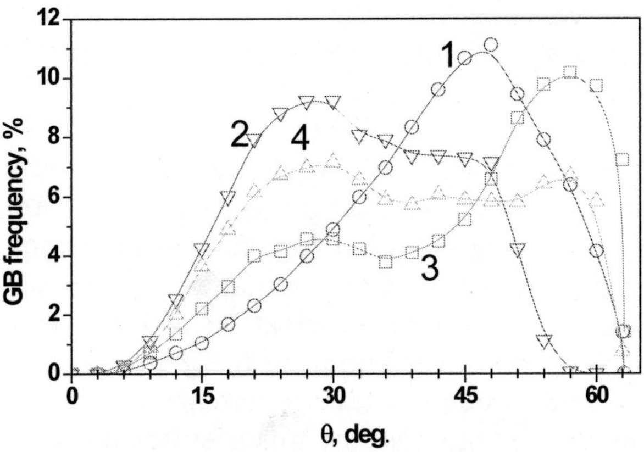

Fig. 2.8. Distribution of the grain boundaries with respect to the misorientation angles obtained for the entire volume of the crystal with the textures: 1) chaotic; 2) $\langle 100 \rangle$; 3) $\langle 110 \rangle$; 4) $\langle 111 \rangle$.

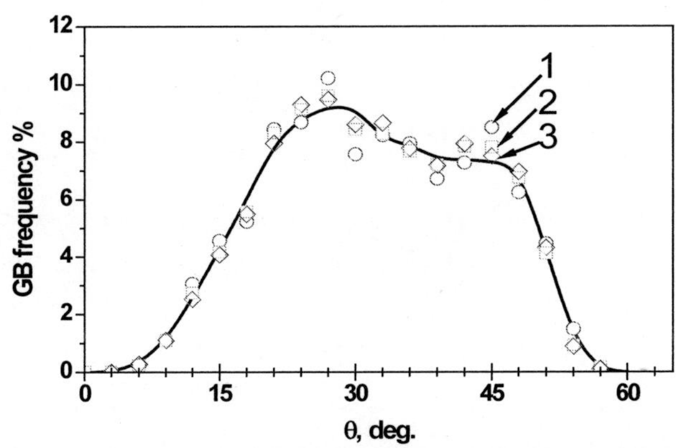

Fig. 2.9. Distribution of the grain boundaries with respect to the angles in the polycrystal with the axial texture $\langle 100 \rangle$ in the entire volume and in the individual cross-sections: 1) $(100)_L$; 2) $(110)_L$; 3) $(111)_L$. The solid line corresponds to the volume distribution.

selected the distribution for the texture $\langle 100 \rangle$ (Fig. 2.9). The solid line in the figure corresponds to the volume distribution and the symbols to the distribution in the flat cross sections. The appropriate distributions with respect to the axes and are shown in Tables 2.5 and 2.6. As indicated by Fig. 2.9, the distributions in all flat cross sections fit quite closely the curve corresponding to the volume distribution, and the scatter of the values is the same for the types of cross-section. The same follows from the comparison of the distributions with respect to the axes Σ for the volume case and flat cross-section (Tables 2.5 and 2.6). Consequently, it can be concluded that the experimental distribution, determined for a relatively large set of the grain boundaries in the

Table 2.3. Distribution of the grain boundaries with respect to the axes, determined for the entire volume for a textureless polycrystal and a polycrystal with three types of texture

Texture type	Zone							
	I	II	III	IV	V	VI	VII	VIII
Chaotic	0.77	2.14	3.56	22.74	53.53	9.36	2.12	5.79
⟨100⟩	2.26	6.44	9.23	38.31	36.55	3.99	0.80	2.41
⟨110⟩	0.47	1.28	2.19	16.13	36.03	12.21	3.17	8.40
⟨111⟩	0.56	1.65	2.45	18.22	52.83	11.01	3.80	9.32

Table 2.4. Distribution of the grain boundaries with respect to Σ obtained from the entire volume in the chaotic polycrystal and the polycrystal with three types of texture

Σ	Chaotic	Texture			Σ	Chaotic	Texture		
		⟨100⟩	⟨110⟩	⟨111⟩			⟨100⟩	⟨110⟩	⟨111⟩
1	2.12	13.04	6.77	10.33	17a	0.18	0.49	0.14	0.16
3	1.63	0.00	3.47	2.50	17b	0.31	0.00	0.60	0.27
5	1.19	3.22	0.29	0.24	19a	0.34	0.54	0.36	0.51
7	0.92	0.06	0.76	1.48	19b	0.22	0.00	0.21	0.27
9	0.98	0.28	0.88	0.62	21a	0.18	0.47	0.27	0.49
11	0.74	0.00	0.79	0.32	25a	0.08	0.43	0.18	0.23
13a	0.27	0.93	0.33	0.60	25b	0.44	0.00	0.60	0.29
13b	0.31	0.49	0.34	0.63	*	3.84	4.24	4.12	4.45
15	0.68	0.48	0.37	0.08	**	85.57	75.33	79.52	76.67

*: $25 < \Sigma < 65$; **: $\Sigma > 65$

Table 2.5. Distribution of the grain boundaries with respect to the axes in the polycrystal with the ⟨001⟩ texture, obtained for the entire volume and for individual cross-sections

Texture type	Zone							
	I	II	III	IV	V	VI	VII	VIII
Volume	2.256	6.444	9.232	38.308	36.551	3.993	0.804	2.412
$(100)_L$	2.034	6.339	8.231	39.688	37.134	3.974	0.663	1.940
$(001)_L$	2.247	6.867	9.272	37.583	36.732	3.934	0.797	2.571
$(110)_L$	2.229	6.625	8.965	37.805	36.987	3.936	0.791	2.663
$(111)_L$	2.195	6.540	8.833	38.031	36.970	4.048	0.779	2.706

Table 2.6. Distribution of the misorientation of the grain boundaries with respect to Σ in the polycrystal with the texture $\langle 100 \rangle$, obtained for the entire volume and in individual cross-sections.

Σ	Volume	Cross section			Σ	Volume	Cross section		
		$(100)_L$	$(110)_L$	$(111)_L$			$(100)_L$	$(110)_L$	$(111)_L$
1	13.04	12.58	12.65	12.80	17a	0.49	0.62	0.60	0.65
3	0.00	0.00	0.00	0.00	17b	0.00	0.00	0.00	0.00
5	3.22	2.60	3.42	3.45	19a	0.54	0.62	0.53	0.54
7	0.06	0.14	0.05	0.06	19b	0.00	0.00	0.00	0.00
9	0.28	0.38	0.28	0.29	21a	0.47	0.47	0.62	0.66
11	0.00	0.00	0.00	0.00	25a	0.43	0.52	0.44	0.41
13a	0.93	1.37	0.96	0.89	25b	0.00	0.00	0.00	0.00
13b	0.49	0.43	0.56	0.57	*	4.24	3.31	4.26	4.24
15	0.48	0.39	0.36	0.41	**	75.33	76.57	75.27	75.21

*: $25 < \Sigma < 65$; **: $\Sigma > 65$

flat cross-section of the specimen, corresponds to the misorientation spectra of the grain boundaries in the entire volume of the polycrystal, regardless of crystallographic anisotropy. It should be mentioned that in this case we are concerned only with the crystallographic anisotropy for the equiaxed grain structure. It is clear that for the anisotropic form of the grains, the misorientation spectra of the grain boundaries in different cross-sections will differ.

References

1. Korn G. and Korn T., A handbook of mathematics, Nauka, Moscow, 1973.
2. Frank F.C., Metall. Trans. A, 1988, vol. 19, No. 3, 403–408.
3. Gertsan V.Yu., Kristallografiya, 1989, vol. 326, No. 3, 588–591.
4. Orlov A.N. (editor), Atomic structure of i/ntercrystalline boundaries, Mir, Moscow, 1978.
5. Rybin V.V. and Perevezentsev V.N., FTT, 1975, vol. 17, No. 11, 3188–3193.
6. Orlov A.N., et al., FTT, 1975, vol. 17, No. 6, 1662–1670.
7. Kronberg M.L. and Wilson F.H., Trans. AIME, 1949, vol. 185, 506–508.
8. Sutton A.P. and Ballufi R.F., Interfaces in crystalline materials, Clarendon Press, Oxford, 1995.
9. Grimmer H., et al., Acta Crystall., 1974, vol. 30A, 197–207.
10. Andreeva A.V. and Fionova L.K., FMM, 1977, vol. 44, No. 2, 395–400.
11. Orlov A.N., et al., Grain boundaries in metals, Metallurgiya, Moscow, 1980.
12. Glymer G. and Chalmers B., High-angle grain boundaries, Mir, Moscow, 1975.
13. Bollman W., Crystal defects and crystalline interfaces, Springer, Berlin, 1970.
14. Zisman A.A. and Rybin V.V., Poverkhnost', 1982, No. 7, 87–90.
15. Zisman A.A. and Rybin V.V., FMM, 1989, vol. 68, No. 2, 264–270.

16. Shvindlerman L.S. and Straumal B.B., Regions of existence of special and non-special grain boundaries, Preprint No. 1068, IFTT, Chernogolovka, 1986.
17. Grimmer H., et al., in: Atomic structure of intergranular boundaries, Mir, Moscow, 1978, 25–54.
18. Mykura H. A., A checklist of cubic coincidence site lattice relations. Grain boundary structure and kinetics, Ohio, ASM. Metals Park, 1980, 445–456.
19. Gertsman V.Yu., et al., Scripta Materialia, 1996, vol. 35, No. 11, 1247–1251.
20. Grimmer H., Scripta metall., 1989, vol. 23, 1407–1413.
21. Gertsman V. Y. , Scripta Metall. Mater., 1992, vol. 27, 291–297.
22. Grimmer H., Mater. Sci. Forum, 1993, vol. 126–128, 269–275.
23. Grimmer H., Acta Cryst., 1976, vol. 32, 783.
24. Brandon D.C., Acta Metall., 1966, vol. 14, 1479–1484.
25. Hirt G. and Lome I., Dislocation theory, Atomizdat, Moscow, 1972.
26. Zhao J. and Adams B.L., Acta Cryst., 1988, vol. A44, 326–336.
27. Frank F.C., Metall. Trans. A, 1988, vol. 19, No. 3, 403–408.
28. Knyazev N.M., Distribution of random misorientations with deviation from the misorientation of special boundaries, Dep. VINITI, 1989, No. 770-B89.
29. Mackenzie J.K., Biometrika, 1958, vol. 45, 229–243.
30. Mackenzie J.K., Acta Met., 1964, vol. 12, No. 2, 223–225.
31. Warrington D.H. and Boon W., Acta Met., 1975, vol. 23, No. 5, 599–607.
32. Rybin V.V., et al., FMM, 1982, vol. 53, No. 3, 544–553.
33. Garbacz A. and Grabskl M.W., Scr. Met., 1989, vol. 23, 1369–1374.
34. Garbacz A. and Grabskl M.W., Arch. Nauki Mater., 1990, vol. 11, No. 3–4.
35. Pshenichnyuk A.I., et al., in: First collection of studies of UTAN Association, UTAM Association, 1990, 60–70.
36. Saltykov S.A., Stereometric metallography, Metallurgiya, Moscow, 1976.
37. Gertsman V.Yu., et al., Simulation of the misorientation distribution of grain in polycrystals, Ed. board Metallofizika, Kiev, 1991, dep. VINITI, No. 636-V91.
38. Dyad'kin I.G., in: Monte-Carlo methods in physics and geography, Ufa, 1973, 4–144.

3

ORIENTATION-DISTRIBUTED PARAMETERS OF THE POLYCRYSTALLINE STRUCTURE

3.1. The distribution function of the grains with respect to crystallographic orientations. Calculation methods

Let us determine a rigorous description of the distribution function of the grains with respect to crystallographic orientations, i.e. orientation distribution function (ODF). Let all the rotations form the space of rotations **G**, and the Euler angles define the coordinate system in **G**. Consequently, $\mathbf{g} = [\alpha, \beta, \gamma]$ determines the rotation in this space. The function $f(\mathbf{g})$ (where $\mathbf{g} \in \mathbf{G}$) of the distribution of the grains with respect to the orientations is determined on the space of rotations **G** and is characterised by the following property. Let $\mathbf{U}\{[\alpha, \beta, \gamma]\}$ be the region in **G**. Consequently, the number of the elementary cells with the orientation belonging to **U**, related to the total number of the cells in the specimen, is:

$$p(U) = N_U / N = \int_U f\left(\left[\alpha,\beta,\gamma\right]\right) d\alpha \sin\beta\, d\beta\, d\gamma. \tag{3.1}$$

The value p (**U**) is often interpreted as the probability of the elementary cell, randomly selected in the specimen as the orientation belonging to **U**. For the direct calculation of the function $f(\mathbf{g})$ it is necessary to measure the volume dV for each orientation $\mathbf{g} = [\alpha, \beta, \gamma]$. In practice, these measurements cannot be taken. Therefore, function $\mathbf{f}(\alpha, \beta, \gamma)$ can be determined only by indirect methods. In this case, the ODF is reconstructed from the pole figures (PF) plotted by x-ray photography. Further mathematical processing of the PF taking into account the crystallographic symmetry of the crystal lattice and the symmetry of the specimen itself makes it possible to determine the three-dimensional distribution function of the grains with respect to the crystallographic orientations. This leads to the main task of quantitative texture analysis:

restoration of the ODF on the basis of the finite number of the PF. The mathematical methods of calculation of the ODF on the basis of PF, proposed by the Russian scientist Viglin [1] and subsequently developed further by Bunge [2] and Roe [3], have been used widely not only abroad [4–7] but also in Russia (see, for example, [8–10]). In addition to this, work is now being carried out to develop and use new methods of calculating the ODF on the basis of the pole figures, but in this work we use the classic methods which will be described briefly.

The pole figure is defined as follows. Let $\mathbf{S}^2$ be the set of the unit vectors, i.e., a sphere. It is assumed that we select a crystallographic plane in the lattice, and the normal to this plane in the crystallographic coordinates system is denoted by **n**. The pole figure $P_h(\mathbf{y})$ (where $\mathbf{y} \in \mathbf{S}^2$) for the vector $\mathbf{h} \in \mathbf{S}^2$ is the positive function, characterised by the following properties. It is assumed that the V-shaped region is situated in $\mathbf{S}^2$. Consequently, the number of the elementary cells whose direction **h** is located in the region $V \cup (-V)$, divided by the total number of such cells, is

$$P(V) = N_{V\cup(-V)} / N = \int_V P_{\mathrm{h}}(\mathbf{y})\,d\mathbf{y} = \int_V P_h(\varphi,\theta)\sin\theta\,d\theta\,d\varphi. \tag{3.2}$$

The number $P(V)$ is usually interpreted as the probability of the cell randomly selected in the specimen having direction **h**, situated in the region $V \cup (-V)$. In other words, $P_h(\mathbf{y})$ is the density of probability of the elementary cell, randomly selected in the specimen, having the vector **h** parallel to the vector **y**. This leads to equality $P_h(\mathbf{y}) = P_h(-\mathbf{y})$. The experiments usually yield the finite number of the PF: $P_{h1}(\mathbf{y}), P_{h2}(\mathbf{y}),\ldots,P_{hN}(\mathbf{y})$.

The distribution functions of the grains with respect to the orientations and the PF are linked together as follows. Let **y** be the unit vector with the spherical coordinates θ and φ, i.e., $\mathbf{y} = (\cos\varphi\sin\theta, \sin\varphi\sin\theta, \cos\theta)$. It is assumed that some number γ is determined in the interval [0, 2π]. Consequently, [**y**, γ] is the rotation with the Euler angles [φ, θ, γ]. All the rotations **g** transforming **y** to **h** ($\mathbf{h}_i = \mathbf{g}\mathbf{y}_i$) are described by the formula

$$g = [\mathbf{h},\varphi]^{-1}[\mathbf{y},0].$$

The following integral relationship can be written

$$P_{hi}(\mathbf{y}) = \frac{1}{4\pi} \int \left\{ f\left([\mathbf{h},\varphi']^{-1} [\mathbf{y},0] \right) + f(-[\mathbf{h},\varphi'][\mathbf{y},0]) \right\} d\varphi'. \tag{3.3}$$

This relationship was used to develop several methods of restoration of the function $f(\mathbf{g})$ with respect to the pole figures.

Ruero and Barro [7] proposed to divide the space of rotations **G** into N_{gr} cells with the volume V_{gr} and regard the function $f(\mathbf{g})$ as a constant inside the i-th cell and equal to f_i. Similarly, the spherical region of determination of the PF, $\mathbf{S}^2$, is divided into N_P cells with the same area. The function $P_h(\mathbf{y})$ is also regarded as constant within the limits of each cell. Thus, the functions $f(\mathbf{g})$ and $P_h(\mathbf{y})$ are replaced by N_{gr} and N_p-dimensional vectors, respectively. The advantage of the vector method is the simple mathematical procedure. In addition, incomplete PF can be used in these cases. A shortcoming is that the ODF, calculated by the given method (as shown by Matthies [11–15]) may have 'ghosts'. However, the vector method has been recently adopted and is used widely in calculations.

Another method was developed by Matthies [11–15] who proposed to find a solution in the form of the sum of some standard ODF. The task of restoration of the ODF can be solved unambiguously if we introduce *apriori* assumptions regarding the type of orientation distribution function of the grains. Matthies proposed to expand the orientation distribution function of the grains as the sum of the functions of the type

$$f(\varepsilon_0, g_0, g) = f(\varepsilon_0, \omega) = s_0 \exp\left(-\frac{\omega^2}{\varepsilon_0} \right), \tag{3.4}$$

where

$$s_0(\varepsilon_0) = \frac{2\sqrt{\pi}}{\varepsilon_0(1 - \exp(-\varepsilon_0^1 / 4))}, \tag{3.5}$$

and ε_0 and ω is the radius and angular distance between $\mathbf{g}_0$ and $\mathbf{g}$, respectively. The function $f(\varepsilon_0, \omega)$ is efficiently approximated by the function which is more suitable for calculations:

$$f(g_0, b, g) = f(s, \omega) = \frac{\exp(s \cos\omega)}{I_0(s) - I_1(s)}, \tag{3.6}$$

where $s = \frac{1}{2}\ln\frac{2}{\sin^2(b/2)}$; $b = 2\sqrt{2\ln\varepsilon_0}$; I_0 and I_1 are the Bessell functions.

The function (3.6) was proposed on the basis of the purely external analogy with the three-dimensional Gaussian normal distribution. The application of the Gaussian distribution must be determined by the effect of the central limiting theorem of the probability theory which gives the statistical substantiation to this distribution. In the studies by Saveleva [16, 17] it is shown that the central limiting theorem on the group of rotation leads to a different type of distribution:

$$f(g) = \sum_{l=0}^{\infty}\sum_{m,n=0}^{\infty}(2l+1)\exp\left\{-\varepsilon^2\left[l(l+1)-m^2\right]-\varepsilon^2\right\}\times \\ \times\exp\left\{-i(n\alpha+m\gamma)\right\}P_l^{mn}(\cos\beta). \tag{3.7}$$

In some partial cases, the PF have relatively large regions in which they convert to zero. This simplifying circumstance can be used in the construction of the ODF, specially when using the discrete method. For the polycrystals consisting of a small number of crystals (coarse-grained material), the PF and ODF are simply the sum of δ-functions. In this exclusive case the problem of determination of the ODF has a unique solution.

In the book, we use the method of calculating the ODF on the basis of the incomplete pole figures proposed in [4, 9, 10]. When calculating the three-dimensional distribution of the crystals with respect to the orientation is it is necessary to consider the fact that the required ODF and the function $A(\mathbf{h}, \mathbf{y})$ of the distribution of the density of the poles (the PF are flat cross sections of the given function) can be represented in the form of series of spherical functions with the expansion coefficient $C_l^{\mu\nu}$ [4]

$$A(h,y) = \sum_{l=0}^{\infty}\sum_{\mu=1}^{M(l)}\sum_{\nu=1}^{N(l)}C_l^{\mu\nu}\left[\frac{4\pi}{2l+1}\right]\dot{\dot{k}}_l^{\mu}(h)\dot{\dot{k}}_l^{\nu}(y); \tag{3.8}$$

$$f(g) = \sum_{l=0}^{\infty}\sum_{\mu=1}^{M(l)}\sum_{\nu=1}^{N(l)}C_l^{\mu\nu}\,\dot{\dot{T}}_l^{\mu\nu}(g). \tag{3.9}$$

The symmetric spherical functions form the total orthogonal systems. The experimental results are used to determine the finite number of the pole figures P_{hi} ($\mathbf{y}$) and not the function $A(\mathbf{h}, \mathbf{y})$. For example, for the metals with the cubic lattice only three or four pole figures can be measured with sufficient accuracy. The coefficients $C_l^{\mu\nu}$ in this case are determined by the method of mean quadratic approximation:

$$R = \sum_{l=0}^{L} \int_{Bi} \left\{ N_{hi} P(h_i, y) - \left[\frac{4\pi}{2l+1} \right] \sum_{\mu=1}^{M(l)} \sum_{\nu=1}^{N(l)} C_l^{\mu\nu} \overset{\vdots}{k}_l^{\mu}(h) \overset{\vdots}{k}_l^{\nu}(y) \right\}^2 dy \to \min, \tag{3.10}$$

Here L is the number of the pole figures, B_i is the integration region of the PF, N_{hi} are the normalisation coefficients for the measured values of the pole figures. In this book, we use the incomplete pole figures and the integration region is determined by the condition $B_i = B(\alpha < 65°)$, where α is the angle of inclination of the specimen. From the condition of the minimum of the functional (3.10), taking into account that according to the normalisation condition $C_0^{11} = 1$, we obtain the system N_{max} of the linear equations in relation to the coefficients $C_l^{\mu\nu}$. For the cubic symmetry of the lattice and the rhombic symmetry of the pole figures the number of equations of the system with the restriction of the series to 22 terms is equal to 124.

A large number of programs have been developed for calculating the ODF on the basis of the experimentally measured pole figures. In this book, we use the programs described in [10] which can be employed to determine the ODF on the basis of incomplete pole figures and are the further development of the Bunge–Roe method.

3.2. Relationship between the grain boundary misorientation distribution and the ODF

Generally speaking, the relationship between the distributions of the crystallographic orientation and the misorientation spectrum of the grain boundaries is ambiguous [18, 19] since the distribution function of the orientations does not contain any information on the neighbourhood of the grains and the misorientations are determined by the nearest neighbours in particular. According to the definition, as already mentioned previously, the ODF is the analogue of the single-particle distribution function – the fraction of the grains with the given orientation of the basis of the crystal lattice, whereas the spectrum of the misorientation of the grain boundaries is an analog of the two-particle function, i.e., the fraction of the pairs of neighbouring grains, with the given misorientation between them. In a general case, these two distribution functions contain different physical information, and from the general physical considerations the two-particle function is reduced to the single-particle function only if the superposition principle is fulfilled, i.e., in this context this means that the orientations of the adjacent grains are independent random functions. Therefore, it is interesting to investigate the extent to which the spectrum of the grain boundaries is determined by the texture and the effect of the correlation in the orientations of the adjacent grains on the misorientation spectrum of the grain boundaries.

We determine the essential and sufficient conditions for the unambiguous restoration of the spectrum of the misorientation of the grain boundaries on the basis of the texture. It is assumed that the orientation E_α of the basis of the crystal lattice of each grain is known, where α is the number of the grain. The basis E_α can be represented either as a matrix (3 × 3), with each column of the matrix specifying the coordinates of the corresponding basis vector in some laboratory coordinate system, or it is parametrised by the system of the Euler angles. Subsequently, for all grain pairs which are in fact neighbours, we calculate the misorientation *R* which transforms the basis E_α to the basis $E_{\alpha+1}$:

$$\widehat{E}_{\alpha+1} = \widehat{R}^{\alpha}_{\alpha+1}\widehat{E}_{\alpha}, \text{ from which } \widehat{R}^{\alpha}_{\alpha+1} = \widehat{E}_{\alpha+1}\widehat{E}^{-1}_{\alpha}. \tag{3.11}$$

It is generally accepted to characterise the rotational matrix in terms of the axis–angle variables (**C**, θ). In this case, the crystal lattice of the grain α, rotated through the angle θ around the axis **C**, coincides with the crystal lattice of the grain α + 1. From the set (C_α, θ_α), determined by this procedure (where α = 1, 2,..., *N,* and *N* is the number of the boundaries in the investigated section) we construct the frequency function *P*(**C**, θ) in such a manner that the quantity *P*(**C**, θ) Δ**C**Δθ determines the fraction of the misorientations in the given element of the phase volume in relation to the total number N_B of the misorientations in the system. The frequency function, obtained by this procedure is regarded as identical with the probability density of the appropriate event and is referred to as the grain boundary misorientation distribution (GBMD). It is assumed that this spectrum characterises not only the section in which it was determined but also the entire specimen of the material as a whole provided that it is spatially uniform. Another section has the same frequency function within the limits of the statistical errors determined by the volume of the sample (i.e., the number of tested boundaries in the section). In turn, this means that the given section is a specific representative of the ensemble and all the statistical characteristics of the material can be determined by averaging with respect to the ensemble, and this procedure is equivalent to the application of the frequency function -

GBMD. The concept of the ensemble makes it possible to introduce immediately the frequency density of probability *N* which contains also the shortened description: one-particle (ODF) and two-particle (GBMD) functions. Let the ensemble be prepared in such a manner as each its representative contains N_{gr} grains. For the unambiguous definition of the probability density it is necessary to determine the procedure for numbering the grains which, after all, can be carried out by several methods. For numeration it is necessary to fulfil the following condition: it is assumed that the grains with the number α must have as its neighbours grains with the numbers α–1 and α+1, i.e., the line connecting the grains in the order of increase of their numbers, is continuous and does not split. Two examples of such numeration for a planar case are shown in Fig. 3.1. Let F ($\mathbf{E}_1$, $\mathbf{E}_2$,..., $\mathbf{E}_N$) be the combined density which determines the probability of the given orientation of each grain. The one-particle function – the density of orientation of the basis of the α-grain – is given by the expression

$$f_\alpha(\widehat{E}) = \int\int F(\widehat{E}_1,...\widehat{E}_{\alpha-1},\widehat{E}_{\alpha+1},...\widehat{E}_N)\times \\ \times\,\delta\widehat{E}_1,...\delta\widehat{E}_{\alpha-1},\delta\widehat{E}_{\alpha+1},...\delta\widehat{E}_N, \tag{3.12}$$

where integration with respect to the basis should be perceived as the integration with respect to the three Euler angles. In averaging of (3.4) with respect to α we obtain the ODF:

$$P_0(\widehat{E}) = \frac{1}{N_{gr}}\sum_{\alpha=1}^{N_{gr}} f_\alpha(\widehat{E}). \tag{3.13}$$

Naturally, this is another form of the equation (3.12). We now examine the restrictions for the misorientation of the grain boundaries, forming the junction (for better understanding, we select a triple junction, Fig. 3.2). The numbers denote the grain boundaries and the appropriate misorientation matrices. If we go completely around the junction and return to the initial point, the crystallographic basis of the grain *A* will be expressed by the following equation:

$$\widehat{A} = \widehat{R}_3\widehat{C} = \widehat{R}_3\widehat{R}_2\widehat{B} = \widehat{R}_3\widehat{R}_2\widehat{R}_1\widehat{A}, \quad \text{or} \quad \widehat{R}_3\widehat{R}_2\widehat{R}_1 = \widehat{I}. \tag{3.14}$$

Thus, knowing the misorientation of the two boundaries in the triple junction, we can calculate the third misorientation. We examine details of the packing shown in Fig. 3.1. Knowing the misorientation of the grain boundaries which intersect the packing lines, and using the relationship (3.14), it is quite easy to calculate all the other misorientations on the mesh (inserts in Fig. 3.1). In accordance with the accepted type of numeration of the grains all the boundaries are divided into two classes: the boundaries distributed along the numeration line are referred to as the given boundaries, and all the other boundaries are referred to as calculated ones. According to this definition, we can also introduce two types of misorientation subspectra: GBMD on the 'given' boundaries – $P_D(\mathbf{R})$ – and GBMD on the 'calculated boundaries' – $P_C(\mathbf{R})$. The principal difference between them is that the misorientations on the 'given' boundaries can be specified completely independently of each other, whereas the misorientations on the 'calculated' boundaries are unambiguously restored from the misorientations at the 'given' boundaries, and also on the basis of the method of packing the numeration lines [19].

According to the general rules of the probability theory [20] $P_D(\mathbf{R})$ is determined by the equation

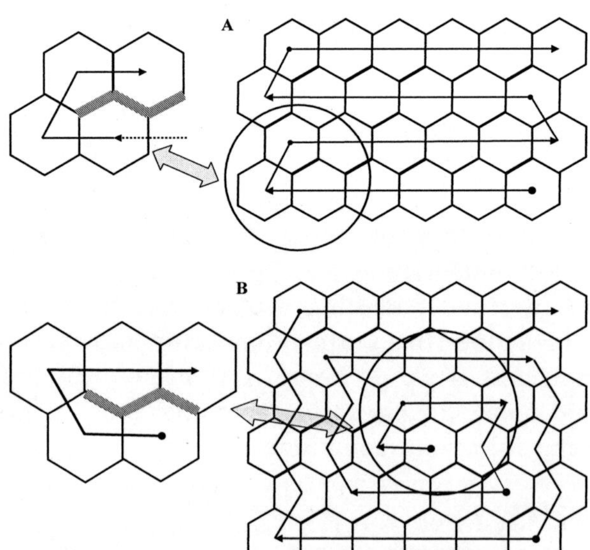

Fig. 3.1. Two possible methods of numeration for the two-dimensional system of the grains. The inserts on the left shows the 'given' and 'calculated' (indicated by the thick line) misorientations of the grain boundaries.

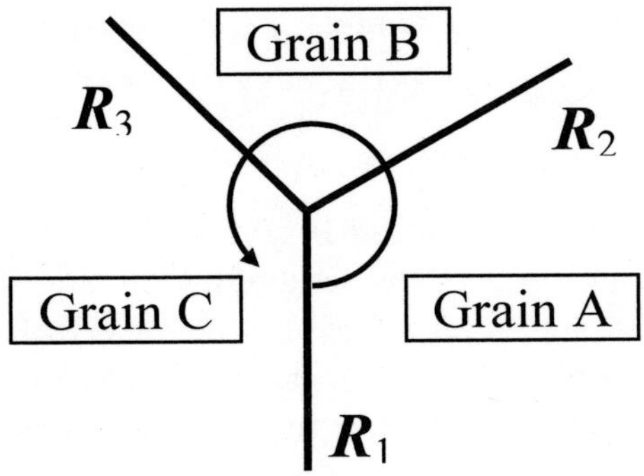

Fig. 3.2. The scheme of the triple junction.

$$P_D(R) = \frac{1}{N_{gr}-1}\sum\int\int\delta(R - E_{\alpha-1}E_{\alpha}^{-1})F(E_1...,E_N)\times$$
$$\times dE_1...,dE_N = \frac{1}{N_{gr}-1}\sum\int\int\delta(R - E_{\alpha-1}E_{\alpha}^{-1})f(E_{\alpha-1},E_{\alpha+1})\times$$
$$\times dE_1...,dE_N. \tag{3.15}$$

If the orientations of the basis of the two adjacent grains are independent quantities and the distribution of each of them is given by the ODF, then

$$f_{\alpha,\alpha+1}(E_{\alpha},E_{\alpha+1}) = P_0(E_{\alpha})P_0(E_{\alpha+1}). \tag{3.16}$$

Consequently, the quantity

$$P_D^*(R) = \int\int\delta(R - E_1E_2^{-1})P_0(E_1)P_0(E_2)dE_1dE_2 \tag{3.17}$$

will be referred to as the non-correlated GBMD. In a general case, the following more general equation should be used instead of (3.16):

$$f_{\alpha,\alpha+1}(E_{\alpha},E_{\alpha+1}) = f_{\alpha}(E_{\alpha})f_{\alpha+1,\alpha}(E_{\alpha+1},E_{\alpha}), \tag{3.18}$$

where f_{α} is determined by expression (3.12) and f $_{\alpha,\alpha+1}$ is the appropriate conditional density of probability. It is not justify to expect that the last probability is determined in particular by the orientations of the basis of the grain α and α+1 since the physical properties of the boundary of the grains are given only by their misorientation and not by the orientation of the grains in the space. Consequently

$$f_{\alpha,\alpha+1}(E_{\alpha+1},E_{\alpha})=g_{\alpha+1,\alpha}(E_{\alpha+1}E_{\alpha}^{-1}). \tag{3.19}$$

With the natural assumption according to which the function g is independent of the position of the boundary in the chain, we come to the conclusion that the function gives the probability of detection of the given misorientation, i.e., it is the spectrum of misorientations $P_D(\mathbf{R})$. Substituting (3.18) and (3.19) into (3.16) gives

$$P_D^*(R)=\int\int\delta(R-E_1E_2^{-1})P_0(E_1)P_D(E_1E_2^{-1})dE_1dE_2. \tag{3.20}$$

Utilising the properties of the δ-function and the unit normalisation $P_0(\mathbf{E})$, it is evident that (3.20) is an identity. This shows that the misorientation spectrum on the numeration line $P_D(\mathbf{R})$ and the texture $P_0(\mathbf{E})$ can be specified independently of each other. However, the actual spectrum $P(\mathbf{R})$ is mean-weighted with respect to the spectra $P_D(\mathbf{R})$ and $P_C(\mathbf{R})$. The weight coefficients are the partial fractions of the 'given' and 'calculated' boundaries, respectively. It is quite easy to estimate the characteristic values of these weights. The well-known Euler relationship between the number of the grains N_{gr}, the number of boundaries N_B and the number of junctions of the grains N_T is fulfilled on the flat network of the boundaries:

$$N_T-N_B+N_{gr}=1. \tag{3.21}$$

In most cases, the junctions of the grains are triple junctions. In this case, three boundaries converge to each junction, and each boundary connects two triple junctions, i.e.

$$3N_T=2N_B. \tag{3.22}$$

The weight coefficient q_D of the 'given' boundaries is equal to the ratio of the number of the boundaries in the numeration line ($N_{gr}-1$) is the total number of the boundaries (N_B).

From (3.21) and (3.22) we obtain

$$q_D=\frac{N_{gr}-1}{N_B}=\frac{1}{3} \text{ and } q_C=1-q_D=\frac{2}{3}. \tag{3.23}$$

Thus, in the proposed method of calculating the misorientation spectrum of the grain boundaries with respect to the texture there are two independent parameters which determine the total spectrum of misorientation of the grains in the polycrystals. This is the texture of the material which determines the type of the spectrum of the misorientation of the grain boundaries on the 'given' boundaries and the nature of packing of these boundaries in the section of the polycrystals.

3.3. Correlation orientation of adjacent grains. The concept of the basis spectra of misorientation of the grain boundaries

In the previous section, attention was given to the presence of two parameters which do not depend on the texture and which can be changed. In this case, it is the way of packing of the 'given' spectrum of the grain boundary and the method of defining the spectrum along the numeration line. Attention will now be given to the effect of the variation of these parameters on the type of resultant misorientation spectrum of the grain boundaries by simulation of the spectrum of the misorientation of the grain boundaries in polycrystals with different modelling textures [19, 21].

Since it was shown in section 2.4 that the flat section is a representative section for the volume specimen, further modelling will be carried out on two-dimensional systems. The calculation of the spectrum of the misorientation of the grain boundaries in the simulation polycrystals has the following algorithm.

1. The modelling structure is given, i.e., ODF, in accordance with which we determine the number N_{gr} of the random orientations (realisations) of the basis. The procedure for pertaining the random orientation of the basis for the modelling axial textures was described in section 2.2.

2. The section of the polycrystalline given – the two-dimensional system consisting of N_{gr} grains. The type of filling of the section by the set of bases, determining the previous stage, is determined. To verify the effect of this parameter, attention will be given to two variants, shown in Fig. 3.1.

3. The sequence of selection of the bases from the N_{gr} realisations in filling of the mesh along the numeration line is determined. This is equivalent to the definition of the type of the spectrum of the misorientation of the grain boundaries on the line of the 'given' grain boundaries. The following variants of the selection and their physical meaning will be discussed:

–the first grain along the numeration line is allocated any of the

N_{gr} bases, for example, the first one. The misorientations between this basis and all other bases are calculated. From the results we select the basis for which the angle of misorientation with the basis of the first grain is minimum. The selected basis is attributed to the second grain. The procedure is repeated for the basis of the second grain and all remaining grains up to complete filling of the section of the polycrystals. In this procedure, the fraction of the low-angle boundaries in $P_D(\mathbf{R})$ is increased to the maximum extent;

–the procedure is identical with the first one but a basis with the minimum misorientation is chosen. The GBMD gives the upper limit of the high-angle random boundaries;

–the procedure is again repeated but now we select the basis with the minimum possible value of the inverse density of the coincident nodes Σ (within the limits of the Brandon deviation). The appropriate spectrum $P_D(\mathbf{R})$ gives the upper estimate of the fraction of the special grain boundaries;

–it is assumed that the orientations of the bases are not correlated along the filling line, i.e., $P_D(\mathbf{R})$ is described by the equation (1.40). This variant is described in section 2.4. Thus, for the same set of the basises, i.e. for the same texture of the specimen, we find four different misorientation spectra of the 'given' boundaries.

In the fully determined polycrystal we calculate the misorientations on the 'calculated' boundaries and construct the complete spectrum of the misorientation of the grain boundaries corresponding to the given texture and the numeration line, and the spectrum of the misorientations the grain boundaries at the 'given' boundaries.

As previously, all the misorientations are reduced to the description with the minimum angle and the axis situated in the standard stereographic angle. The modelling system has the form of a square filled with cubic lattice grains. The grains are represented by regular hexagons. On the whole, the calculation block contains 625 grains and 1776 boundaries between them. The third of all the boundaries is related to the 'given' boundaries. On these boundaries we specify the misorientation spectrum which forms when the previously described procedure is applied, i.e., the set of bases corresponding to the given texture, is ordered in accordance with one of the previously mentioned variants (A, B and C).

In the case of the chaotically disoriented ensemble (CDE) of the grains for the three types of the axial texture ($\langle 100\rangle$, $\langle 110\rangle$, $\langle 111\rangle$) we obtain three sets of ordered bases in accordance with the variants A, B and C. Using this procedure, we calculate the spectrum of the misorientation of the grain boundaries for 12 sets of the ordered bases in the structure of the polycrystals. Figure 3.3 and Tables 3.1 and 3.2 show as an example the distribution with respect to the angles, axes and Σ for the CDE for two packing variants shown in Fig. 3.1. The distributions correspond to the three types of ordering of the set of the basis for the CDE (A, B, C). Identical results for the texture $\langle 100 \rangle$ are shown in Fig. 3.4 and also in Tables 3.3 and 3.4. The solid line on the histograms shows the 'non-correlated' spectra of the grain boundaries for the chaotic polycrystals and the texture $\langle 100 \rangle$ (Fig. 2.6) with the weight coefficient 2/3. The non-correlated spectrum is taken with this weight since the contribution of the 'calculated' boundaries to the total spectrum is proportional to the number of the boundaries and they equal 2/3 of the total number of the boundaries for the given method of packing of the 'given' boundaries. Detailed analysis of the results indicates that for any type of simulation texture the total spectrum of misorientation of the grain boundaries does not depend, within the limits of the statistical error, on the type of packing of the 'given' boundaries. The type of correlation has a controlling effect on the calculated spectrum of the misorientation of the grain boundaries and greatly increases the fraction of the boundaries of the appropriate type.

Thus, it has been established that for the given texture we can unambiguously restore four types of the spectrum of the misorientation of the grain boundaries which differ from each other by the correlation in the orientation of the axes along the packing line of the 'given' boundaries. In the case of the correlation of type *A* in the GBMD the fraction of the low-angle boundaries is increased with the maximum extent, in the case of the correlation of type *B* – the fraction of high-angle arbitrary grain boundaries, and in the case of the correlation of type C – the fraction of the special (or close to special) grain boundaries. The fourth spectrum (type D) corresponds to the absence of any correlation the orientations of the adjacent grains along the packing line of the 'given' boundaries.

All four spectra are the boundary distributions in the region of determination of the actual spectrum of misorientation of the grain boundaries and can be referred to as the basis spectra of the grain boundaries. The actual GBMD can be represented as the mean-weighted sum of the basis spectra:

$$P(\theta) = \sum_{i=1}^{4} q_i P_i(\theta), \tag{3.24}$$

where $P_i = \{P_A, P_B, P_C, P_D\}$ are basic GBMD; q_i are the expansion coefficients satisfying the normalisation condition .

Table 3.1. Distribution of the axes in the textureless crystal for three types of ordering of the bases (A, B, C) and two packing methods (*a*, *b*)

Zone		I	II	III	IV	V	VI	VII	VIII
A	a	1.24	3.15	4.45	26.85	50.23	7.15	1.86	5.07
	b	1.01	2.93	4.22	27.53	49.10	8.62	1.97	4.62
B	a	0.34	1.13	2.37	14.47	38.74	32.10	2.42	8.45
	b	0.34	1.30	2.08	16.50	38.12	32.26	1.69	7.71
C	a	0.79	2.20	2.53	15.65	35.64	4.90	13.96	24.32
	b	0.79	2.14	1.91	14.13	37.00	4.26	13.75	23.03

Table 3.2. Distribution with respect to in a textureless polycrystals for three types of ordering of the basis (A, B, C) and two packing methods (a, b)

Σ	*A*		*B*		*C*	
	a	*b*	*a*	*b*	*a*	*b*
1	34.29	35.70	1.80	0.85	2.20	1.52
3	1.01	1.46	4.39	3.89	32.43	32.49
5	1.01	0.56	0.51	0.51	1.97	1.86
7	0.34	0.23	0.73	0.39	0.90	0.79
9	0.85	0.56	0.73	1.01	0.85	0.73
11	0.56	0.39	0.23	0.39	0.51	0.62
13a	0.11	0.39	0.17	0.11	0.34	0.17
13b	0.17	0.23	0.45	0.34	0.45	0.28
15	0.34	0.29	0.34	0.56	0.62	0.51
17a	0.11	0.06	0.28	0.11	0.11	0.17
17b	0.23	0.34	11.60	11.60	0.17	0.28
19b	0.11	0.17	0.28	0.28	0.28	0.23
21a	0.06	0.00	0.11	0.00	0.11	0.23
25a	0.06	0.06	0.00	0.00	0.00	0.00
25b	0.00	0.51	0.28	0.28	0.11	0.11
*	3.32	3.44	2.93	2.59	2.76	2.31
**	57.26	55.52	75.28	76.96	56.14	57.49

*: $25 < \Sigma \leq 65$; **: $\Sigma > 65$

The texture of the polycrystals determines the type of basic spectra $P_i(\theta)$. The coefficients q_i describe the contribution of each of the basic GBMDs to the total spectrum and depend on the prior history of the thermomechanical effect on the material. The values of the weight fractions can be determined only by comparing the experimental spectrum with the expansion (3.24). If the considerations are based on the mean quadratic approximation, the coefficient q_i can be determined by minimising the sum of the mean quadratic deviations of the calculation spectrum from the experimental one, i.e., minimising

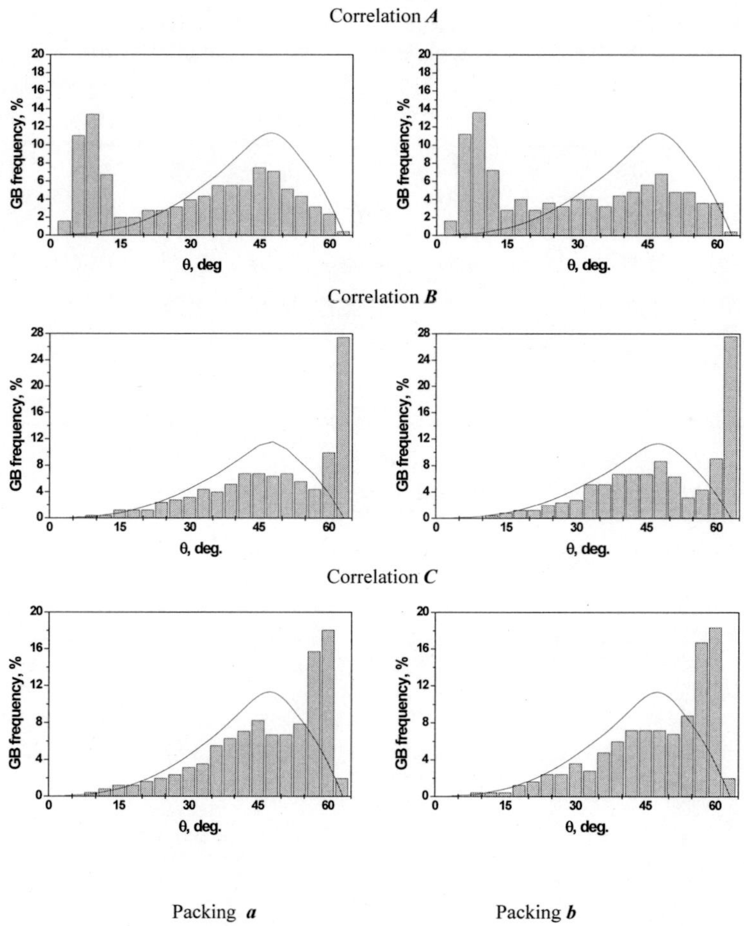

Fig. 3.3. Distribution of the misorientation angles in a textureless polycrystals for three types of correlation in the orientation of the adjacent grains and two packing methods: *a* (left) and *b* (right), see Fig. 3.1.

the equation

$$F(q_i) = \sum (P(q_i) - P_{eq})^2. \tag{3.25}$$

Taking into account the normalisation condition, q_i is the solution of the system of linear equations:

$\partial F(q_i)/\partial q_k = 0$, where $k = 1, 2, 3$.

Table 3.3. Distribution of the axes in the polycrystals with the texture ⟨100⟩for three types of ordering of the basis (*A*, *B*, *C*) and two packing methods (*a*, *b*)

Zone		I	II	III	IV	V	VI	VII	VIII
A	*a*	1.07	3.38	4.84	27.31	48.76	8.45	1.80	4.39
	b	1.35	4.00	6.87	29.67	44.26	6.64	1.91	5.29
B	*a*	2.98	5.63	8.56	43.98	33.84	32.53	0.96	1.52
	b	2.82	6.59	9.63	46.06	30.41	32.14	0.56	1.80
C	*a*	11.37	22.97	6.59	25.85	28.38	2.14	10.85	21.86
	b	12.22	23.99	7.77	26.69	24.27	3.15	10.56	21.35

Table 3.4. Distribution with respect to for a polycrystals with a texture <100> for three types of ordering of the basis (A, B, C) and two packing methods (a, b)

Σ	*A*		*B*		*C*	
	a	*b*	*a*	*b*	*a*	*6*
1	42.57	59.57	9.97	13.63	8.22	8.50
3	0.00	0.00	0.00	0.00	0.00	0.00
5	0.62	0.34	2.59	3.38	33.11	33.39
7	0.11	0.00	0.00	0.00	0.17	0.17
9	0.23	0.00	0.28	0.17	0.73	0.51
11	0.00	0.00	0.00	0.00	0.00	0.00
13a	0.34	0.73	0.68	0.34	0.79	1.30
13b	0.90	1.07	0.51	0.45	0.68	0.45
15	0.06	0.11	2.20	2.42	0.56	0.45
17a	0.17	0.17	0.79	0.23	0.00	0.56
17b	0.00	0.00	0.00	0.00	0.00	0.00
19b	1,30	0.06	0.23	0.23	0.85	0.56
21a	1.13	0.73	0.73	0.17	0.39	0.28
25a	0.11	0.34	0.34	0.11	0.23	0.23
25b	0.00	0.00	0.00	0.00	0.00	0.00
*	2.65	2.87	3.72	3.60	2.93	2.76
**	49.83	33.78	77.98	75.28	51.35	50.85

*: $25 < \Sigma \leq 65$; **: $\Sigma > 65$

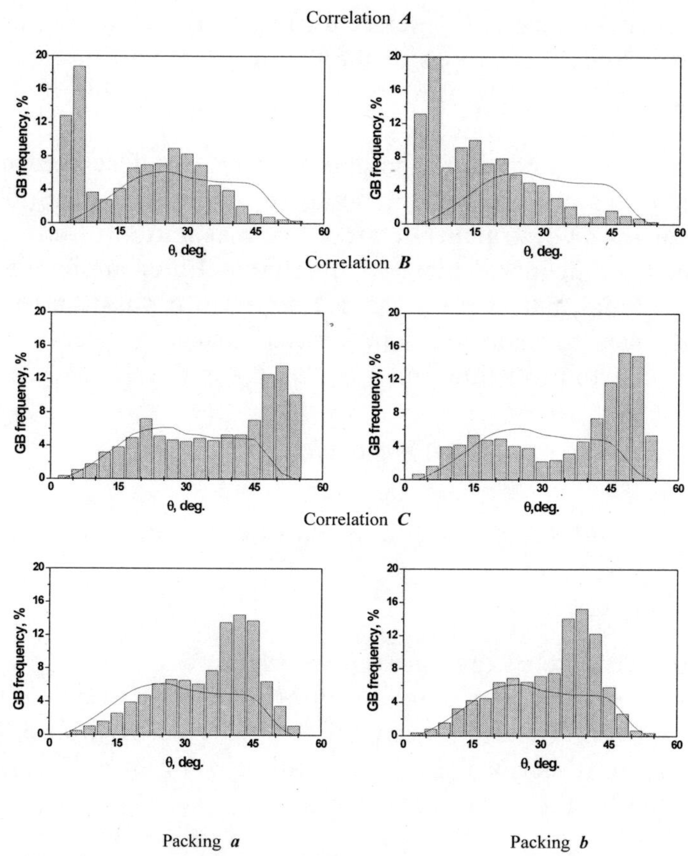

Fig. 3.4. Distribution of the misorientation angles in the polycrystals of the modelling texture ⟨100⟩ for the three types of correlation in the orientation of the adjacent grains and two packing methods: *a* (left) and *b* (right) (see Fig. 3.1.

3.4. Modelling the misorientation spectra of the grain boundaries in the FCC crystals with modelling ODF

As shown in the previous section, the calculation spectra for the axial textures, corresponding to different correlation types, are characterised by considerable differences. It is now necessary to verify this conclusion for the modelling textures similar to the actual texture [22]. At the same time, the orientation distribution function of such a texture should be less difficult to analyse than the ODF for the real polycrystals. The investigations were carried out on two types of texture, greatly differing from each other. The first type of texture corresponds to the texture of pure metals, rolled at room temperature. This is the so-called copper texture. The second term corresponds to the sharp cubic texture formed in pure FCC metals during recrystallisation annealing. In further sections, these types of the texture state will be referred to as DEF and RX, respectively. Both types of texture are represented by the orientation distribution functions. The quantitative characteristic of the texture was determined using the approach briefly described in section 3.1. We now examine in greater detail the algorithm of modelling of the ODF for the two structural states. The orientation of any grain can be described by means of three parameters – Euler angles (φ_1, φ, φ_2), and the rotational matrix **g** (φ_1, φ, φ_2) defines all possible rotations of the crystal basis of each grain in space. Thus, the three-dimensional orientation distribution function **f**(φ_1, φ, φ_2) is described as follows

$$f(q)dg = f(\varphi_1, \varphi, \varphi_2)(\sin\varphi / 8\pi^2)d\varphi_1 d\varphi d\varphi_2 = dV / V. \tag{3.26}$$

Here dV is the volume of the grains with the orientation in the interval

$$\left[\varphi_1 + d\varphi_1, \varphi + d\varphi, \varphi_2 + d\varphi_2\right], \tag{3.27}$$

and V is the volume of the entire polycrystal.

The Euler angles and ideal orientations (hkl) [uvw], describing the planes (hkl) situated parallel to the rolling plane, and the crystallographic directions [uvw], coinciding with the rolling direction, are connected by the relationships

$$\begin{aligned}
\varphi_1 &= \arcsin\left(w / \sqrt{u^2 + v^2 + w^2}\sqrt{(h^2 + k^2 + l^2)(h^2 + k^2)}\right); \\
\varphi &= \arccos\left(l / \sqrt{h^2 + k^2 + l^2}\right); \\
\varphi_2 &= \arccos\left(k / \sqrt{h^2 + k^2 + l^2}\right).
\end{aligned} \tag{3.28}$$

To obtain the modelling ODF of the rolled state, it is necessary to solve the inverse problem – construct the distribution function on the basis of the ideal orientations. For this purpose, the ODF is regarded as a set of N textural components with the given Euler angles (φ_{1i}, φ_i, φ_{2i}, where i = 1, 2,..., N), having the total width Ψ_{0i} have the half of the maximum and the normalisation coefficients M_i. Part of the grains of the polycrystal may have an arbitrary orientation (the so-called background). These orientations are selected with the normalisation multiplier. The total normalisation condition gives the equality $\sum_{i=1}^{N} M_i + F = 1$. In practice, it is convenient to calculate initially the coefficient of expansion of the ODF into a series:

$$C_l^{uv} = F_{0l} + \sum_{i=1}^{N} M_i \frac{\exp\left[-(l\psi_{0i})^2/4\right] - \exp\left\{-\left[(l+1)\psi_{0i}\right]^2/4\right\}}{1-\exp(-\psi_{0i}^2)} \times \tag{3.29}$$

$$\times T_l^{\mu\nu}(\varphi_1, \varphi, \varphi_2),$$

where $F_{0l} = F\delta_{0l}$; δ_{0l} is the Kronecker symbol. Consequently, the orientation distribution function of the grains is calculated from the formula (3.29), presented in a slightly different form:

$$f(\varphi_1, \varphi, \varphi_2) = \sum_{l=0}^{L} \sum_{\mu=1}^{M(l)} \sum_{\nu=1}^{N(l)} C_l^{\mu\nu} T_l^{\mu\nu}(\varphi_1, \varphi, \varphi_2), \tag{3.30}$$

where the truncation radius of the polynomials L is selected equal to 22. The appropriate values of (*hkl*) [*uvw*] for the recrystallisation texture are equal to (100) [001]. Consequently, we can use formally the same procedure of modelling of the ODF as in the case of the rolling texture. The values of the positions of the textural components, their half width and the weight multipliers are presented in Table 3.5. All the parameters were selected similar to those detected in the experiments [23, 24]. Figures 3.5 and 3.6 show the cross-section of the calculated ODF for the texture of the type of copper and recrystallisation texture. The intensity and shape of the peaks correspond to the well-known experimental data [23–25].

The misorientation spectra of the grain boundaries were calculated using the two-dimensional model described in previous sections and consisted of 625 grains and 1776 boundaries between them. The ODF of the data were used to determine 635 crystallographic bases using the following procedure. The generator of random numbers was used to simulate three angles φ_1, φ and φ_2, corresponding to the region of determination of the ODF. Subsequently, the fourth random number $\delta(0 \leq \delta \leq f_{max}(\varphi_1, \varphi, \varphi_2))$ was selected in the region of the variation of the orientation distribution function. If the condition $\delta \leq f(\varphi_1, \varphi, \varphi_2)$ was fulfilled, the resultant coordinates were accepted and three angles were used to construct the basis corresponding to the given ODF:

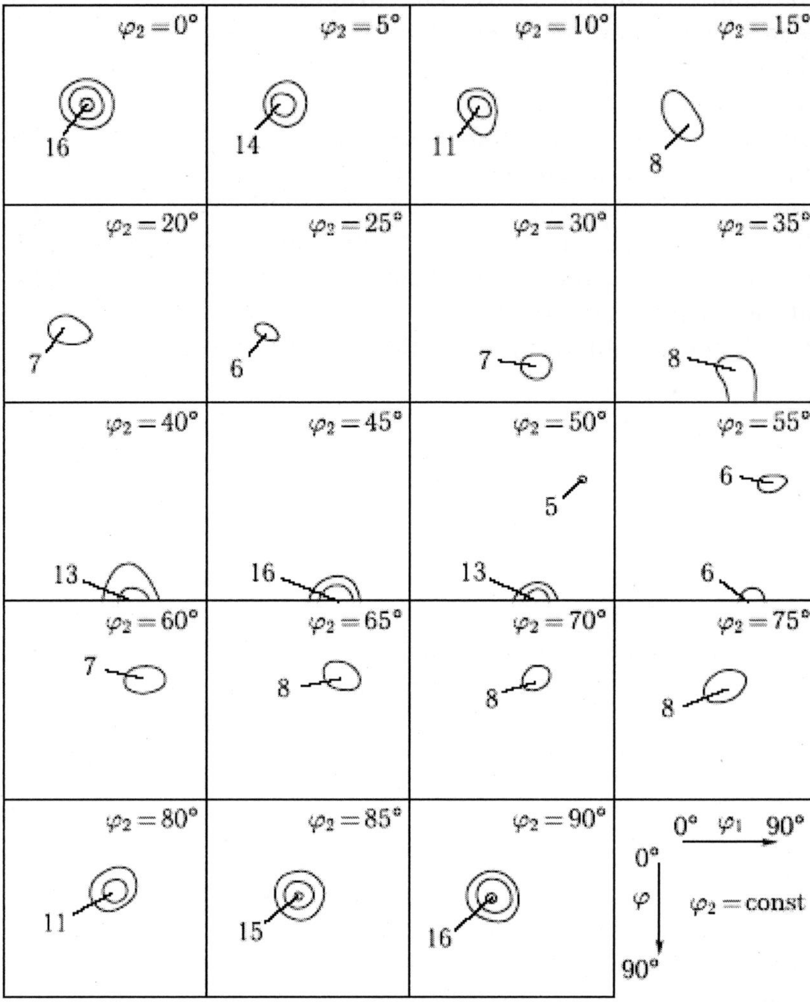

Fig. 3.5. Cross-section of the modelling of the ODF for the rolling texture (DEF).

$$B = \begin{pmatrix} \cos\varphi_1\cos\varphi\cos\varphi_2 - \sin\varphi_1\sin\varphi_2 & -(\cos\varphi_1\cos\varphi\sin\varphi_2 + \sin\varphi_1\cos\varphi_2) & \cos\varphi_1\sin\varphi \\ \sin\varphi_1\cos\varphi\cos\varphi_2 + \cos\varphi_1\sin\varphi_2 & -\sin\varphi_1\cos\varphi\sin\varphi_2 + \cos\varphi_1\cos\varphi_2 & \sin\varphi_1\sin\varphi \\ -\sin\varphi\cos\varphi_2 & \sin\varphi\sin\varphi_2 & \cos\varphi \end{pmatrix}. \quad (3.31)$$

Using all the determined bases, the procedure described previously was used to calculate the basis spectra of the misorientation of the grain boundaries (types A–D). As already mentioned, to analyse the misorientation spectra of the grain boundaries, it is necessary to select the standard GBMD. The standard can be represented by the chaotic spectrum of the misorientation of the grain boundaries, for example, the theoretically calculated Mackenzie misorientation spectrum [18]. In this case, the standard spectrum was represented by the basis spectra of the type *D* determined in the previous section. The GBMD, modelled for the rolling texture, was compared with the *D*-spectrum of the chaotic polycrystal, and for the recrystallisation texture with the GBMD of the

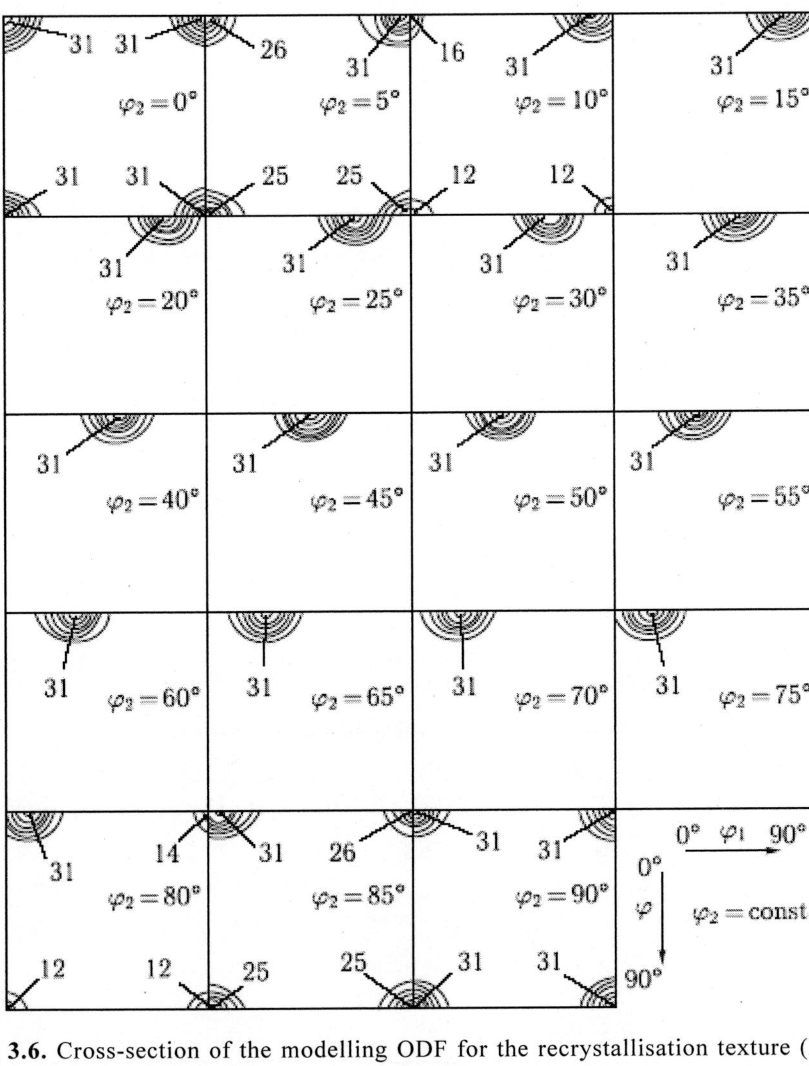

Fig. 3.6. Cross-section of the modelling ODF for the recrystallisation texture (RX).

Table 3.5. Values of the parameters used for modelling of the ODF

Texture type	Ideal orientations		Euler angles			Half width	Weight multiplier
	{*hkl*}	⟨*uvw*⟩	ρ_1	ρ	ρ_2		
DEF	011	211	35.3	45.0	0.0	10	0.32
	123	634	59.0	36.7	63.4	10	0.32
	112	111	90.0	35.3	45.0	10	0.06
	Background						0.30
RX	001	100	0.0	0.0	0.0	10	0.31
	Background						0.69

Table 3.6. Distribution of the axes in the textureless polycrystal for the four types of ordering of the bases (*A*, *B*, *C*, *D*) for the rolling texture (DEF) and the textureless state (TL).

Zone		I	II	III	IV	V	VI	VII	VIII
A	DEF	7.1	3.8	4.5	25.8	44.0	9.5	1.1	4.2
	TL	1.2	3.2	4.5	26.9	50.2	7.2	1.9	4.9
B	DEF	1.0	1.8	2.8	16.0	40.7	27.9	2.5	7.5
	TL	0.3	1.1	2.4	14.5	38.3	32.3	2.4	8.5
C	DEF	1.7	5.4	2.6	17.6	33.7	6.1	12.6	20.3
	TL	0.8	2.2	2.5	15.7	35.6	4.9	14.0	24.3
D	DEF	1.0	3.0	5.0	22.6	50.7	9.3	2.6	5.8
	TL	0.8	2.1	3.6	22.7	53.5	9.4	2.1	5.8

Table 3.7. Distribution with respect to in the textureless polycrystal for four types of bonding of the bases (A, B, C, D) for the rolling texture (DEF) and the textureless state (TL).

Σ	*A*		*B*		*C*		*D*	
	DEF	TL	DEF	TL	DEF	TL	DEF	TL
1	33.9	34.3	4.8	1.8	2.3	2.2	3.1	2.1
3	0.6	1.0	4.6	4.4	28.3	32.4	1.8	1.6
5	2.0	1.0	0.7	0.5	5.2	2.0	2.3	1.2
7.65	9.9	7.4	23.3	18.0	13.0	8.3	14.5	9.5
>65	53.6	57.3	66.6	75.3	51.2	56.1	78.3	85.6

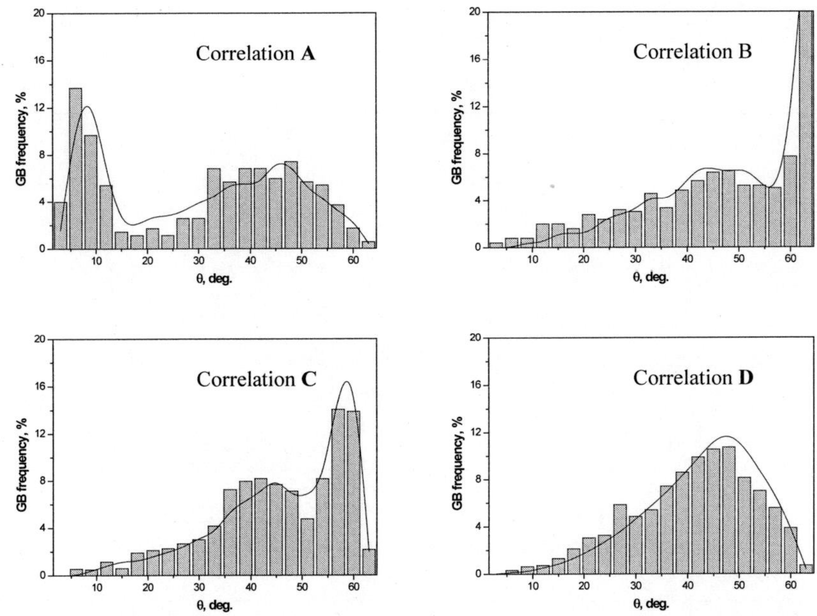

Fig. 3.7. Distribution of the misorientation angles for the deformation texture. The solid line shows the appropriate distribution, calculated for the textureless state.

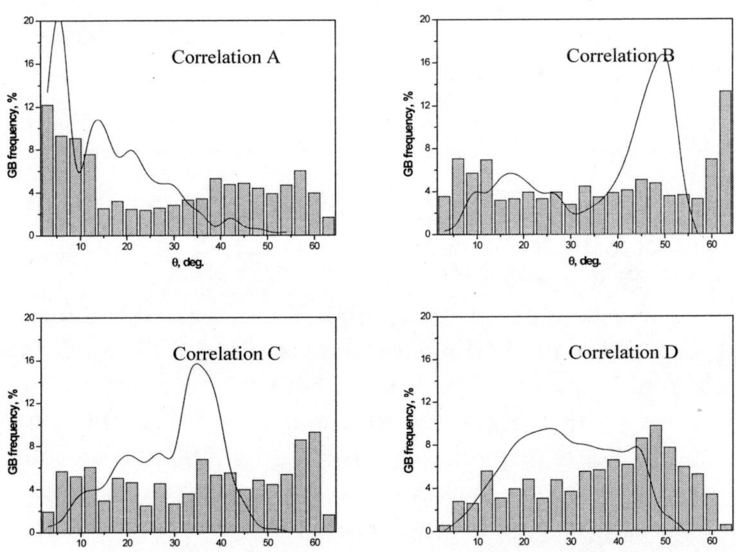

Fig. 3.8. The distribution of the misorientation angles for the recrystallisation texture. Solid line indicates the appropriate distribution, calculated for the axial texture.

Table 3.8. Distribution of the axes for four types of ordering of the bases (A, B, C, D) in the polycrystal with the recrystallisation texture (RX) and the modelling axial texture (100) AX

Zone		I	II	III	IV	V	VI	VII	VIII
A	RX	14.0	3.4	4.8	25.5	37.3	10.0	1.4	3.6
	AX	1.1	3.4	4.8	27.3	48.8	8.5	1.8	4.3
B	RX	6.8	2.9	4.2	22.9	32.9	22.7	1.8	3.8
	AX	3.0	5.6	8.6	14.0	33.8	32.5	1.0	1.5
C	RX	8.6	8.4	3.7	17.8	29.9	10.1	8.0	13.5
	AX	11.4	23.0	6.6	15.9	28.4	2.1	10.9	21.7
D	RX	5.3	3.3	5.1	24.6	44.8	11.1	1.2	4.6
	AX	2.3	6.4	9.2	38.3	36.6	4.4	0.8	3.4

Tables 3.9. Distribution with respect to Σ for the four types of ordering of the bases (A, B, C, D) in the polycrystal with the recrystallisation texture (RX) and the modelling axial texture ⟨100⟩ (AX)

Σ	*A*		*B*		*C*		*D*	
	RX	AX	RX	AX	RX	AX	RX	AX
1	42.9	42.6	25.6	10.0	19.5	8.2	13.3	13.0
3	0.3	0.0	2.5	0.0	17.5	0.0	1.4	0.0
5	1.5	0.6	0.8	2.6	5.6	33.1	2.0	3.2
7.65	9.5	7.0	17.8	19.4	16.7	17.3	14.0	15.5
>65	45.8	49.8	53.3	78.0	40.7	51.4	69.3	75.3

type *D* determined for the modelling axial texture ⟨100⟩. The calculated spectra (together with the standard spectra) are shown in Fig. 3.7 and 3.8 and also in Tables 3.6–3.9.

Detailed comparison of all the results shows that all the three types of basic spectrum for the rolling texture are similar to the GBMD for the textureless polycrystal. This relates both to the distribution of the angles and axes of the misorientation and to the distribution of the grain boundaries with respect to (inverse density of the coincident angles). For example, all the basis of misorientation spectra of the grain boundaries for the rolling texture have a maximum in the angle range 45–40° which corresponds to the maximum of the Mackenzie distribution full of however, all the three types of the basic spectrum greatly differ from each other, confirming the conclusion made in the previous chapter. This conclusion indicated that the type of distribution of the misorientation of the grain boundaries depend both on the texture and on the type of correlation in the orientation of the adjacent grains. The effect of the last factor is very strong. Only the very sharp texture (which is very difficult to obtain in practice) can suppress the effect of correlation and fully determine the type of the misorientation spectrum of the grain boundaries. For example, in comparison of the basic spectra for the two investigated types of the texture (DED and AX) shows that they greatly differ for different textures. This conclusion confirms the conclusion made in the previous section on the strong effect of the textural state on the misorientation spectrum of the grain boundaries the polycrystalline materials. Analysis of the basic GBMD for the recrystallisation texture show the following special feature: for all types of basic spectra is the typical feature is the large fraction of the low-angle grain boundaries and the high density of the misorientation axes of the grain boundaries in the vicinity of the $\langle 100 \rangle$ pole. The latter is spent by the fact that the [100], [010] and [001] axes are equivalent for the cubic lattice, and the selected recrystallisation texture is sharper in comparison with the modelling texture $\langle 100 \rangle$, used as the reference value.

In this section we described the method of calculating special misorientations in the crystals with the monochromatic lattice, and also the model and algorithm of calculating the misorientation spectrum of the grain boundaries on the basis of the textural analysis results. The results of the calculations of the coincident-site lattice in the monochromatic lattice were used to compile a table of special boundaries in crystals having the given lattice symmetry. The main conditions, essential for the restoration of the spectra of the grain boundaries with respect to the orientation distribution function were formulated, and the concept of the basic spectra, which can be used to determine the actual spectrum has the mean-weighted some of the basic spectra, was introduced. The results show that the misorientation distribution of the grain boundaries depend greatly on the crystallographic texture. However, the texture itself does not determine unambiguously the spectrum of the grain boundaries because the same texture may correspond to different distributions of the angles and axes of misorientation, depending on the correlation in the orientation of the adjacent grains. According to the experimental results, this correlation does not depend on the method of defining the acting lines of the 'given' grain boundaries but is controlled by their misorientation spectrum.

Using this approach it was shown for the first time that regardless of the type of texture, the misorientation distribution of the grains in any flat cross-section is representative of the entire volume specimen. Consequently, we can draw conclusions regarding the volume fraction of the distribution of the angles and axes of misorientation on the basis of the experimentally verified statistics of the grain boundaries on the flat sections in scanning electron microscopy and in foils in transmission electron microscopy.

The report approach has been tested on a polycrystal with the given modelling 'rolling' and 'recrystallisation' texture. The results of the analysis of the calculated basic spectra of the misorientation of the grain boundaries for the case of the texture close to the experimental texture, it can be concluded that both factors – the sharpness of the texture and the type of correlation – fully determine the type of the spectrum of the grain boundaries in the investigated materials.

References

1. Viglin A.C., FTT, 1960, vol. 2. No. 10, 2463–2476.
2. Bunge H.J., Z. Metallk., 1965, vol. 56, No. 12, 872–874.
3. Roe R.J., J. Appl. Phys., 1965, vol. 36, No. 6, 2024–2031.
4. Bunge H.J. Matematische Methoden der Texturanalyse. Berlin, Acad. Verlag, 1969.
5. Matthies S. Actuelle Problem der Quantativen Texturanalyse. Berlin, 1982.
6. Pospieshe J., Luche K., Acta Metall., 1975, vol. 23, No. 8, 997–1007.
7. Papirov I.I. (editor), New methods of investigating the texture of polycrystalline materials, Metallurgyia, Moscow, 1985.
8. Vishnyakov Ya.D., et al., Theory of formation of textures in metals and alloys, Nauka, Moscow, 1979.
9. Aleksandrov I.V. and Kaibyshev O.A., Zavod. Lab., 1978, No. 3, 286–289.
10. Strartseva T.V., et al., ibid, 1978, 334–338.
11. Matthies S., Krist. und Techn., 1980, vol. 15, No. 4. 431–444.
12. Matthies S., Krist. und Techn., 1980, vol. 15, No. 5, 601–614.
13. Matthies S., Krist. und Techn., 1980, vol. 15, No. 7, S. 823–835.
14. Matthies S., Krist. und Techn., 1980, vol. 15, No. 10, 1189–1195.
15. Matthies S., Krist. und Techn., 1980, vol. 15, No. 11, 1323–1328.
16. Savelova T.I., Zavod. Lab., 1989, vol. 55, No. 9, 57–60.
17. Nikolaev D.I. and Savelova T.I., Izv. AN SSSR, Metally, 1989, No. 6, 165–169.
18. Gertsman V. Y., Zhilyaev A.P., Pshenichnyuk A.I., Materials Forum, 1990, vol. 62–64, Pt.II, 41–43.
19. Gertsman V. Y., et al., Acta Met., 1992, vol. 40, 1433–1439.
20. Korn G. and Korn T., A handbook of mathematics, Nauka, Moscow, 1973.
21. Zhilyaev A.P., et al., Strength of Materials. Proc. of the 10th Int. Conf. ICSMA 10, Sendai: The Japan Institute of Metals (Ed. by H. Oikawa et al.), 1994, 887–890.
22. Alexandrov I.V., et al., Modelling Simul. Mater. Sci. Eng., 1995, vol. 3, 149–159.
23. Schmidt U. and Lucke K., Texture Cryst. Solids, 1979, vol. 3, 85–112.
24. Pospiech J. and Lucke K., Acta Metall. Mater., 1975, vol. 23, 997–1007.
25. Alexandrov I.V. and Davies G.J., Mater. Sci. Eng., 1985, vol. 75, L1–L4.

4

EXPERIMENTAL INVESTIGATIONS OF GRAIN BOUNDARY ENSEMBLES IN POLYCRYSTALS

4.1. Diffraction methods of measuring misorientation

The successes in the determination of the crystal geometry parameters of the grain boundaries, in particular, the most important parameter – the misorientation of the adjacent grains – are associated with the development of experimental techniques [1, 2]. The efficiency of the experimental techniques is determined by two main parameters: the accuracy of measuring misorientation and the localisation of the procedure, i.e., the minimum size of the area from which it is still possible to obtain the diffraction pattern without superposition from the adjacent grains. Twenty years ago the application of electron microscopes, using auto-emission cathodes (field emission gun) and generating the electron beam with the size of up to 20 nm and with high intensity (sufficient for producing the diffraction pattern) was only in the initial stage [1, 2]. It was therefore necessary to develop conventional methods [3], in particular, the method of determination of the maximum error of determination of the misorientation in statistical investigations. It should be noted that the traditional methods are also used extensively at the moment. This is an essential tool for the precision measurement of the misorientation of the grain boundaries. In addition, similar methods are used when it is not possible to use the automated methods. Because of these circumstances, the original results obtained in the investigation of the experimental errors and described in the present chapter are still valid at the present time.

4.1.1. Methods of measuring the misorientation of two adjacent grains

In transmission electron microscopy, there is a relatively large number of diffraction methods [3, 4] used for determining the misorientation of two adjacent grains. Two of these methods will be discussed as an example.

The first is the method using the Kikuchi lines and characterised by the maximum accuracy of determination of the misorientation (~0.1°). All the methods based on the application of the Kikuchi lines are in fact equivalent. They differ only in the selection of the quantities measured on the microdiffraction pattern. The Ryder–Pitch method [5, 6] uses three reflections which do not belong to the same zone, and three pairs of corresponding Kikuchi lines. Figure 4.1 shows the diagram of measurement using this procedure for a single reflection and the corresponding Kikuchi lines. The deviation from the accurate Bragg position is represented by parameter a_i. The angle between the direction of the beam **Z** and the direction to the reflection $\mathbf{g}_i$ is determined by the expression

$$\cos\left(\Phi_i\right) \approx \frac{x_i + 2a_i}{2L}, \tag{4.1}$$

Here L is the effective length of the diffraction chamber. For the small diffraction angles $x_i / L = \lambda\,|\mathbf{g}_i|$, where λ is the wavelength of the electrons. Consequently

$$\cos\left(\Phi_i\right) \approx \frac{x_i + 2a_i}{a_i}\frac{\lambda}{2}, \tag{4.2}$$

On the other hand,

$$\cos\left(\Phi_i\right) = \frac{\left(\mathbf{Z}_i\mathbf{g}_i\right)}{\left|\mathbf{Z}_i\right|\left|\mathbf{g}_i\right|}. \tag{4.3}$$

Combining (4.2) and (4.3) we obtain the expression

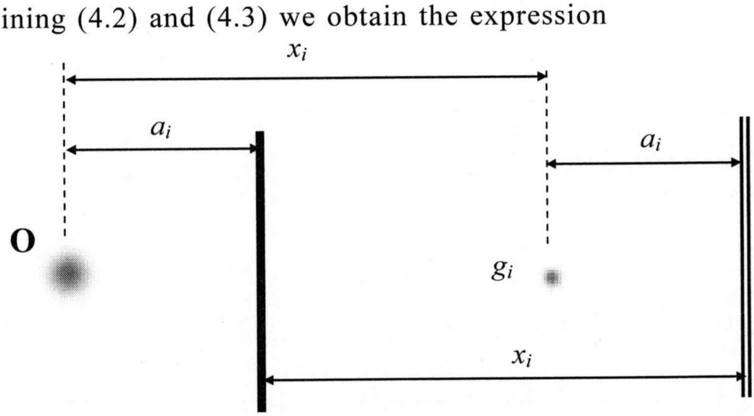

Fig. 4.1. Measurement of distances in the Ryder–Pitch method.

$$\left(\mathbf{Z}_i\mathbf{g}_i\right)=\frac{\lambda}{2}\frac{x_i+2a_i}{a_i}\left|\mathbf{Z}_i\right|\left|\mathbf{g}_i\right|^2, \quad \text{where } i=1, 2, 3. \tag{4.4}$$

The direction of the beam **Z** in this case is determined by the general solution of the three equations of the type (4.4):

$$\mathbf{Z}=a_1\left|\mathbf{g}_1\right|^2\left[\mathbf{g}_2\times\mathbf{g}_3\right]+a_2\left|\mathbf{g}_2\right|^2\left[\mathbf{g}_3\times\mathbf{g}_1\right]+a_3\left|\mathbf{g}_3\right|^2\left[\mathbf{g}_1\times\mathbf{g}_2\right] \tag{4.5}$$

where **Z** is the axis of the zone of the vectors $\mathbf{g}_1$ and $\mathbf{g}_2$, with the ends of the vectors situated on the Ewald sphere. In the Ball method [7] it is proposed to exclude the point defects and determine the vectors **g** by measurements of only of the position of the Kikuchi lines. Since this method is used for the determination of the experimental error of measurement of the misorientation of the two grains, it will be described in greater detail. The Kikuchi lines were discovered in 1928; they are named after the scientist who observed them for the first time. The mechanism of the formation of these lines may be described as follows. In inelastic and non-coherent scattering in thick foils, the electrons can be subjected to secondary coherent scattering, if the Bragg law is satisfied at a specific orientation of the reflecting planes. Radiation is scattered in cones. If the electron beam falls symmetrically on the reflecting plane scattering cones of the same intensity form on both sides of the plane. For the normal conditions, the traces of these reflections are seen on photoplates (or a screen) as straight lines (dark and light), corresponding to the planes (*hkl*) and –(*hkl*). The distance between the lines for each Kikuchi pair is proportional to 2θ, i.e., $p_i \sim 2\theta$, or $p_i \sim 1/d_{hkl}$. Having three pairs of the Kikuchi lines (i.e., knowing the width p_i and angles α_i between them), it is possible to index the diffraction pattern:

$$p_i / p_j = \frac{\sqrt{h_i^2+k_i^2+l_i^2}}{\sqrt{h_j^2+k_j^2+l_j^2}};$$

$$\cos\alpha_{ij} = \frac{h_ih_j+k_ik_j+l_il_j}{\sqrt{h_i^2+k_i^2+l_i^2}\sqrt{h_j^2+k_j^2+l_j^2}}.$$

This expression holds for cubic lattice crystals. Knowing the indexes of the Kikuchi lines, we can calculate the appropriate indexes of the Kikuchi poles and find the crystallographic coordinates of the electron beam.

The methods of calculating the misorientation of the grains on the basis of the microdiffraction data are in fact reduced to two main methods. The first method uses only the directions of the beam. However, in this case it is necessary to record electron diffraction patterns at at least two positions of the goniometer. The second method uses electron diffraction diagrams taken from the adjacent grains only in one position of the goniometer but in addition to the direction of the beam the orientation of the grain is calculated using some direction situated in the plane of the electron diffraction diagram (for example, the edge of the photoplate). With other conditions being equal, the first method is characterised by an order of magnitude higher accuracy of calculating the orientation of the grain. The algorithm of calculating the misorientation using the first method may be described as follows. If the crystallographic direction in the grain A (direction of the beam), given by the vector $\mathbf{A}$ (A_1, A_2, A_3), is parallel to the appropriate vector in the grain B (and $|\mathbf{A}| = |\mathbf{B}| = 1$), the misorientation $\mathbf{R}$ between the grains A and B corresponds to equality $\mathbf{B} = \mathbf{RA}$. Consequently, it is quite easy to determine the misorientation matrix: $\mathbf{R} = \mathbf{BA}^{-1}$. Knowing the matrix $\mathbf{R}$, it is quite easy to determine the misorientation angle and axis.

This method will be examined in greater detail. If the electron diffraction pattern shows three pairs which do not belong to the same zone of the Kikuchi lines, then there are three Kikuchi poles. To calculate the beam direction, it is necessary to know the mutual position of the triangle, formed by the Kikuchi lines, and the 'trace' of the primary beam. Different measurements can be used for the calculations. The distances a (Fig. 4.2) correspond to the angular deviations α_i from the direction of the primary beam of the crystal planes generating the appropriate Kikuchi lines ($\mathbf{g}$ are the unit vectors of the normals to these planes (h,k,l)). In this case, the main system of the equations for calculating the direction of the electron beam $\mathbf{Z}$ has the following form

$$(\mathbf{Zg}_i) = \sin\alpha_i. \tag{4.6}$$

The angles α determined from the angular scale of the electron diffraction pattern or using the equation

$$\sin\alpha_i = \frac{\alpha_i}{\sqrt{\alpha_i^2 + L^2}} \tag{4.7}$$

where L is the effective length of the diffraction chamber.

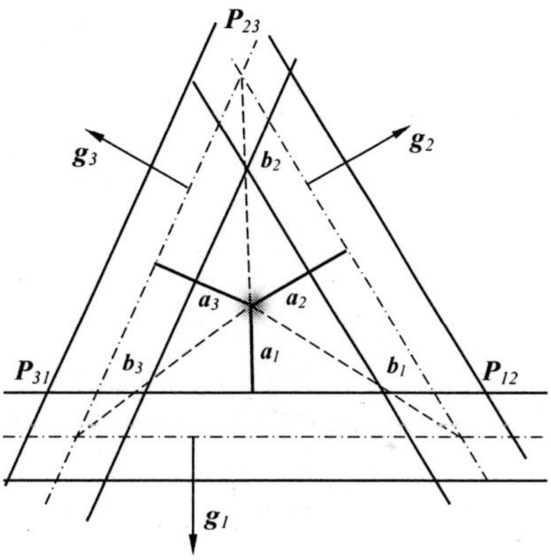

Fig. 4.2. Diagram showing the measurement of the distances for the determination of the crystallographic coordinates of the electron beam on the Kikuchi diagram.

The beam direction can also be calculated using the distance bi as to the Kikuchi poles P_j (Fig. 4.2) [8]. In this case, the equation used to calculate the crystallographic coordinates of the electron beam is:

$$\left(\mathbf{ZP}_{ij}\right) = \sin\beta_{ij}. \tag{4.8}$$

Like the angles α_i, the angles β_{ij} are determined using the distances b on the angular scale or using the equation

$$\cos\beta_{ij} = \frac{L}{\sqrt{b_i^2 + L^2}} \tag{4.9}$$

If the diffraction pattern contains only one Kikuchi pole, the beam direction can be calculated using the system consisting of the equations of the type (4.6) and (4.8), and the normalisation condition: $|\mathbf{Z}| = 1$.

The misorientation of the two grains is calculated from the known orientations of both grains. These orientations can be determined either in the absolute form, i.e., on the basis of the ratio to the laboratory system of the coordinates, or in the relative form, i.e., using the relationship, for example, in the coordinate system connected with the crystallographic lattice of one of the grains. In order to determine unambiguously the position of the grain, it is necessary to determine at least two non-parallel directions in the grain. These directions can then be used to determine the orientation of the crystal in relation to the external laboratory coordinate system. However, if such pairs of directions are selected in the adjacent grains and their mutual position is known, then data on the similar orientation are sufficient for calculating the misorientation of these grains.

As already mentioned, the advantage of the method using the Kikuchi lines is a relatively simple procedure for obtaining experimental information and high accuracy of the method. A shortcoming is the fact that the electron diffraction patterns with distinctive Kikuchi lines can be obtained only from relatively perfect crystals (with the dislocation density not exceeding 10^9 m^{-2}) in thick sections of the foil, and also with the localisation of the method, restricted to several microns [9].

The second method, characterised by high localisation (up to 0.1 μm), is the single reflection method [9, 10]. The method is based on the direct determination of the coordinates of the vector **g**. Consequently, only a single diffraction reflection can be seen clearly on the diffraction pattern. The method is based on the determination of the orientation of the individual grains in relation to the laboratory coordinate system. The grain orientation is determined using the orientation matrices which can be constructed having only three non-complanar reference vectors, expressed in both the laboratory coordinate system, connected with the diffraction pattern, and in the crystallographic system of the coordinates connected with the unit vectors of the reciprocal lattice. The reference vector is represented by the direction to the node of the reciprocal lattice, situated on the Ewald sphere. The resultant orientation matrices of two adjacent grains are used to calculate the misorientation matrix which is used to determine the minimum misorientation taking into account the operators of the rotational symmetry of the crystal lattice. The advantage of the single reflection method is the high localisation and the possibility of working in elastically loaded regions with the dislocation density up to 10^{12} m^{-2}. Shortcomings of the method include the low accuracy (in comparison with the Kikuchi line method) of determination of the misorientation (~0.3–0.5°) and a high labour content of the calculations.

In addition to the previously examined diffraction methods, used in transmission electron microscopy, the misorientation of the grain boundaries was also determined using the channelling patterns of the electrons (in a scanning electron microscope) [11] and also the diffraction x-ray methods [12].

4.1.2. The experimental measurement error

The main problem in the experimental verification of the ensembles of the grain boundaries in polycrystalline materials is the problem of calculating the error of experimental determination of the misorientation of two adjacent grains [13, 14]. In particular, the accuracy of determination of the misorientation depends on the accuracy of determination of the directions of the beam and the error of determination of the misorientation. The main sources of the errors in the determination of the beam orientation are [13, 14]:

– distortion of the foil as a result of thermal stresses, generated by heating with the electron beam, the nonuniform compression of the foil in the holder, sagging of the foil and the effect of its own weight, etc;

– the blurriness of the reflection and Kikuchi lines on the electron diffraction pattern (the accuracy of determination of the central reflection, i.e., the trace of the primary beam, is especially important);

–the error of measurement of the distances on the electron diffraction pattern;

–the errors of the values of the wavelength of the electrons, the effective length of the diffraction chamber, the crystal lattice spacing of the investigated material.

The error of determination of the misorientation depends on the angle of inclination of the goniometer between two positions in which the electron diffraction patterns are recorded. In addition, a contribution to this error is also provided by the fact that the microdiffraction section cannot be repeated in different positions of the goniometer. Thus, the accuracy of determination of the misorientation depends on a large number of factors. Therefore, it is not possible to obtain the analytical expression suitable for the resultant error. Taking this into account, the accuracy was verified by experiments. To determine the accuracy of determination of the beam direction, one of the electron diffraction patterns is used to calculate several variants of some direction using different Kikuchi lines as the initial data. The angular scatter characterises the required accuracy. The value of the error of determination of the misorientation is necessary to calculate the misorientation of the same boundarie using, as the reference vectors, different pairs of beam directions into the grains determined for different positions of the goniometer. The scatter of the calculated misorientation values characterises the accuracy of determination of misorientation. In early studies [15, 16], the accuracy of determination of the beam directions was estimated using single examples without sufficient statistical data. The relationship of the accuracy of calculation of misorientation with the error of determination of the beam direction was not determined. To solve these problems, the following experiments were formulated. A JEM-2000EX microscope with the accelerating voltage of 160 kV ($\lambda = 0.00285$ nm) was used to investigate the accuracy of determination of the misorientation of a twin boundary (the type of the boundary was determined on the basis of the external morphological features) in annealed Ni–Cr alloy (lattice spacing a = 0.3562 nm). The electron diffraction patterns were recorded at the effective length of the chamber of $L = 610 \pm 3$ mm for nine positions of the goniometer in the inclination angle range ±60° around a constant axis. The calculations, starting with the indexing stage, were carried out in a computer. The data on the position of the Kikuchi lines and the trace of the primary beam were fed into the computer memory using a digitiser. The mean error of measurement of the distances was 0.2 mm, and this accuracy was restricted by the extent of blurring of the Kikuchi lines and not by the accuracy of the measurements.

As an example, Fig. 4.3 shows the photographs of a typical electron diffraction pattern and its scheme with the indexed Kikuchi lines and Kikuchi poles. The presence of several Kikuchi poles makes it possible to use different methods for calculating the beam direction. Table 4.1 shows the values of the beam directions, calculated using different procedures [6, 15, 17, 18]. 167 variants of the beam directions were found for 20 recorded electron diffraction patterns. The histogram in Fig. 4.4 shows the distribution of deviations from the mean values according to these calculations. It can be seen that the maximum distribution is found in the range 0.02–0.03°, and the majority (>75%) of the values belongs in the range 0–0.04°. These data are in agreement with the estimates of the accuracy carried out in other investigations. For example, in [19] the beam direction was determined with the accuracy of ~0.06° (at the accuracy of measurement of the distances of 0.5 mm). The results obtained in [16] also show the scatter in relation to the mean order expressed in hundredths of a degree. The data presented in Table 4.1 also shows that the accuracy decreases in the mean if the calculations are carried out using the poles far away

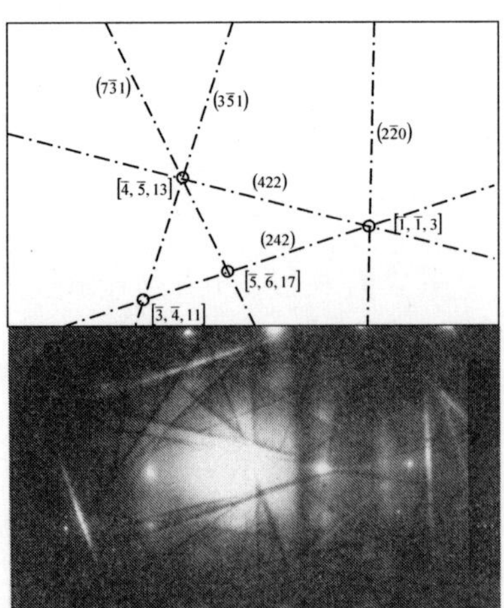

Fig. 4.3. Scheme (a) and photograph (b) of an electron diffraction pattern with equal to lines. To simplify understanding, only the mean lines for each pair of the Kikuchi lines are given.

Table 4.1. Beam directions, calculated using different methods

No.	Kikuchi lines	Beam direction	Deviation, deg
	Calculated on the basis of three poles		
1	(4–40) (–351) (–4–2–2)	[–0.29299 –0.33245 0.89645]	0.121
2	(4–40)(–351)(–2–4–2)	[–0.29362 –0.33135 0.89666]	0.139
3	(4–40)(–351)(7–31)	[–0.29485 –0.33361 0.89541]	0.025
4	(–4–2–2)(7–31)(242)	[–0.29530 –0.33366 0.89525]	0.044
	Calculated on the basis of three lines		
5	(4–40)(–351)(–4–2–2)	[–0.29479 –0.33340 0.89551]	0.011
6	(4–40)(–351)(–2–4–2)	[–0.29444 –0.33314 0.89572]	0.021
7	(4–40)(–351)(7–31)	[–0,29506 –0,33326 0,89548]	0,018
8	(_4_2–2)(7–31)(242)	[–0.29514 –0.33316 0.89549]	0.023
9	(_4_2–2)(–351)(242)	[–0.29461 –0.33330 0.89561]	0.010
	Calculated on the basis of two lines and a pole		
10	(4–40) (–351)	[–0.29383 –0.33275 0.89607]	0.066
11	(–35l)(–4–2–2)	[–0.29494 –0.33331 0.89550]	0.012
12	(4–40)(–351)	[–0.29474 –0.33365 0.89544]	0.026
13	(–440) (3–5–1)	[–0.29331 –0.33239 0.89637]	0.106
14	(–35l)(–2–4–2)	[–0.29486 –0.33316 0.89558]	0.007
15	(4–40) (242)	[–0.29437 –0.33344 0.89064]	0.026
16	(4–40) (–351)	[–0.29525 –0.33314 0.89546]	0.029
17	(–351)(–73–l)	[–0.29534 –0.33338 0.89534]	0.037
18	(4–40) (–73–1)	[–0.29499 –0.33358 0.89538]	0.027
19	(422) (–73–1)	[–0.29536 –0.33303 0.89546]	0.037
20	(242) (–73–1)	[–0,29568 –0.33318 0.89530]	0.055
21	(–35l)(–2–4–2)	[–0.29506 –0.33353 0.89537]	0.028
22	(3–5–l)(422)	[–0.29449 –0.33322 0.89568]	0.017
23	(3–5–l)(242)	[–0.29462 –0.33329 0.89561]	0.009
24	(–351) (242)	[–0.29462 –0.33339 0.89551]	0.011

Table 4.1. (Cont.)

No.	Kikuchi lines	Beam direction	Deviation, deg
	Calculated using the Ryder–Pitch methods		
25	(4–40) (–351) (–4–2–2)	[–0.29505 –0.33344 0.89541]	0.023
26	(4–40)(–351)(–2–4–2)	[–0.29483 –0.33329 0.89554]	0.006
27	(–4–2–2)(–351)(242)	[–0.29526 –0.33331 0.89539]	0.031
	Calculated using the Pumprey–Bowkett and Ball methods		
28	(4–40) (–351)	[–0.29441 –0.33301 0.89578]	0.026
29	(4–40) (422)	[–0.29501 –0.33362 0.89536]	0.030
30	(4–40) (242)	[–0.29486 –0.83341 0.89551]	0.012
31	(–351)(–4–2–2)	[–0.29515 –0.33338 0.89540]	0.027
32	(–351)(–2–4–2)	[–0.29500 –0.33330 0.89548]	0.015
33	(422) (242)	[–0.29530 –0.33313 0.89545]	0.032
	Calculated using the Gertsman–Valiev and Skakova–Golub'–Orlova methods		
34	(4–40) (–351)	[–0.29435 –0.33301 0.89580]	0.029
35	(4–40) (422)	[–0.29481 –0.33351 0.89547]	0.018
36	(4–40) (242)	[–0.29447 –0.33347 0.89559]	0.022
37	(–351)(–4–2–2)	[–0.29507 –0.33321 0.89519]	0.018
38	(–351)(–2–4–2)	[–0.29490 –0.33316 0.89557]	0.009
39	(422) (242)	[–0.29505 –0.33354 0.89538]	0.027
	Mean	[–0.29476 –0.33322 0.89559]	

from the centre of the electron diffraction pattern. The effect of the errors in the determination of the individual quantities, used in the calculations, on the accuracy of determination of the beam direction was verified by experiments. The variation of one of these quantities λ, L or *a*, by 1% leads to changes in the direction of the beam $\Delta\mathbf{Z}$ by $\leq 0.1°$. Usually, the crystal lattice spacing *a* is determined with a considerably higher accuracy (at least 0.01%) and this error causes that

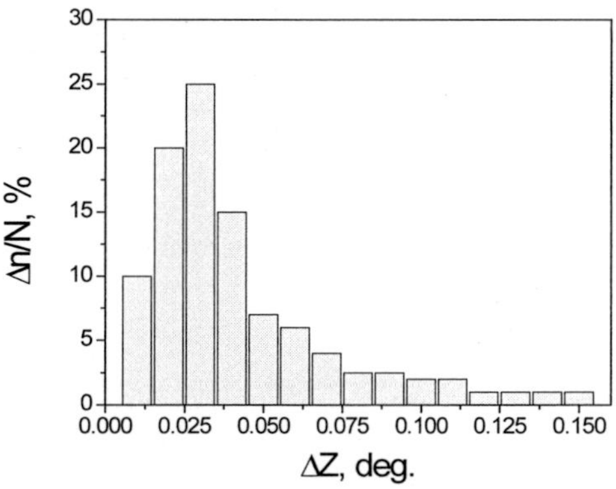

Fig. 4.4. Distribution of deviations from the mean beam direction determined by calculations using Kikuchi lines.

the $\Delta \mathbf{Z}$ values are expressed in thousandths of a degree. The wavelength of the electrons and the effective length of the diffraction chamber in every specific case can also be determined with the accuracy better than 1%, and the indeterminacy finding these values results in the error in the direction of the beam of the order of hundredths of a degree. Therefore, one can agree with the conclusions made in [13]: the accuracy of determination of the misorientations is restricted by indeterminacies in the measurement of the diffraction pattern and the error in the diffraction constant of the device. The mean values of the beam direction for all nine pairs of the electron diffraction patterns were used to calculate 45 values of the misorientation of the investigated twin boundary. The misorientation matrix was determined for each variant on the basis of the two pairs of directions, and the misorientation matrix was then used to calculate the misorientation axis and angle. All the 45 misorientation matrices were orthonormalised with the accuracy of at least 10^{-7}. In Fig. 4.5 the main deviations of the misorientation values are plotted in relation to the difference in the inclination angles of the goniometer. Each point on the graph is related to 3–7 misorientations for which the mean misorientation and the mean quadratic deviations were calculated. For example, the misorientation at $\Delta\varphi = 90°$ was calculated on the basis of the beam directions, determined for the goniometer positions of –60° and +30°, –45° and +45°, –30° and +60°, i.e., these are independent measurements for the same boundary. Fig. 4.5 shows the following. Firstly, as expected, the error depends strongly on the angle between the positions of the goniometer. At $\Delta\varphi = 45°$ the mean square scatter of the misorientations is almost completely independent of $\Delta\varphi$. Secondly, the error of determination of the misorientations is almost an order of

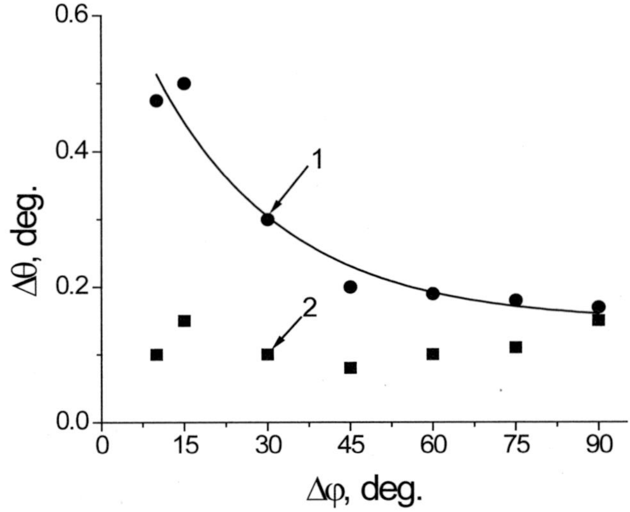

Fig. 4.5. Dependence of the mean quadratic error of determination of misorientation on the position of the goniometer: deviation from the mean (1) and ideal (2) misorientation.

magnitude higher than the mean error of determination of the beam direction. The latter was verified for many other cases, and in all cases the accuracy of determination of the misorientation was at least a half an order of magnitude lower than the accuracy of determination of the beam directions. Figure 4.5 also shows the deviations of the mean misorientations for every $\Delta\varphi$ from the ideal misorientation $\Sigma 3$. It may be seen that they are smaller than the mean quadratic deviations, i.e., the misorientation of the investigated boundary did not differ from the ideal misorientation within the range of the experimental error. The misorientation error obtained in the investigated example (approximately 0.15°) is evidently close to the maximum accuracy of determination of the misorientation attainable in the given experiment. Generally speaking, this accuracy is at the moment obtained only in the investigations of the individual boundaries because in investigating the ensemble of the boundaries it is almost impossible to realise simultaneously all optimum conditions ensuring this type of accuracy. Therefore, the accuracy of determination of the misorientation in statistical investigations is always at least an order of magnitude lower.

It should be mentioned that in many cases it is not possible to achieve the maximum accuracy because this accuracy may not simply correspond to the actual situation. The point is that even if the microdiffraction patterns for different positions of the goniometer are obtained strictly from the same areas of the foil in each grain (this is highly problematic), the calculated misorientation characterises the misorientation of certain sections of the crystals in the vicinity of the boundary and not the misorientation of the boundary. In addition to this, there are almost always differences in the orientation of the crystals in different sections as a result of distortion of the foil and simply as a result of its sagging in the holder and of the effect of its own weight, not mentioning the cases in which the investigated section contains dislocations (even single ones). Generally, as reported in [16], the final error is determined mainly by the residual deformation of the foil and distortions of the diffraction pattern and not by the errors of the method of determination of orientation and misorientation.

In conclusion, we present the results of identical investigations of the accuracy of determination of the orientation and misorientation of the adjacent grains with twin orientation, in the same material, but with the smaller grain size. The appropriate diffraction patterns were obtained by diffraction in a converging beam [20]. It should be mentioned that when using diffraction in the converging beam some error sources provide a considerably smaller contribution. This method can be used almost in all cases to obtain efficiently resolved Kikuchi lines. In addition, the condition of non-repeatability of the diffraction section in the submicrocrystalline materials is not so strict. The accuracy of determination of the beam direction was evaluated using the previously mentioned procedure, calculating several variants of the beam direction on the same electron diffraction pattern and using different Kikuchi lines. The experiments were carried out using a foil of submicrocrystalline Ni–Cr alloy (d <0.1 μm), and a twin boundary was again tested. To obtain diffraction in the converging beam, experiments were carried out using a stationary non-coherent beam with the convergence angle α_0 = 3mrad and the diameter of several tens of nanometres [20, 21].

The typical electron diffraction pattern, obtained by diffraction in the converging beam, is identical with that shown in Fig. 4.3. The calculation results of the beam direction are presented in Table 4.2. This angle scatter indicates the deviation from the mean angle for each calculation variant and characterises the accuracy of locating the beam. It can be seen that the accuracy of determination of the beam direction is 0.01–0.02°. Table 4.3 shows the characteristics of the diffraction patterns, produced in adjacent grains in two positions of the goniometer. The results were used to calculate the misorientation of the grain boundary:

Table 4.2. Calculation of the beam direction using different Kikuchi lines

No.	Kikuchi lines	Beam direction	Deviation, deg
1	(113) (–5–1–3) (3–1 –1)	[–0.10651 –0.91801 0.38199]	0.016
2	(113) (–5–13) (2–2–4)	[–0.10658 –0.91782 0.38242]	0.014
3	(113) (—3—1—3) (3–1–1)	[–0.10635 –0.91801 0.38202]	0.021
4	(113)(–3–1–3)(2–2–4)	[–0.10656 –0.91787 0.38231]	0.007
5	(204) (4–2–4) (–422)	[–0.10690 –0.91796 0.38201]	0.019
6	(424) (3–13) (20–4)	[–0.10684 –0.91781 0.38238]	0.016
7	(424) (–3–15)(–1–1–5)	[–0.10670 –0.91784 0.38233]	0.009
Mean from 7 measurements		[–0.10664 –0.91790 0.38221]	

Table 4.3. Calculation of two parallel directions of the beam in adjacent grains

Grain	Position of goniometer	Beam direction
A	1	[–0.10734 –0.91873 0.380010]
A	2	[0.53992 –0.78529 0.303000]
B	1	[–0.41446 0.90163 –0.132362]
B	2	[0.82281 –0.26453 –0.503300]

$$R = \begin{pmatrix} -0.66683 & -0.33455 & 0.66590 \\ 0.66673 & -0.66699 & 0.33256 \\ 0.33289 & 0.66573 & 0.66782 \end{pmatrix}. \tag{4.10}$$

and also the angle of misorientation (59.95°) and the misorientation axis ([0.5767–0.5770–0.5784]). The determined misorientation differs from the special boundary Σ3 (60° [111]) by 0.097°.

The investigated examples of the determination of the beam orientation and grain misorientation show that the application of diffraction in the converging beam makes it possible to determine the misorientation of two grains with the error which at least is not greater than the error obtained when using diffraction from the selected region. Thus, the method can be used to determine with high accuracy the crystal geometry parameters of the grain boundary in submicrocrystalline materials and also investigate local changes of the misorientation in coarse-grained materials.

4.2. Experimental spectra of the grain boundaries in FCC polycrystals

We discuss the results of several experimental investigations of the misorientation spectra of the grain boundaries carried out in recent years [22–26], paying special attention to FCC metals. To summarise the results obtained in individual investigations they are presented in tables.

In [22] the method of Laue x-ray diffraction was used to measure the misorientation of 22 grain boundaries in pure nickel. The microstructure with the mean grain size of 1000 μm was produced after annealing at 1573 K for 10 min with preliminary deformation of the specimens (ε = 5%).

The grade AISI 304L stainless steel [24] with the microstructure with the mean grain size of ~100 μm was produced by the following heat treatment: the specimen was subjected to treatment to a solid solution at T = 1093°C (30 min) and subsequently aged at T = 593°C for 1000 h.

The investigated Ni–Cr alloy specimens were produced from the hot rolled strip of industrial purity rolled at room temperature with the reduction of ε = 70% followed by annealing at T = 993 K (2 h). Consequently, the primary recrystallisation took place in the alloy. The mean grain size of the produced microstructure was 6 μm (or 2 μm taking into account the annealing twins). In this condition, measurements were taken of the misorientation of 134 grains [26]. The specimens were then annealed at 1273 K (30 min). Selective recrystallisation resulted in the formation of a microstructure with the mean grain size of 13 μm (~4 μm taking into account the twins). A total of 162 misorientations of the grain boundaries was measured for this condition [27].

The stainless steel (Kh16N15M3B) investigated in [28] had a microstructure with the mean linear grain size of the order of 40–60 μm, produced by annealing at 1150°C (30 min). The authors of this book measured the misorientations of 133 grain boundaries.

In [29] the specimens of AISI 304L and s were produced by rolling at room temperature (ε = 90 %), followed by annealing. The 304L steel was annealed at T = 1123 K, and steel 316L at 1423 K for 1 h. This resulted in the formation of a microstructure with the mean grain size of 11.4 μm (304L) and 12.2 μm (316L). 172 and 169 misorientations of the grain boundaries were measured for the specimens, respectively.

Analysis of the data, presented in Table 4.3, shows the obvious identity of the GBMD in the FCC materials, susceptible to twinning in annealing. The common feature of all the materials included in the table is the fact that they have a recrystallised structure: fine-grained Ni–Cr alloy showed primary recrystallisation, whereas in the other materials the microstructure formed in individual stages of selective recrystallisation. Thus, in all the cases the structure formed under the effect of internal driving forces – the deformation energy, stored in the material, and the tendency for the decrease of the total energy of the grain boundaries. Consequently, on the basis of these results the authors of [28–30] concluded that there is some stable misorientation distribution of the grain boundaries in all the materials, susceptible to twinning in annealing. The controlling role in the spectrum of the grain boundaries was played by the boundaries of the type $\Sigma 3^n$ formed as a result of multiple twinning. The length of the grain boundaries was also measured in [29]. The resultant distribution of the relative length $\Sigma 3^n$ is shown in Table 4.4. Evidently, the boundaries $\Sigma 3$ in the steels 304L and 316L are dominant not only as regards the number but also the total length of the interface. The considerably smaller length of the grain boundaries of the types $\Sigma 9$ and $\Sigma 27$ shows that they are not advantageous from the viewpoint of energy and form in the structure of the polycrystals as a result of topological restrictions in contact of the boundaries $\Sigma 3$. In [30, 31] it was attempted to generalise the observed experimental results from the viewpoint of the percolation theory of the ensemble of the grain boundaries. It is clear that as regards the macroscopic properties (for example, corrosion resistance of polycrystalline materials) the important factor is not only the number

Table 4.4. Distribution of grain boundaries with respect to σ in the FCC crystals

Σ	Ni	AISI 304L	NiCrl	NiCr2	Stainless steel	AISI 304L	AISI 316L
3	41.0	42.0	34.5	34.6	34.6	35.5	33.1
9	9.0	15.0	7.5	7.4	6.0	5.2	4.7
27	6.0	4.0	5.0	3.1	5.2	2.4	2.4
81	2.0	—	3.0	—	3.8	—	—
Others	42.0	39.0	50.0	54.9	50.4	56.9	59.8
$\langle d \rangle$, μm	1000	100	6	13	40–60	11.4	12.2
Number of boundaries	222	118	134	162	133	172	169
Reference	[33]	[35]	[25]	[26]	[28]	[29]	[29]

of special boundaries but also the method by which they are combined in the ensemble, i.e., whether there is a percolation cluster of 'weak' grain boundaries spreading from one surface to another, or whether there are special grain boundaries distributed on the network acting as an obstacle in the path of propagation of corrosion. In a recent study [32] using an OIM automatic attachment [1, 2] experiments were carried out to measure GBMD in two nickel-based alloys (alloys 600) and iron (type AISI 304L and 316L alloys) subjected to corrosion tests. The spectrum of the misorientation of the grain boundaries of these materials is typical with the dominance of the twin grain boundaries 3. The verification results are presented in Table 4.5. Analysis of the distribution of the corrosion cracks in the network of the grain boundaries shows that the degree of corrosion damage in the grain boundaries Σ3 is the lowest. Other boundaries are subjected to corrosion failure with the probability proportional to the frequency of finding these boundaries in the ensemble of the grain boundaries. The low-angle grain boundaries are characterised by higher resistance to failure than the high-angle arbitrary grain boundaries. The orientation of the plane of the grain boundary is very important for the propagation of the corrosion cracks. However, in some cases, the high-angle boundaries of

Table 4.5. Distribution of the grain boundaries with respect to Σ in the alloys 600 and AISI, subjected to the corrosion test [32]

	Alloy 600-I			Alloy 600-II			Alloy 600-III		
Σ	All GBs (527)	Corros-ion (64)	Rel. length	All GBs (578)	Corros-ion (24)	Rel. length	All GBs (889)	Corros-ion (119)	Rel. length
LAGB	3.2	1.6	—	2.6	—	—	4.4	2.5	—
3	29.7	—	43.5	27.7	4.2	44.0	24.2	2.5	52.8
5–29	7.0	6.4	8.,9	10.1	4.2	8.4	6.3	10.1	4.2
HAGB	60.1	92.0	47.6	59.6	91.6	47.6	65.1	84.9	43.0

	AISI 304L			AISI 316L		
Σ	All GBs (161)	Corrosion (66)	Rel. length	All GBs(204)	Corrosion (30)	Rel. length
LAGB	1.2	1.5	—	3.4	3.3	—
3	31.7	3.0	40.1	28.4	—	35.4
5–29	9.2	12.1	10.3	22.1	30.0	21.9
HAGB	57.9	83.4	49.6	46.1	66.7	42.7

the grains did not fail even if the orientation of the grain boundaries was favourable for the propagation of corrosion.

The statistics of the misorientation of the grains in the materials with the FCC lattice not susceptible to twinning in annealing will be investigated. In the group of pure metals one such metal is aluminium, and the distribution of misorientation and aluminium has been studied in a large number of investigations. Only some of them will be mentioned [33–36]. Pure aluminium (99.999 %) was studied in these investigations. The specimens were annealed at $T = 0.9T_M$ (500 h). This was followed by measurements of 440 grain boundary misorientations. The verification results are presented in Table 4.6. In [34] the authors studied the effect of the annealing time and temperature on the changes in the misorientation spectrum of the grain boundaries. Pure aluminium was annealed at $T = 0.95\ T_M$ (50 h) and at $T = 0.6\ T_M$ (500 h). 110 grain boundary misorientations were measured in each condition. The results show that the maximum fraction in the group of the special boundaries was represented by the Σ3 grain boundaries characterised by the most favourable energy parameters; the number of these boundaries does not change during heating. It should be mentioned that the fraction of all special boundaries in the GBMD is more than a quarter of all the measured boundaries. The statistics of the misorientation of the grain boundaries in aluminium was measured in greater detail in [35–37], where the attention was given to the effect on GBMD of not only annealing temperature and time but also purity of the material. The investigations were conducted on commercial purity aluminium (AO) and high purity aluminium (A999; 99.999 %), the experimental results are also presented in Table 4.6. The specimens of AO and A999 were

Table 4.6. Statistics of the misorientation of the grain boundaries in aluminium [36, 48]

Σ	Commercial aluminium			Pure aluminium			
1(LABG)	26.1	17.6	14.1	16.3	12.9	13.7	5.4
3–27	7.8	5.9	2.6	7.4	20.5	16.4	33.8
HAGB	66.1	76.5	83.3	76.3	76.6	69.9	60.8
No. of measured boundaries	115	115	78	431	124	146	74
Annealing conditions	823 K, 3 h	823 K, 3 h + 853 K, 10 h	683 K, 4 h	684 K, 0.5 h	684 K, 1.5 h	723 K, 1 h + 773 K, 1 h	773 K, 3 h

rolled with the reduction of 90 and 60%, respectively. They were subjected to initial heat treatment at T = 623 K (30 min; A999) and at T = 823 K (3 h, AO). Subsequent annealing was carried out in accordance with the procedure shown in Table 4.6. The misorientation of the grains was measured by the method of local x-ray diffractometry [38]. The number of the measured misorientations of the grain boundaries for each state is also shown in the table. Aluminium is characterised by the strong dependence of the fraction of the grain boundaries 3 (and, correspondingly, Σ9) on the purity of the material [37]. As the purity of the material increases, the fraction of the grain boundaries Σ3 becomes greater. It should be mentioned that this tendency was not detected in the materials susceptible to twinning in annealing. Another special feature by which the aluminium differs is the high fraction of the low-angle boundaries, reaching 20% of the total number of the grain boundaries. Evidently, the latter is associated with the fact that as a result of the small degree of splitting of the dislocations in aluminium deformation is accompanied by easy climb of the dislocations and, consequently, the formation of low-anlge sub-boundaries which are subsequently inherited in the material during recrystallisation [39]. A similar phenomenon takes place in the Ni–Cr alloy, subjected to hot deformation. Regardless of the fact that the spectrum of the grain boundaries after static recrystallisation is characterised by a high fraction of the special grain boundaries (especially, Σ3), after hot deformation with ε = 20%, the fraction of the low-angle boundaries reaches 15.4%. In this case, the fraction of the twin boundaries Σ3 is approximately 20% [14].

4.3. Orientation distribution function in Ni–Cr alloy. Experimental and modelling GBMDs

4.3.1. Orientation distribution function in Ni–Cr alloy and stainless steels

In previous chapters, it was found that the texture in the polycrystals with the modelling texture does not control unambiguously the GBMD. Generally speaking, the same texture may be associated with different spectra depending on the correlation and orientation of the adjacent grains whose nature is determined by the type of specific material and its thermomechanical prior history. For analysis of the relationships governing the formation of the spectrum of the grain boundaries of real materials it is possible to determine the GBMD, using the quantitative characteristic of the texture – ODF – by computer modelling, with the GBMD and in different types of correlation with orientation of the adjacent grains, and it is possible to compare them with the experimentally detected GBMD. This is possible as a result of the investigations carried out in [25–28] in which measurements were taken of the GBMD for a Ni–Cr alloy in two recrystallised conditions and for AISI 304L and 306L stainless steels. The resultant orientation distribution functions and the spectra of the misorientation of the grain boundaries calculated on the basis of these functions will be examined in greater detail [41].

The material for the investigation of the relationship of the texture and the spectrum of the grain boundaries was a Ni–Cr alloy(thermomechanical treatment of this material was described in detail in [25, 26, 41]), and grade AISI stainless steels [28, 41]. It should be mentioned that the initial hot-rolled Ni–Cr alloy strip of industrial purity was rolled at room temperature with the reduction of 70% and this was followed by annealing at 993 K for 2 h. Primary recrystallisation of the alloy resulted in the formation of a fine-grain microstructure with the mean grain size of ~6 μm (or ~2 μm taking into account the annealing twins). A similar state is denoted as Ni–Cr alloy A. Subsequently, the specimens were subjected to additional annealing at 1273 K for 30 min. Selective recrystallisation increased the mean grain size to ~13 μm. This condition is denoted as Ni–Cr alloy B. Details of the experimental measurements of the grain misorientation were published in [25, 26] and in a dissertation by V.N. Danilenko [42]. It should only be mentioned that 134 boundaries were tested in condition A and 162 grain boundaries in condition B.

To describe and analyse the textural condition of the investigated steels, experiments were carried out with the standard method which makes it simple to determine the quantitative characteristic of the texture – the orientation distribution in the space of the Euler angles – on the basis of the experimentally determined incomplete pole figures. The following pole figures were selected for the Ni–Cr alloy: (200), (111), (220) and (311). The ODF were determined using the internal regions of the pole figures restricted by the radial angle $\alpha = 65°$. X-ray diffraction investigations were carried out in a DRON-3 diffractometer with an automatic attachment [43]. The pole figures were recorded with the step of variation of the radial and azimuthal angles (α and β) of 5°. The cross sections of the ODF, obtained on the basis of the results of calculations using the incomplete pole figures (φ_2 = const) are presented in Fig. 4.6 and 4.7. Figure 4.6 shows the ODF of Ni–Cr alloy A after the first annealing (993 K, 2 h), and Fig. 4.7 – after the second annealing treatment (1273 K, 30 min). Analysis of the ODF data shows

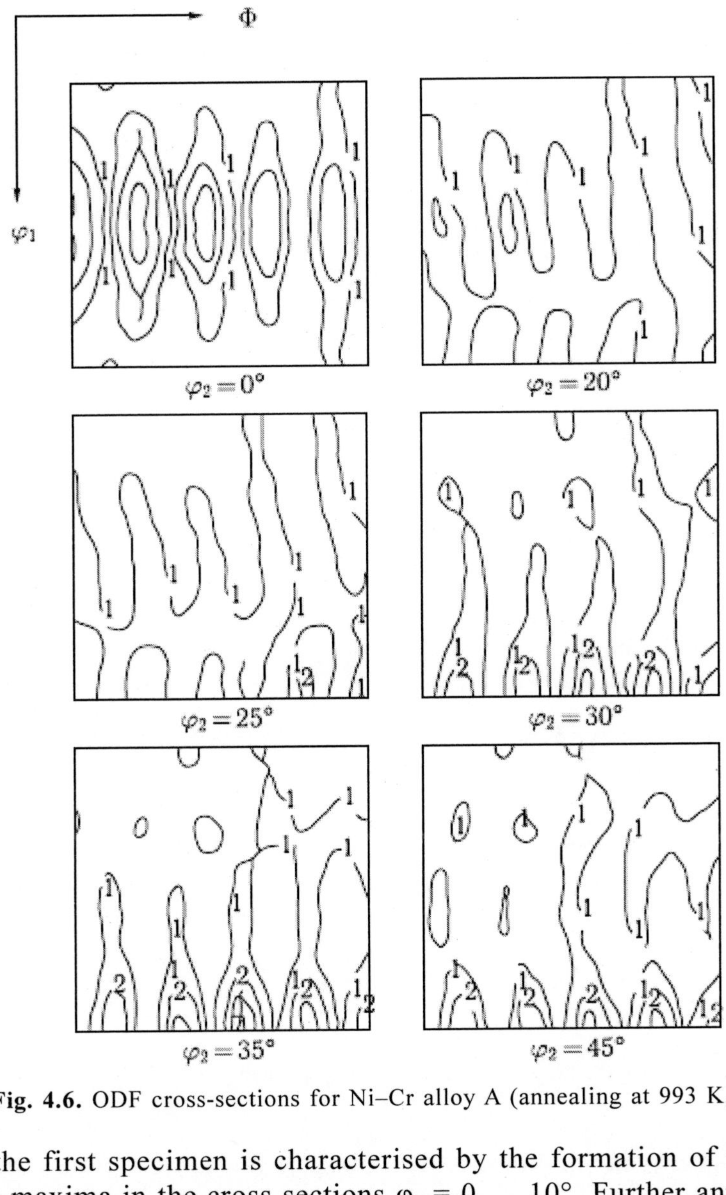

Fig. 4.6. ODF cross-sections for Ni–Cr alloy A (annealing at 993 K, 2 h).

that the first specimen is characterised by the formation of relatively sharp maxima in the cross sections φ_2 = 0,..., 10°. Further annealing is associated with the scattering of the texture, and the form of the ODF is typical of the polycrystals with a weak preferential orientation of the crystals. The first type is connected with the rolling texture. The maxima $\{011\}\langle 211\rangle$ and $\{112\}\langle 111\rangle$ were found in Fig. 4.6. However, the analysis of the ODF pattern is made more complicated by the presence of additional maxima. Additional annealing results in the formation of the peak $\{100\}\langle 001\rangle$in the corners of the cross sections.

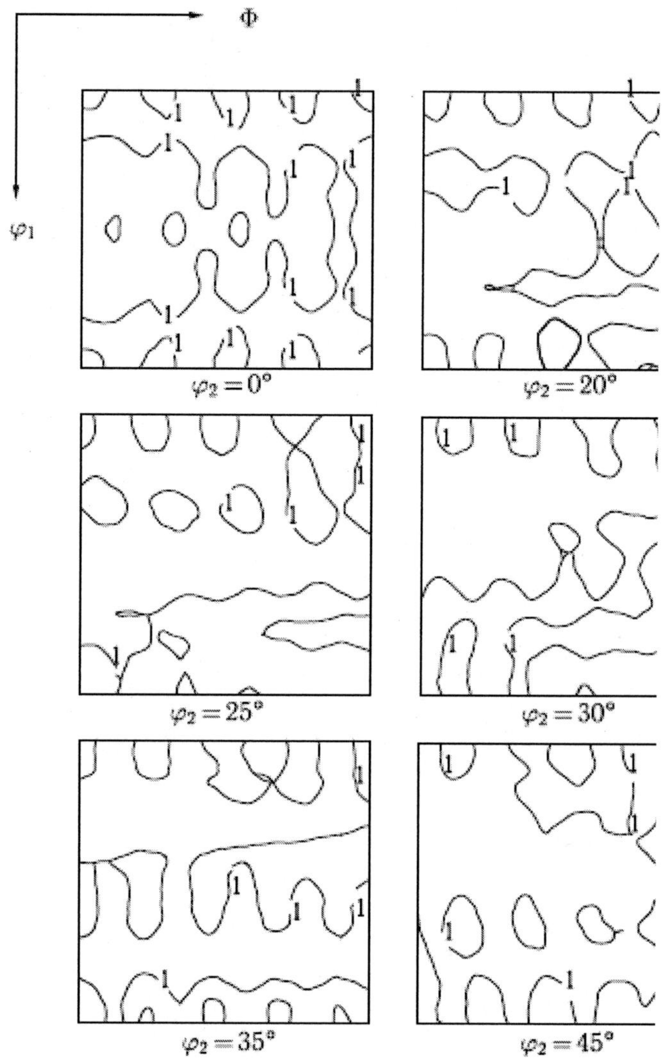

Fig. 4.7. ODF cross-sections for Ni–Cr alloy B (annealing at 1273 K, 30 min).

This peak is associated with the beginning of recrystallisation. Identical changes in the texture maxima were observed in [44] with the increase of the annealing temperature of the Fe–50% Ni alloy up to 900°C. The stoichiometric iron–nickel alloy shows the transition from the rolling texture to the recrystallisation texture, probably as a result of the insufficiently long holding time of the specimens of the Ni–Cr alloy. The recrystallisation texture did not manage to form completely and this was also reflected in the sharpness of the texture. Evidently, the latter resulted in small differences in the fraction of the special boundaries,

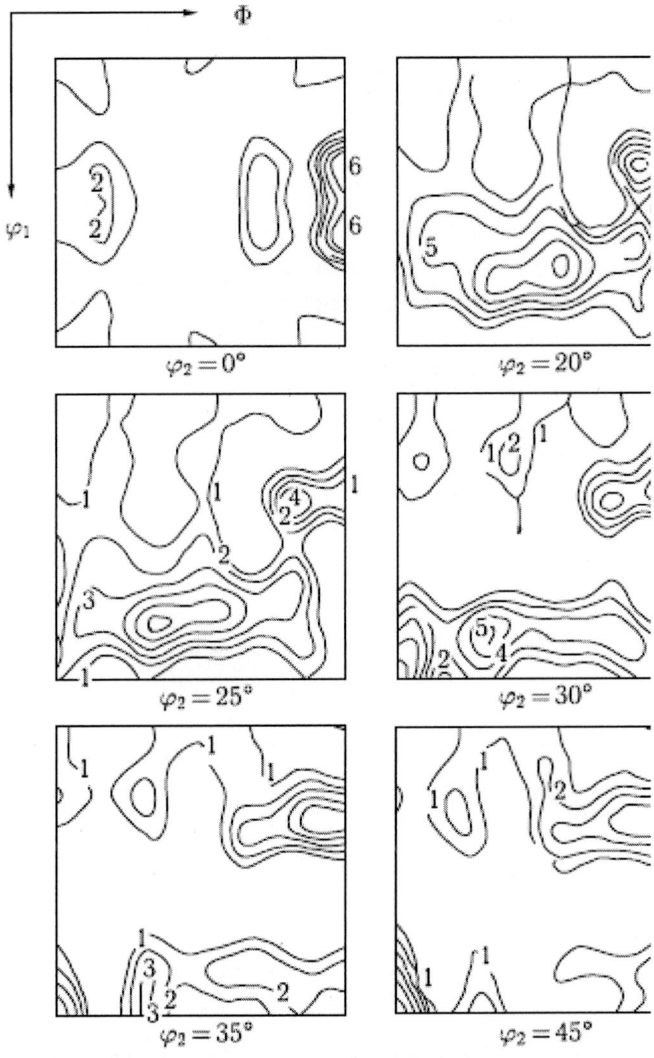

Fig. 4.8. ODF cross-sections for AISI 304L stainless steel.

determined by experiments and the calculations. Even in this case, the application of the correlation methods results in rational agreement of the experimental results with the calculated data.

Figures 4.8 and 4.9 shows the cross sections of the distribution function of the grains in the crystallographic orientation for the AISI 304L and AISI 316L stainless steel. In both cases, the pole figures of the type (200), (220) and (311) are characterised by a relatively high intensity maximum. The texture of the AISI 304L is characterised by two orientation tubes. The skeleton tube of one of them extends from the ideal orientation (131) [112] to the ideal orientation (110) [111].

The second tube is associated with the blurring of the maximum (011) [011]. The orientation distribution function of the AISI 316L steel is also characterised by two orientation tubes. The skeleton tube of the first of these tubes links the ideal orientation (011) [100] with the ideal orientation (121) [321]. The second tube extends from the ideal orientation (120) [001] to the orientation (110) [001]. The comparison of the ODF of Ni–Cr alloy and the stainless steels shows a large difference in the texture of these materials. The experimentally determined orientation distribution function is used for the analysis of the crystal basis in the modelling polycrystal.

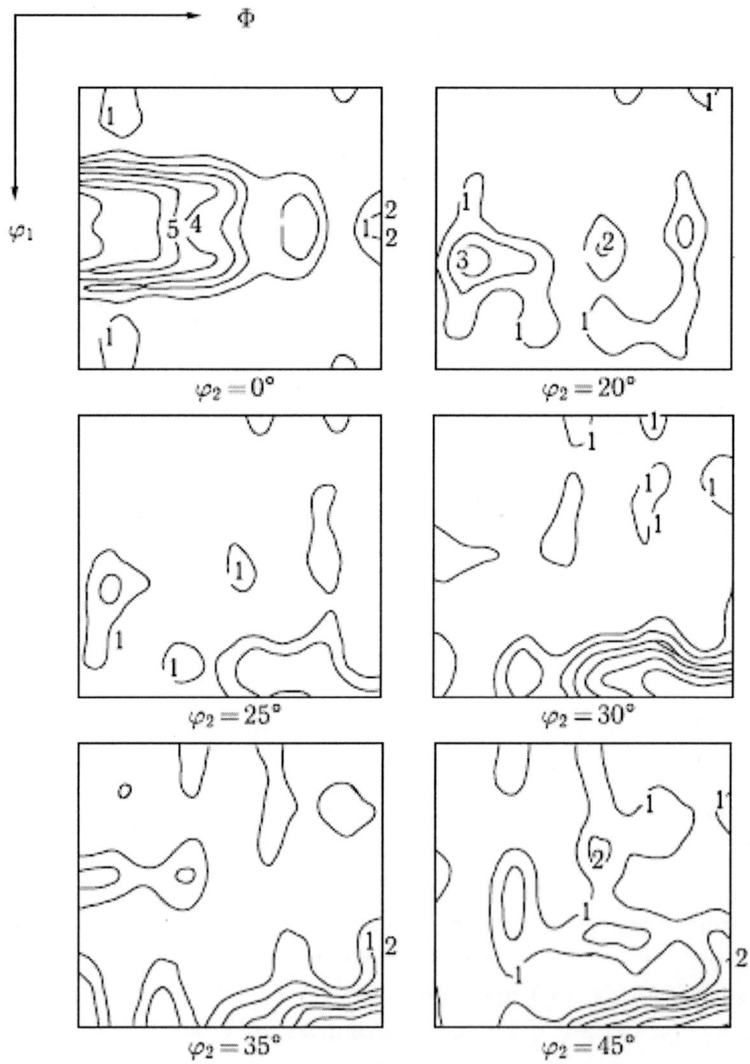

Fig. 4.9. ODF cross-sections for AISI 316L steel.

4.3.2. Modelling spectra of the misorientation of the grain boundaries in Ni–Cr alloy and AISI stainless steels. Comparison with the experimental results

Figures 4.10 and 4.11 shows the calculated and experimental distributions with respect to the misorientation angle for the Ni–Cr alloy in the first and second state with different types of correlation neighbours. The appropriate distributions in the axes and Σ are shown in Tables 4.7–4.10. The identical data for the AISI stainless steels are presented in Fig. 4.12 and 4.13 and also in the Tables 4.11–4.14.

Analysis of the GBMDs produced by modelling and of the appropriate different types of correlation in the orientation of the

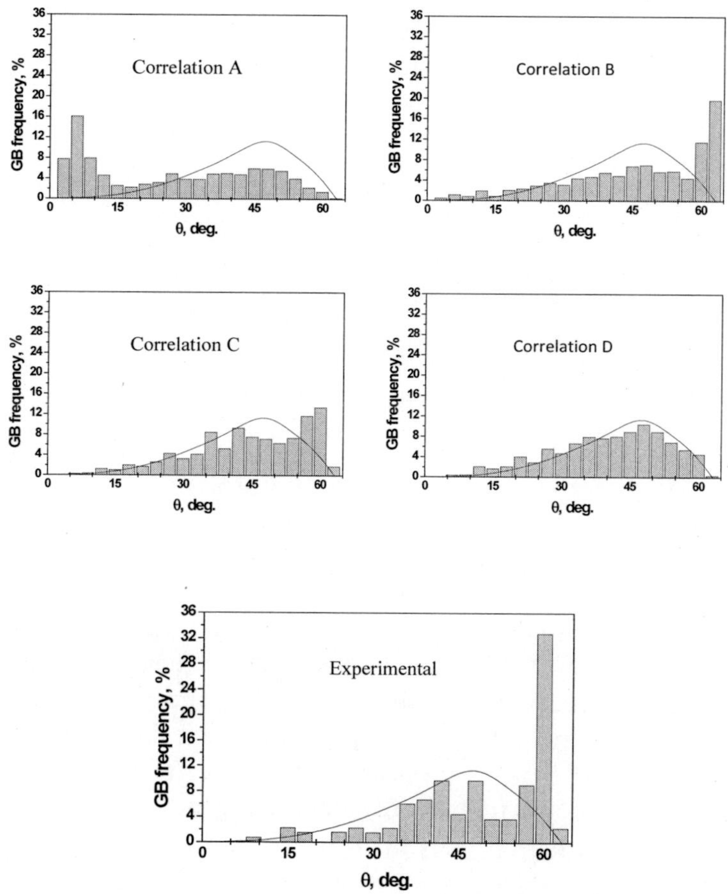

Fig. 4.10. The distributions of the misorientation angles in Ni–Cr alloy A (annealing at 993 K, 2 h), calculated on the basis of the actual ODF and experimental distributions [25, 42].

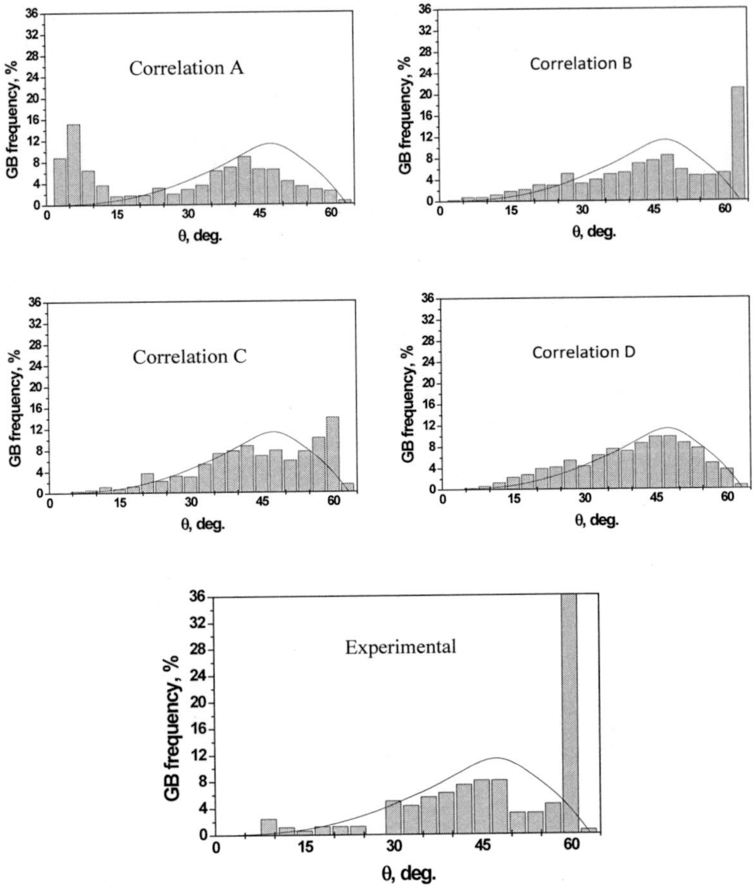

Fig. 4.11. The distributions of the misorientation angles in Ni–Cr alloy B (annealing at 1273 K, 30 min), calculated on the basis of the actual ODF and experimental distributions [26, 42].

adjacent grains (type *A–D*) shows the large differences in GBMD for the same specimen. This applies to both Ni–Cr alloy and the stainless steels. Comparison of the distributions of the misorientation of the grain boundaries with respect to Σ shows that the spectrum modelled for the correlation of the neighbours of type *C* at which the special boundaries prevail in the GBMD, is closest to the experimental spectrum. The calculations of the weight coefficients using equation (3.24) give the identical results: $q_1 = q_2 = q_4 = 0$; $q_3 = 1$. It should be mentioned that the distributions of the same type for different investigated materials are in good agreement with each other. Consequently, it may be concluded that the twinning process is the main factor of the formation of the GBMD in annealing in the materials with the low stacking fault energy. Thus, the number of the boundaries Σ3 in the experimental spectrum

Table 4.7. Calculated and experimental distributions of the misorientation axes in Ni–Cr alloy A after annealing at T = 993 K (2 h)

Zone	I	II	III	IV	V	VI	VII	VIII
Type A	12.84	3.04	4.39	26.07	40.77	8.45	1.30	3.15
Type B	1.01	1.46	3.21	18.19	40.60	26.35	1.80	7.38
Type C	2.37	5.91	1.52	17.51	37.22	5.01	19.69	20.78
Type D	0.90	2.53	4.67	23.09	51.41	10.08	1.63	5.69
Experiment	0.00	0.75	2.24	16.42	31.34	7.46	27.61	14.18

Table 4.8. Calculated and experimental distributions of the grain boundaries with respect to Σ in Ni–Cr alloy A after annealing at T = 993 K (2 h)

Σ	A	B	C	D	Experiment
1	35.64	3.83	2.82	4.28	2.99
3	0.96	3.43	24.94	1.35	37.31
5	1.41	0.68	6.14	1.41	0.75
7	0.62	0.73	2.14	0.84	1.49
9	1.13	0.62	1.86	0.96	3.73
11	0.56	0.28	0.56	0.68	0.75
13a	0.11	0.45	0.56	0.23	0.00
13b	0.17	0.06	0.56	0.62	0.00
15	0.90	0.17	0.96	0.56	0.75
17a	0.00	0.06	0.17	0.11	0.00
17b	0.11	9.40	0.17	0.23	0.00
19a	0.06	0.11	0.23	0.17	0.75
21a	0.06	0.06	0.17	0.23	0.00
25a	0.06	0.00	0.11	0.06	0.00
25b	0.11	0.11	0.34	0.39	0.00
*	2.96	2.70	2.34	3.60	5.22
**	54.96	77.31	55.35	83.33	46.26

*: 25 < Σ < 65; **: Σ > 65

is greater than the number of the boundaries of the same type in the modelled GBMD with the C-correlation. Other same time, in the experiments the boundaries Σ5 are not detected. These differences are associated mainly with the fact that the simplest model of the grain structure of the polycrystal, consisting of regular hexagons, were selected in this case. Generally speaking, this representation does not correspond to the topology of the grain structure in the actual materials, subjected to twinning in annealing. In [45] it was reported that the

Table 4.9. Calculated and experimental distributions of the misorientation axes in Ni–Cr alloy *B* after annealing at T = 1273 K (30 min)

Zone	I	II	III	IV	V	VI	VII	VIII
Type A	12.22	3.27	4.56	24.10	43.64	8.56	0.90	2.76
Type B	1.41	1.35	2.25	18.98	41.05	26.07	2.08	6.81
Type C	2.31	4.90	2.53	16.84	35.98	7.10	10.25	20.10
Type D	1.01	2.53	3.66	23.76	52.70	8.90	2.14	5.29
Experiment	0.00	1.85	2.47	9.88	38.90	5.56	33.95	7.41

Table 4.10. Calculated and experimental distributions of the grain boundaries with respect to Σ in Ni–Cr alloy B after annealing at T = 1273 K (30 min)

Σ	A	B	C	D	Experiment
1	38.23	4.56	3.10	4.56	4.32
3	0.23	4.28	25.06	1.52	35.80
5	0.90	0.90	5.74	1.24	0.62
7	0.73	0.28	1.91	0.62	0.00
9	0.11	0.62	2.03	0.96	6.79
11	0.39	0.23	0.96	0.45	1.23
13a	0.17	0.23	0.56	0.34	0.00
13b	0.23	0.34	0.56	0.34	0.00
15	0.28	0.51	1.07	0.39	0.00
17a	0.06	0.06	0.11	0.11	0.00
17b	0.06	8.90	0.23	0.23	0.00
19b	0.28	0.45	0.28	0.45	0.00
21a	0.11	0.23	0.23	0.28	0.00
25a	0.28	0.11	0.11	0.06	0.00
25b	0.11	0.39	0.34	0.28	0.00
*	2.71	2.80	2.53	3.65	3.09
**	54.96	75.11	54.84	83.67	48.14

*: $25 < \Sigma < 65$; **: $\Sigma > 65$

fraction of the boundaries Σ3 in the GBMD calculated only on the basis of the geometrical considerations corresponds to the boundaries formed during recrystallisation as a result of contact of two grains with twin orientation, whereas there are other mechanisms of the formation of

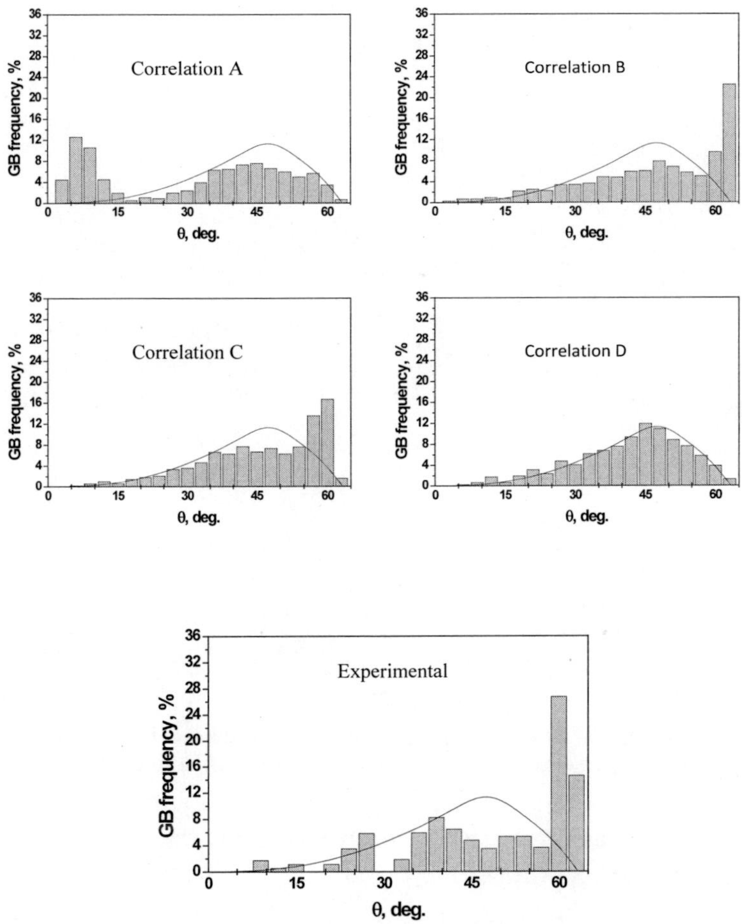

Fig. 4.12. The distribution of the misorientation angles in the AISI 304L stainless steel calculated on the basis of the actual ODF and experimental values [27, 28].

twins (for example, splitting of the grain boundaries into two parts). In this case, the difference in the fractions of the grain boundary's of the type Σ3 corresponds to the part of the grain boundaries which formed by a mechanism other than the contact of two grains with twin orientation. It should be mentioned that, as shown by recent EBSD investigations for Ni–Cr alloy A in [46], the fraction of the grain boundaries Σ3 is approximately 24%, which is statistically equivalent to the fraction of the twins obtained in our calculations (25%).

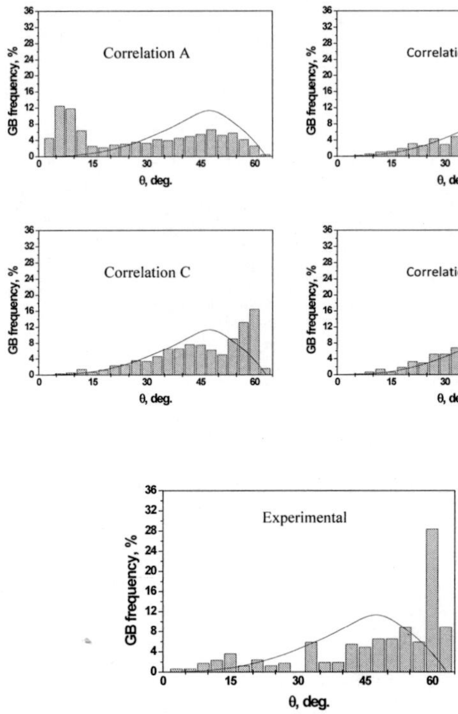

Fig. 4.13. The distribution of the misorientation angles in the AISI 316L stainless steel calculated on the basis of the actual ODF and experimental values [29, 41].

4.4. Special features of the grain boundaries in the FCC materials with a high stacking fault energy

4.4.1. Rolling and annealing texture of aluminium

In the previous section, for basic spectra of the misorientation of the grain boundaries were determined by computer modelling of GBMD on the basis of the actual ODF in the twinning materials. Comparison with the experimental GBMD values shows that the calculated spectrum, characterised by the type *C* correlation, correspond to the actual GBMD in the twinning FCC polycrystals. It is interesting to solve the identical task for the materials with a high stacking fault energy, i.e., not susceptible to twinning in annealing. The important problem is also the prediction of the GBMD in the same material but with the microstructure with a different grain size. These tasks have been solved for aluminium.

The chemical composition of the investigated material was as follows: Ar (argon) 0.42%; Cu (copper) 0.09%; Mn (manganese) 0.06%;

Table 4.11. Calculated and experimental distributions of the misorientation angles in the AISI 304L stainless steel

Zone	I	II	III	IV	V	VI	VII	VIII
Type *A*	6.53	3.15	4.79	24.10	42.79	10.64	1.80	6.19
Type *B*	1.18	2.08	2.48	14.64	41.39	27.59	2.98	7.66
Type *C*	1.69	4.84	2.87	16.39	34.74	6.53	12.44	20.50
Type *D*	1.75	2.31	4.17	21.62	52.98	10.02	2.14	5.01
Experiment	0.59	1.17	1.17	14.62	34.50	5.85	37.42	4.67

Table 4.12. Calculated and experimental distributions of the grain boundaries with respect to Σ in the AISI 304L stainless steel

Σ	*A*	*B*	*C*	*D*	Experiment
1	33.95	2.65	2.48	3.60	2.92
3	1.46	5.24	28.94	1.58	36.84
5	0.62	1.07	4.84	1.86	1.17
7	1.24	0.73	0.96	0.90	0.58
9	0.34	0.56	1.07	0.96	5.26
11	0.45	0.28	0.84	0.62	0.58
13a	0.11	0.45	0.17	0.23	0.00
13b	0.34	0.17	0.06	0.34	1.17
15	0.79	0.28	1.06	0.79	0.00
17a	0.06	0.17	0.11	0.11	0.00
17b	0.17	11.15	0.06	0.33	1.17
19a	0.06	0.11	0.45	0.45	0.00
19b	0.11	0.00	0.11	0.00	0.00
21a	0.11	0.11	0.06	0.23	0.00
21b	0.56	0.28	0.34	0.56	1.17
23b	0.62	0.33	0.28	0.73	0.00
25a	0.11	0.17	0.11	0.23	0.00
25b	0.39	0.23	0.17	0.51	1.75
*	5.86	8.17	5.18	9.95	80.21
**	52.65	67.85	52.70	76.01	39.18

*: $25 < \Sigma < 65$; **: $\Sigma > 65$

Ni (nickel) 0.07%; Ti (titanium) 0.03%; the balance Al (aluminium). The materials prepared by the following procedure. After equal channel angular pressing (ECAP) [25], the produced cylinder was rolled at room temperature into a 1 mm thick strip. Blanks for further heat treatment

Table 4.13. Calculated and experimental distributions of the misorientation of axes in AISI 316L stainless steel

Zone	I	II	III	IV	V	VI	VII	VIII
Type *A*	7.55	3.15	3.94	24.72	43.19	11.37	1.80	4.28
Type *B*	0.98	1.30	3.38	16.78	40.99	26.91	1.58	8.11
Type *C*	2.02	4.96	3.04	15.09	34.01	6.42	11.99	22.47
Type *D*	1.24	2.70	4.22	21.90	53.27	19.53	2.14	4.90
Experiment	0.00	0.59	0.00	10.65	37.87	9.47	33.73	7.69

Table 4.14. Calculated and experimental distributions of the grain boundaries with respect to Σ in AISI 316L stainless steel

Σ	*A*	*B*	*C*	*D*	Experiment
1	33.95	2.65	2.48	3.60	2.92
3	1.46	5.24	28.94	1.58	36.84
5	0.62	1.07	4.84	1.86	1.17
7	1.24	0.73	0.96	0.90	0.58
9	0.34	0.56	1.07	0.96	5.26
11	0.45	0.28	0.84	0.62	0.58
13a	0.11	0.45	0.17	0.23	0.00
13b	0.34	0.17	0.06	0.34	1.17
15	0.79	0.28	1.06	0.79	0.00
17a	0.00	0.17	0.06	0.11	0.00
17b	0.23	11.60	0.28	0.28	0.59
19a	0.23	0.34	0.45	0.23	0.00
19b	0.06	0.11	0.28	0.06	1.18
21a	0.11	0.17	0.28	0.17	0.59
21b	0.23	0.51	0.39	0.39	0.00
23b	0.17	0.34	0.17	0.28	0.00
25a	0.00	0.11	0.00	0.11	0.00
25b	0.34	0.23	0.34	0.56	0.59
*	5.51	8.38	5.23	6.99	5.32
**	51.63	67.23	50.90	80.29	42.01

*: $25 < \Sigma < 65$; **: $\Sigma > 65$

were cut by electrospark cutting from the central part. Annealing at T = 433 K for 45 min resulted in the formation of a fine-grained structure with the mean grain size of approximately 1–2 μm (Fig. 4.14). Annealing of the initial material at T = 583 K for 1 h resulted in the formation of a grain structure with $d \sim$ (20–40) μm (Fig. 4.15).

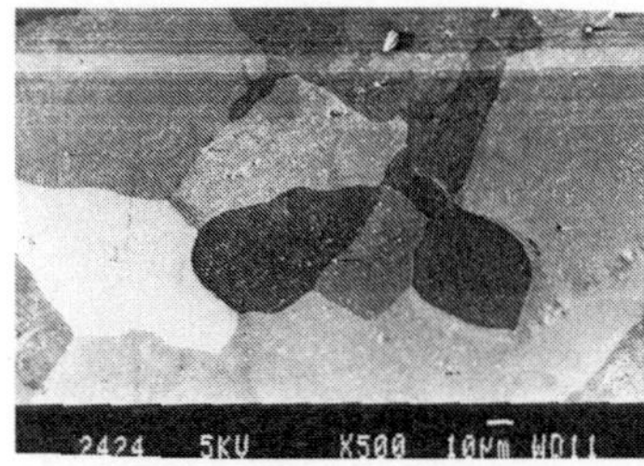

Fig. 4.14. Microstructure of aluminium after annealing at T = 433 K (45 min).
Fig. 4.15 (right). Microstructure of aluminium after annealing at T = 583 K (1 h).

Incomplete pole figures were produced for these two conditions in the investigated material. In the case of fine-grained aluminium pole figures were recorded using an automatic texture attachment. The pole figures for coarse-crystalline aluminium were produced in the non-automatic regime. To average out the reflected signal, the specimens oscillated in the plane-parallel direction with the amplitude of approximately 1 mm. The step of the variation of the radial and azimuthal angle was 5° in both cases. The orientation distribution function was restored on the basis of the three incomplete pole figures: (111), (200) and (220). The mathematical method of determination of the grain distribution function on the basis of the orientations was described in detail previously. The ODF cross-sections φ_2 = const for aluminium in the first and second state are shown in Fig. 4.16 and 4.17, respectively. The ODF, shown in the figures, are of approximately the same type with a relatively high intensity maximum and are characterised by two orientation tubes. The skeleton part of the first of them extends from the ideal orientation (131) [112] to the ideal orientation (230) [001]. The second skeleton tube forms at a point between the ideal orientations (001) [110] and (233) [331] and extends to the ideal orientation (113) [110] (Fig. 4.16). The main difference between the ODF of coarse- crystalline aluminium and that of fine-crystalline aluminium is the scattering of the main texture maxima and the formation of a new maximum in the range of ideal orientation (011) [111]. The appearance of the new maximum is associated in all likelihood with the development of recrystallisation, i.e., with the formation of grains with a new orientation in the material.

4.4.2. Grain boundary ensembles in aluminium: experiments and modelling

Experimental verification was carried out on annealed aluminium with the mean grain size of ~20–40 μm. The foils for the investigations were prepared from the same specimens for which the pole figures were

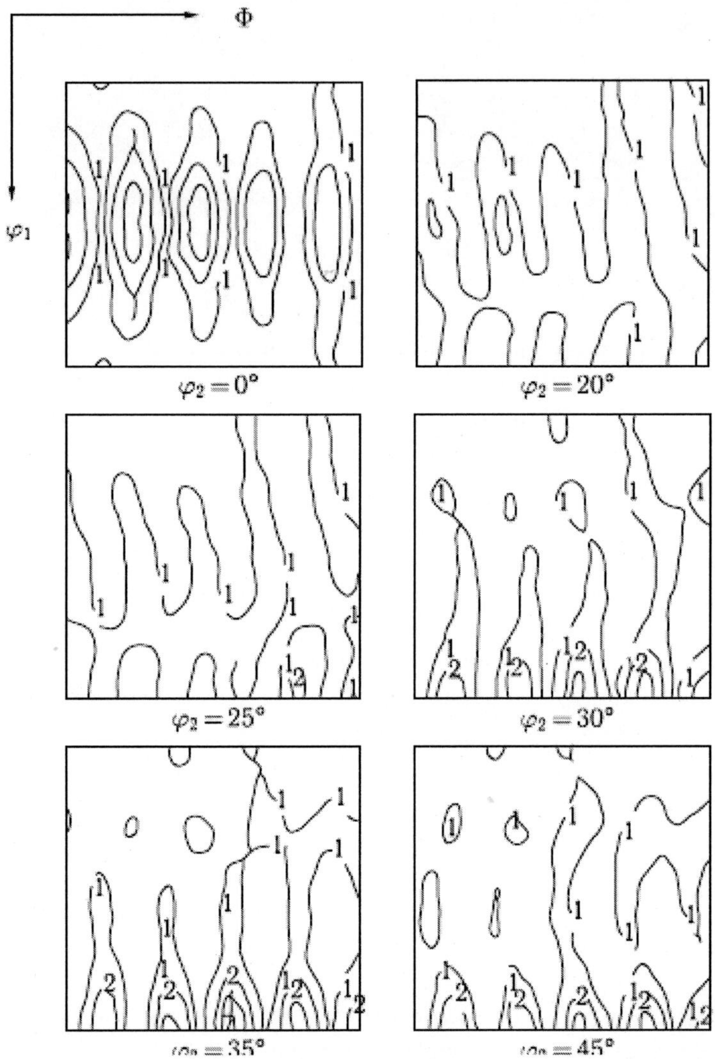

Fig. 4.16. ODF cross sections for aluminium (annealing at 433 K, 45 min).

recorded. The corresponding characteristic sections are shown in Fig. 4.18 and 4.19 and the Tables 4.15 and 4.16 show the parameters of the grain boundaries in the form of the angle–axis dependence. A total of 100 boundaries were tested. The results of experimental verification are presented in the standard form as three distributions (with respect to angles, axes and inverse density of the coincident angles Σ) in Fig. 4.20 and 4.21 and also in Tables 4.17–4.20. A large fraction of all the boundaries in the experimental GBMDs is represented by the low-angle boundaries (30%). Of the four boundaries considered which

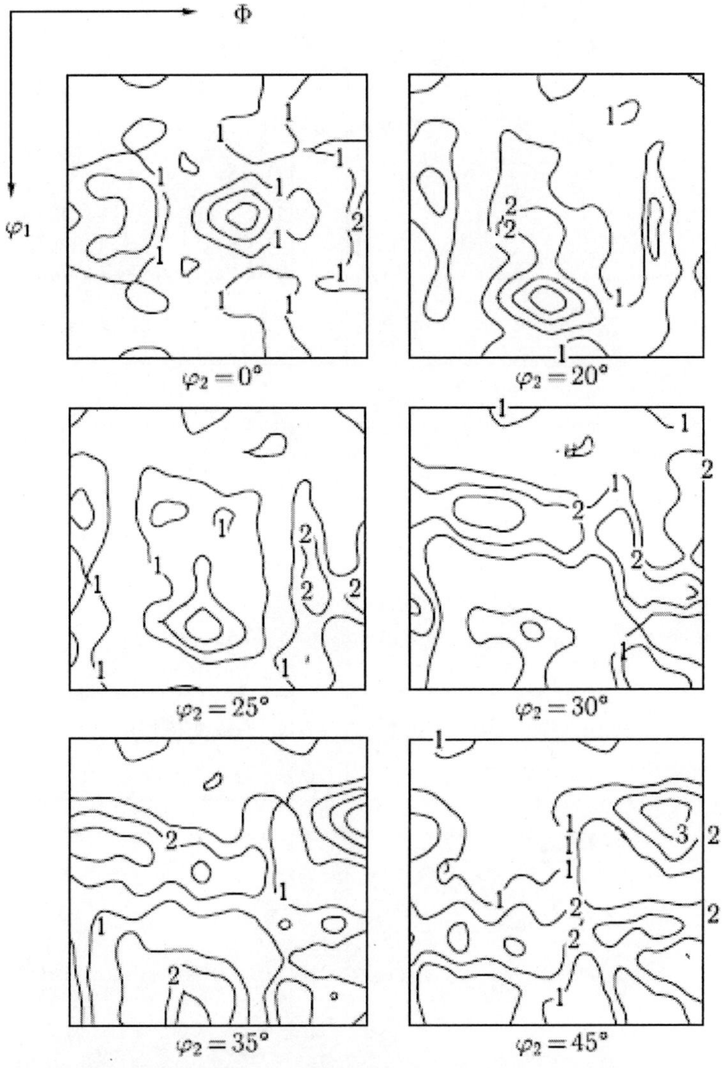

Fig. 4.17. ODF cross sections for aluminium (annealing at 583 K, 1 h).

are close to the special boundary Σ3 (60°/[111]), only one satisfies the more stringent Brandon criterion: $\Delta\theta \leq 8\Sigma^{-1/2}$ deg. The majority of the misorientation axes (60%) are concentrated in the zone 6 (see Fig. 2.4), situated between the directions $\langle 110 \rangle$ and $\langle 111 \rangle$. Further 30% of the misorientation axes are found in the adjacent zones (5 and 8). Thus, the distribution of both the misorientation axes and of the angles in coarse-grained aluminium greatly differs from the distributions in the textureless material. A large part of the low-angle

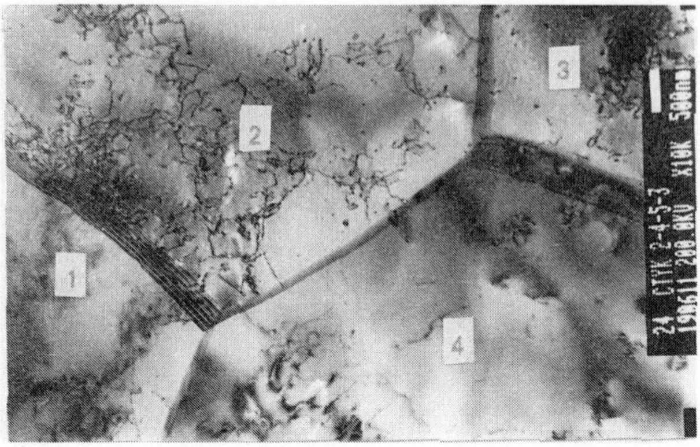

Fig. 4.18. Photograph of area I of coarse-grained aluminium foil (annealing at T = 583 K, 1 h).

Table 4.15. Parameters of the grain boundaries tested in the area I (see Fig. 4.18)

No.	Boundary	Angle, deg	Axis	GB type
1	1–2	54.4	–0.384 –0.494 0.779	General
2	1–3	58.0	0.441 0.534 –0.722	General
3	1–4	57.6	–0.582 0.534 –0.456	General
4	2–3	10.7	–0.479 0.848 –0.228	LAB
5	3–4	7.2	0.274–0.945 0.178	LAB

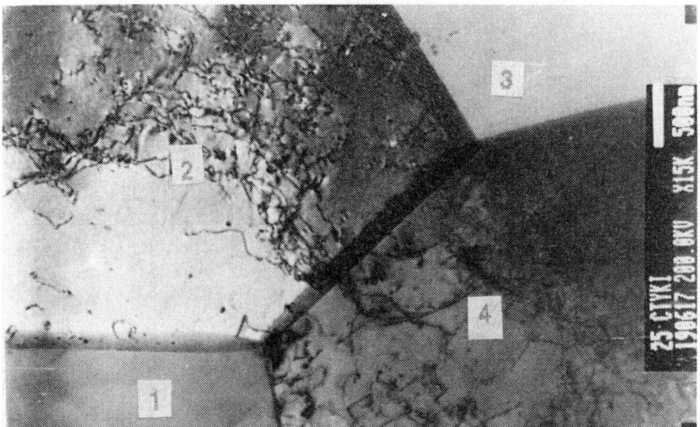

Fig. 4.19. Photograph of area II of coarse-grained aluminium foil (annealing at T = 583 K, 1 h).

boundaries (10–20%) was found in almost all experiments with the certification of the ensemble of the grain boundaries in aluminium, described in the literature [34, 35, 47]. As already mentioned, the latter

Table 4.16. Parameters of the grain boundaries, tested in the area II (Fig. 4.19)

No.	Boundary	Angle, deg	Axis	GB type
1	1–2	44.9	0.393 –0.392 0.832	Special
Close to Σ21, 44.414°/[112] ($\Delta\theta$ = 1.257°; $\Delta\theta_{Br}$ = 3.273°)				
2	1–4	15.8	0.645 –0.715 –0.271	LAB
3	2–3	55.4	0.714 0.699 0.053	General
4	2–4	57.8	0.424 –0.541 –0.726	General
5	3–4	6.0	0.010 0.035 0.999	LAB

is associated with the fact that because of the low degree of splitting of the dislocations in aluminium deformation is accompanied by relatively easy climb and, consequently, by the formation of low-angle boundaries which are inherited by the material in subsequent heat treatment.

Modelling of the misorientation spectra of the grain boundaries in aluminium was carried out using the method described previously in detail and the actual ODFs. The basic spectra differing in the type of correlation along the packing line of the 'given' GBMD (chapter 3) were determined for both states of aluminium. Figures 4.20 and 4.21 and Tables 4.17–4.20 show the results of numerical modelling of the basic spectra of the four correlation types (*A–D*). For aluminium in both states the calculated spectra of the same type are identical within the statistical error range. The highest fraction of the low-angle grain boundaries (Σ1) is detected in GBMD with the type A correlation, and the largest number of the misorientation axes in this case is situated in zone 5 of the standard stereographic angle.

The true spectrum for coarse-grained aluminium was presented in the form of the linear combination of the basic spectra:

$$P(\theta)=\sum_{i=l}^{4} q_i P_i(\theta) , \qquad (4.11)$$

where $P_i = \{P_A, P_B, P_C, P_D\}$ are the basic GBMDs; q_i are the expansion coefficients satisfying the normalisation condition. The coefficients q_i can be determined minimising the sum of the mean quadratic deviations of the calculated spectra from the experimental one, i.e., minimising the expression (2.9). Similar calculations give the following values for aluminium in the second state (annealing at T = 583 K, 1 h):

$q_1 = 0.8;\ \ q_2 = 0.1;\ q_4 = 0.1.$

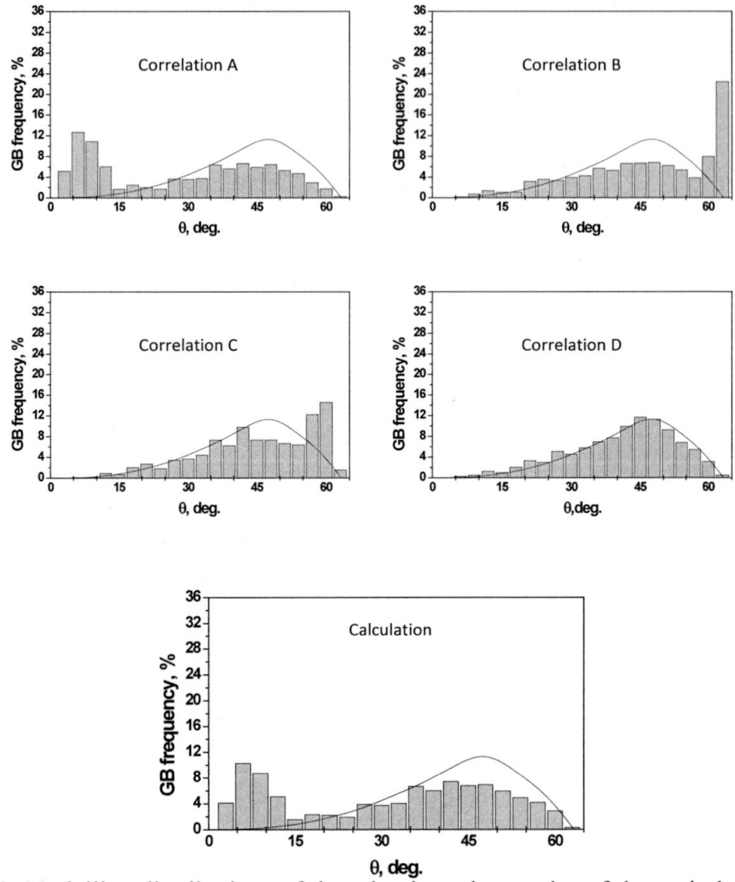

Fig. 4.20. Modelling distributions of the misorientation angles of the grain boundaries in aluminium (433 K, 45 min) and the calculated GBMD.

The coefficients are determined for the basic spectra *A–D* were used to determine the calculation spectrum for fine-grained aluminium. Figure 4.20 show the distribution of the misorientation angles of the calculation spectrum, and Tables 4.17 and 4.18 give the distribution with respect to the misorientation axes and the inverse density of the coincident sites.

Comparison of the calculation and experimental spectra for coarse-grained aluminium (Table 4.20) shows satisfactory agreement for the functions of the low-angle boundaries (27.4 and 30%), the fractions of the twin boundaries $\Sigma 3$ (4 and 3.7%), and also the fractions of the high-angle boundaries (54 and 55.7%). Consequently, it may be assumed that the spectrum of the misorientation of the grain boundaries of fine-grained aluminium (see the column 'calculated' in Table 4.18) also contains a high number of the low-angle boundaries (26.6%).

Table 4.17. Distribution of the misorientation axes in aluminium after annealing at 433 K (45 min)

Zone	I	II	III	IV	V	VI	VII	VIII
Type A	7.83	3.32	4.51	25.62	42.06	9.52	2.31	4.84
Type B	0.90	1.80	3.83	17.06	40.77	27.31	1.91	6.42
Type C	2.53	7.04	3.38	17.17	32.94	5.86	10.81	20.27
Type D	1.35	2.53	4.34	22.35	51.58	10.14	1.97	5.74
Calculated	6.65	3.61	4.38	24.45	42.10	9.21	23.13	16.4

Table 4.18. Distribution of the grain boundaries with respect to Σ in aluminium after annealing at 433 K (45 min)

Σ	A	B	C	D	Calculated
1	35.13	3.04	1.80	3.98	28.59
3	0.68	3.60	26.80	1.58	3.38
5	1.69	1.46	7.55	1.41	2.25
7	0.79	0.73	0.79	0.73	0.78
9	0.79	0.90	0.62	1.30	0.82
11	0.28	0.62	0.68	0.68	0.36
13a	0.28	0.34	0.39	0.45	0.31
13b	0.62	0.17	0.45	0.39	0.58
15	0.23	0.39	0.34	0.84	0.30
17a	0.00	0.06	0.11	0.06	0.02
17b	0.11	10.41	0.34	0.17	0.14
19a	0.11	0.23	0.51	0.28	0.17
19b	0.11	0.06	0.06	0.06	0.10
21a	0.28	0.00	0.06	0.11	0.24
21b	0.17	0.28	0.39	0.28	0.20
23	0.22	0.28	0.28	0.45	0.25
25a	0.11	0.00	0.28	0.11	0.13
25b	0.39	0.23	0.34	0.45	0.39
*	4.98	9.29	5.62	7.36	6.31
**	52.03	67.91	52.59	79.28	54.81

*: $25 < \Sigma < 65$; **: $\Sigma > 65$

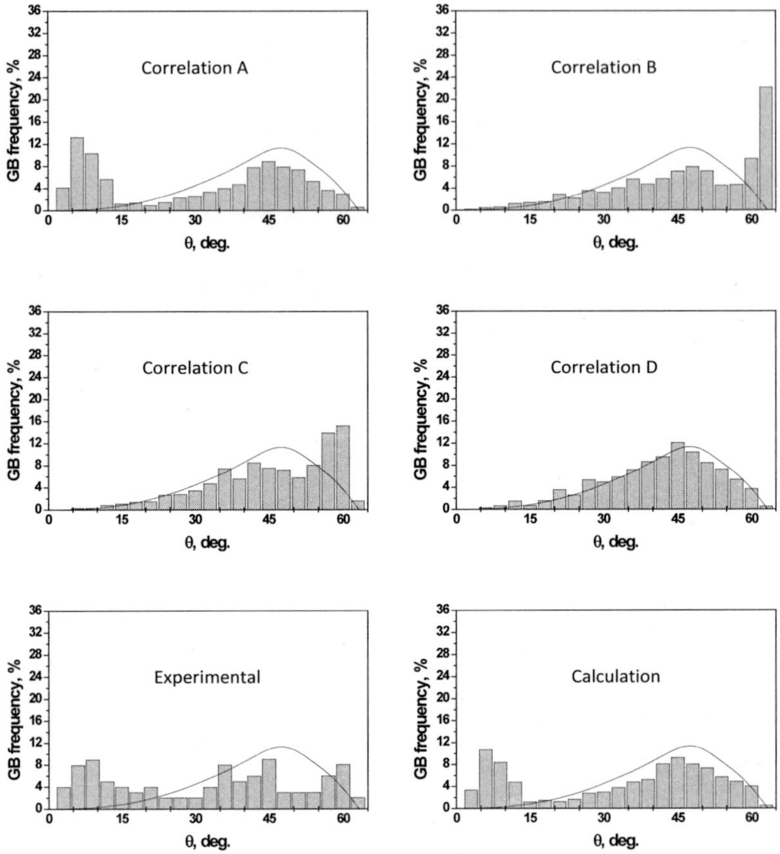

Fig. 4.21. Modelling distributions of the misorientation angles of the grain boundaries in aluminium (433 K, 45 min) and the experimental GBMD.

This chapter includes the results of experimental investigation of the accuracy of measurement of the direction of the electron beam on the diffraction pattern. These measurements were taken for the case of the twin boundary Σ3 in the Ni–Cr alloy using both commercial diffraction and diffraction in a converging beam. The latter method, combined with the single reflection method, can be used to measure the misorientation of the grain boundaries in the materials with the ultrafine-grained structure. It is shown that the maximum possible accuracy of determination of the grain orientation in both cases is approximately 0.01°. This conclusion holds for almost all methods of calculating the direction of the electron beam in a crystal described in the literature [9]. The accuracy of calculation of the misorientation of two adjacent grains is an order of magnitude lower and equals 0.1°. The accuracy of determination of the misorientation of two grains in

Table 4.19. Distribution of misorientation axes in aluminium after annealing at 583 K (1 h)

Zone	I	II	III	IV	V	VI	VII	VIII
Type *A*	7.54	2.37	3.60	19.99	47.75	11.88	1.69	5.18
Type *B*	1.13	1.24	2.59	16.84	40.09	28.77	2.31	7.04
Type *C*	1.69	5.69	3.04	15.54	33.45	6.81	11.99	21.79
Type *D*	0.85	2.20	4.79	23.87	50.39	9.91	1.24	6.76
Experiment	0.00	0.00	1.00	5.00	16.00	60.00	3.00	15.00
Calculated	6.29	2.68	3.67	19.93	46.58	11.18	2.67	7.00

Table 4.20. Distribution of grain boundaries with respect to Σ in aluminium after annealing at 583 K (1 h)

Σ	*A*	*B*	*C*	*D*	Experiment	Calculated
1	33.56	3.43	2.31	2.93	30.00	27.37
3	0.84	4.17	29.00	1.24	4.00	3.70
5	0.62	0.90	5.97	1.30	0.00	1.22
7	1.07	0.79	1.30	1.24	2.00	1.11
9	1.01	0.62	0.90	1.13	0.00	1.01
11	1.30	0.56	0.56	0.68	0.00	1.16
13a	0.11	0.06	0.45	0.28	0.00	0.16
13b	0.28	0.28	0.45	0.45	0.00	0.32
15	0.17	0.56	0.56	0.56	0.00	0.25
17a	0.06	0.11	0.11	0.17	0.00	0.00
17b	0.45	11.09	0.06	0.17	0.00	0.38
19a	0.11	0.11	0.23	0.23	0.00	0.14
19b	0.00	0.06	0.00	0.00	0.00	0.00
21a	0.00	0.06	0.00	0.23	0.00	0.00
21b	0.34	0.17	0.23	0.68	2.00	0.36
23	0.11	0.28	0.28	0.39	0.00	0.16
25a	0.06	0.11	0.06	0.11	0.00	0.00
25b	0.23	0.17	0.28	0.34	0.00	0.24
*	6.36	8.73	6.24	8.70	8.00	6.74
**	53.32	67.74	51.01	79.17	54.00	55.68

*: 25 < Σ < 65; **: Σ > 65

statistical investigations is almost an order of magnitude lower than the maximum possible accuracy because it is not simple to realise simultaneously the optimum conditions ensuring maximum accuracy.

Analysis of the literature data obtained in the investigations of the ensemble of the grain boundaries in the FCC material shows a tendency for the formation of the stable spectrum of the misorientation of the grain boundaries in the materials with low and mean stacking fault energies. Similar materials are susceptible to twinning in annealing and have a high percentage of twin boundaries ($\Sigma 3$), characterised by the lowest excess energy. The fraction of the boundaries $\Sigma 3$ may exceed 30%. In addition, the $\Sigma 9$ and $\Sigma 27$ boundaries, topologically linked with these boundaries, are also characterised by a higher frequency characteristic. At the same time, the FCC polycrystals with a high stacking fault energy (for example, aluminium) show the tendency for inheriting the ensemble of the grain boundaries formed as a result of the preliminary thermomechanical effect (for example, rolling). Long-term annealing (of pure aluminium) is required to increase greatly the fraction of the special boundaries [37]. These special features of the FCC materials have been used as a test in the verification of the proposed procedure for calculating the spectra of the grain boundaries for the given texture, described in chapter 1. The twinning materials have the grain boundary ensemble which can be described with sufficient accuracy by the modelling spectrum with the type *C* correlation with the maximum possible fraction of the special grain boundaries. For example, for the Ni–Cr alloy annealed at 993 and 1273 K (2 h and 30 min, respectively), the fraction of the twins in the experiments was 37.3 and 35.8%. The calculated fractions of the grain boundaries $\Sigma 3$ were equal to 25% in both cases. A small difference may be caused by the insufficient size of the experimental sample. It should be mentioned that 100–200 boundaries for the specimens were measured in the experiments (in comparison with more than 1000 grain boundaries in modelling). In addition to this, in actual cases the topology did not correspond to the selected model of the hexagonal grains. Quaternary or even quinary junctions were found in the experiments and this was not included in the computer model. The ratio of the axes of the twin boundaries in the AISI stainless steels was approximately the same, although in the case of AISI 316L steel, the calculated fraction of the twins (29%) was very close to the experimental value (33%). The new data, obtained using the automatic methods of measuring the spectra of the grain boundaries in Ni–Cr alloy [48] showed excellent agreement between the experimental and calculated spectra.

The misorientation of 100 grain boundaries was studied by experiments in ultrafine-grained aluminium. The results show that the rolled and subsequently annealed specimens 'inherit' the high fraction of the low-angle boundaries (30%). This is in excellent agreement with the data published in previous studies [33]. Comparison of the experimental and modelling spectra was carried out to determine the expansion coefficient of the real spectrum of the misorientation of the grain boundaries into basic spectra. Subsequently, these weight multipliers were used to calculate the actual spectrum of aluminium, produced by rolling the ECAP cylinder.

Thus, the proposed method makes it possible to model the spectra of the misorientation of the grain boundaries in polycrystalline materials and can also be used for materials with lattices other than cubic. This problem will be studied later on the example of zirconium oxide with the monoclinic lattice.

References

1. Dingley D.J. and Field D.P., Mater. Sci. Technol., 1996, No. 12. 1.
2. Field D.P., Ultramicros., 1997, vol. 7, No. 1.
3. Utevskii L.M., Diffraction electron microscopy in metals science, Metallurgiya, Moscow, 1973.
4. Thomas G. and Goringe M.J., Transmission electron microscopy of materials, Nauka, Moscow, 1983.
5. Ryder P.L. and Pitsch W., Phil. Mag., 1967, vol. 15, 437–446.
6. Ryder P.L. and Pitch W., Phil. Mag., 1968. vol. 18, 807–816.
7. Ball C.J., Phil. Mag. A., 1981, vol. 44, No. 6. 1307–1317.
8. Goringe M.J., Loberg B. and Smith D.A., Phys. Stat. Sol. (a), 1979, vol. 55. 569–572.
9. Vergazov A.N., Rubtsov A.S. and Rybin V.V., Proc. 10th National conference on electron microscopy, vol. 1, VINITI, Moscow, 1976, 199–202.
10. Vergazov A.N. and Rybin V.V., Crystallogeometrical analysis of the structure of metals and alloys in electron microscopy, LDNTP, Leningrad, 1984.
11. Watanabe T., Phil. Mag. A, 1983, vol. 47, No. 1, 141–146.
12. Utenkova O.V., et al., Zavod. Lab., 1985, No. 3, 32–35.
13. Pumprey P.M. and Bowkett K.M.. Phys. Stat. Sol. (a), 1970, vol. 3, 375–381.
14. Gertsman V.Yu. and Zhilyaev A.P., Zavod. Lab., 1990, vol. 56, No. 1, 30–34.
15. Pumprey P.M. and Bowkett K.M., Phys. Stat. Sol. (a), 1970, vol. 2. 339–346.
16. Kozubowski L.A., Phys. Stat. Sol. (a), 1977, vol. 43, 535–539.
17. Skakova T.Yu., et al., FMM, 1980, vol. 50, No. 1, 213–215
18. Gertsman V.Yu. and Valiev R.Z., Zavod. Lab., 1981, No. 11, 57–61.
19. Pumprey P.M. and Bowkett K.M., Microscopic Electronique, Proc. Int. Congress, Grenoble, 1970, vol. 2. 189–190.
20. Danilenko V.N., et al., Zavod. Lab., 1991, vol. 57, No. 9, 41–43.
21. Yaichkov I.N. and Maksimov S.K., ibid, 1990, vol. 56, No. 6, 58–65.
22. Lim L.C. and Raj R., Acta Met., 1984, vol. 32, No. 8, 1177–1181.
23. Don J and Majumdar S., Acta Met., 1980, vol. 34, No. 5. 961–967.
24. Sukhomlin G.D. and Andreeva A.V., Phys. Stat. Sol. (a), 1983, vol. 78, No 1, 33–341.

25. Gertsman V.Yu., et al., FMM, 1989, vol. 68, No. 2, 348–352.
26. Gertsman V.Yu., et al., Metallofizika, 1990, vol. 12, No. 3, 120–122.
27. Gertsman V.Yu., et al., Colloque de Physique, 1990, vol. 51, 1151–1154.
28. Gertsman V.Yu., et al., Metallofizika, 1990, No. 2, 113–115.
29. Gertsman V.Yu. and Tangri K., Phil. Mag. A, 1991, vol. 64, No. 6. P. 1319–1330.
30. Gertsman V.Yu., et al., Acta Metall. Mater., 1994, vol. 42, 1785–1794.
31. Gertsman V.Yu., et al., Acta Mater., 1996, vol. 44, No. 7, 2869–2882.
32. Gertsman V.Yu. and Bruemmer S.M., Acta Mater., 2001, vol. 49, 1589–1598.
33. Fionova L.K., FMM, 1979, vol. 48, No. 5, 998–1003.
34. Fionova L.K., Poverkhnost', 1982, No. 5, 43–46
35. Rybin V.V., et al., ibid, 1984, No. 10, 107–116.
36. Rybin V.V., et al., ibid, 1984, No. 9, 107–111.
37. Kozlov A.L., Dissertation, FTI, Izhevsk, 1985.
38. Rybin V.V., et al., Zavod. Lab., vol. 46, No. 7, 600–604.
39. Valiev R.Z., Crystallo-geometrical analysis of intercrystalline boundaries, Nauka, 1991.
40. Danilenko V.N., et al., Phil. Mag. Lett., 1995, vol. 71, No. 1, 39–44.
41. Zhilyaev A.P., et al., Acta Metall. Mater., 1993, vol. 41, No. 9, 2657–2665.
42. Danilenko V.N. Dissertation, IPSM, Ufa, 1992.
43. Gubeev N.A., et al., Zavod. Lab., 1991, No. 12, 31–32.
44. Caleyo F., et al., Scripta Mater., 1999, vol. 41, No. 8, 847–853.
45. Kurzydlowski K.J., Scripta Metall., 1991, vol. 25, 1099–1105.
46. Danilenko V.N., Private communication, 2002.
47. Fionova L.K., Poverkhnost', 1982, No. 5, 43–46.
48. Danilenko V.N., et al., Zavod. Lab., 2005, vol. 71, No. 10, 25–28.

5

GRAIN BOUNDARY SLIDING IN METALLIC BI- AND TRICRYSTALS

After discussing the methods of verification of the individual grain boundaries and ensembles of the grain boundaries, we can find the answer to the question of how the properties of the grain boundaries are manifested in deformation of the materials. We begin with the basic element – the bicrystal.

5.1. Dislocation nature of grain boundary sliding (GBS)

In almost all experiments, the bicrystal is tested in tension under creep conditions. The dependence of the extent of sliding S on loading time t, applied force $\sim\sigma$, temperature T and the set of the parameters R, specifying the misorientation of the grain boundary and orientation of the plane of the boundary in relation to the loading axis, is recorded in experiments. A large number of experiments have been carried out and they make it possible to describe the dependence $S = S(t, \sigma, T, R)$ to the extent sufficient for the formulation of modelling considerations.

The characteristic time dependence of the extent of sliding is shown in Fig. 5.1 [1, 2]. The investigated materials (Cd, Zn) were characterised by the HCP lattice. Consequently, experiments can be set up to ensure that the bicrystal with the given misorientation is oriented in relation to the loading axis in such a manner that in one case the grain boundary sliding is accompanied by intragranular dislocation activity ('stimulated' GBS), and in the other case the shear stresses on the basic planes of both grains are equal to zero ('pure' GBS). The time dependences of these two types of sliding greatly differ. In the case of pure GBS sliding takes place in the given period with the almost constant rate (without hardening). The rate of stimulated GBS at the initial moment of time is an order of magnitude (or more) higher than the pure sliding rate. This rate decreases monotonically with increasing time and at some moment (in Cd ~600 s; in Zn ~200 s) reaches the

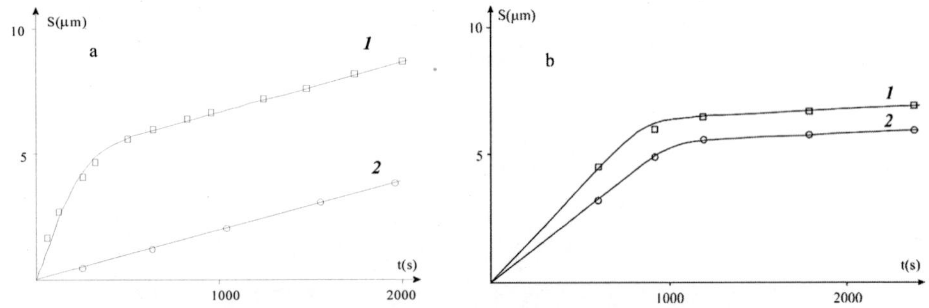

Fig. 5.1. Time dependence of the extent of sliding in bicrystals: a) Cd (T = 421 K; σ = 0.36 MPa; 1) stimulated GBS; 2) pure GBS) [1]; b) Zn (T = 323 K; 1) σ = 1.2 MPa; 2) σ = 0.9 MPa) [2], the curves are the results of theoretical calculations.

level of pure GBS.

The plots of the set of curves for different values of σ and T, similar to those shown in Fig. 5.1, after appropriate treatment [1, 2] shows that the following relationships are fulfilled with satisfactory accuracy:

$$\dot{S}_0 \sim \sigma_B^2 \exp(-E_V / kT); \quad (5.1)$$

$$\dot{S} \sim \sigma_B \exp(-E_B / kT), \quad (5.2)$$

where $\dot{S}$ is the rate of pure GBS; $\dot{S}_0$ is the rate of stimulated GBS at the initial moment; σ_B is the shear stress in the plane of the boundary of the bicrystal; E_B, E_V is the activation energy of grain boundary and of volume diffusion, respectively.

The dependence on the orientation parameters is quite difficult to show in a suitable form because the symbol R represents in fact seven independent variables (two planes and rigid translation). The experimentally determined qualitative relationships will now be discussed. It was shown in [3] that the sliding rate in two mutually perpendicular directions, belonging to the boundary, is anisotropic. Depending on the misorientation, the values of anisotropy $S_{\|}/S_{\perp}$ can differ by an order of magnitude. There are data [2] indicating the high value of the sliding rate in the twisting boundaries in comparison with the tilted boundaries.

Combined investigations of sliding and migration have been carried out in a number of experiments. The results show that the ratio $S(t)/M(t)$ (where $M(t)$ is the time dependence of the extent of migration) does not depend on time. The migration direction coincides with the direction of the perpendicular plane of the boundary of the component of the Burgers vector of the mobile grain boundary dislocations (GBD). In addition, $S/M = b_{\|}/b_{\perp}$, where $b_{\|}$ is the component of the Burgers vector parallel to the boundary plane [2, 5, 6].

It was established in almost all experiments that the sliding in the grain boundary plane is nonuniform. Thus, in addition to the previously mentioned quantities, the extent of sliding also depends on the spatial coordinates in the boundary plane: $S = S(x)$. The dependences shown in Fig. 5.1 were determined by averaging S with respect to the length of the boundary along thc sliding direction.

At least three of the previously mentioned experimental factors – nonuniformity of sliding, the linear relationship between sliding and migration, the anisotropy of sliding on the plane of the boundary – indicate the dislocation nature of GBS. Theoretical models [7] based on the assumption on the 'liquid-like' state of the boundary reproducing the rheology of viscous sliding can be used in describing the GBS in ceramic materials where the melting of the boundary as a result of deviation from the stoichiometric in the vicinity of the grain boundary is indeed detected in a number of cases at high temperatures. However, in the case of the metallic materials all the results indicate the dislocation nature of the GBS.

Analysis of the experimental results show that the adequacy of the theoretical models used for the description of the grain boundary sliding process is not controlled by the reproduction of the individual numerical values characterising the process (this is quite easy to achieve in the conditions of high indeterminacy of the input parameters) and is controlled by the description of the functional dependences on time, applied force and temperature.

One of the first theoretical models treating the grain boundary sliding as the process of movement of grain boundary dislocations, was developed by Gates [8] who derived an equation for the speed of movement of structural grain boundary dislocations:

$$V_B = b_B \sigma_B \frac{\delta D_B}{kT} C, \tag{5.3}$$

where b_B is the modulus of the Burgers vector of the grain boundary dislocations; σ_B is the shear stress in the boundary plane; δ is the effective width of the boundary; D_B is the grain boundary diffusion coefficient; T is temperature; C is the dimensionless coefficient whose numerical value is determined by the nature of the sliding set of the grain boundary dislocations. For example, in the case of sliding GBDs with short pinned sections (the points of exit of the lattice dislocations to the plane of the grain boundary)

$$C = \frac{2\pi\lambda\Omega}{L^2 b_L^2 \cos^2\gamma \ln(X / r_0)} \tag{5.4}$$

where λ is the mean distance between the pinned sections along the dislocations; L is the length of the pinned sections; γ is the angle between the plane of the grain boundary and the axis of rotation (in the case of a clean tilted boundary the system of structural dislocations does not contain the sliding set); X is the diffusion path length equal to half the distance between the pinned sections; Ω is the atomic volume; b_L is the modulus of the Burgers vector of the lattice dislocations. If the pinning points of the sliding dislocations are represented by steps at the boundary then

$$C = \frac{\Omega(1 + \mathrm{tg}\,\alpha\,\mathrm{ctg}\,\gamma)}{b_B^2 X\,\mathrm{tg}\,\alpha\,\mathrm{ctg}\,\gamma}, \tag{5.5}$$

where tg α is linked with the fraction of the steps in the total length of the dislocations. In the case of a non-conservative motion when the dislocation emits (or absorbs) vacancies along the entire length of the dislocation line:

$$C = \frac{\Omega \sin\gamma}{b_B^2 X \cos^2\gamma}. \tag{5.6}$$

Evaluating the rate of grain boundary sliding on the basis of the Orowan ratio:

$$\dot{S} = b_B \rho_B V_B, \tag{5.7}$$

Gates expressed the density ρ_B (x, t) of the grain boundary dislocations by the density of the structural grain boundary dislocations connected with the misorientation at the investigated boundary. This shows that this model cannot be used for describing the stimulated grain boundary sliding because the experiments carried out on the bicrystals show unambiguously the strong effect of the intragranular processes on the density of the grain boundary dislocations and the weak relationship of the density with the density of the structural grain boundary dislocations. To describe pure grain boundary sliding, it is also insufficient to use only information on the sliding as on the movement of the structural grain boundary dislocations because this process would lead to the removal of screw dislocations from the boundary (transformation of an arbitrary boundary to the tilted boundary) and subsequently to the monotonic change of the misorientation. No such evolution was observed in the experiments. Thus, the model of the pure grain boundaries sliding sgall be a model of the source of the grain boundary dislocations formed at the boundary without the participation of intragranular processes.

In [9] it was attempted to describe stimulated GBS taking into account the intragranular dislocation activity. The model is based on the following assumptions. The lattice dislocations from both grains travel to the boundary (from the opposite side) under the effect of the applied stress and, remaining lattice dislocations, move along the boundary in the opposite directions ensuring the corresponding transfer of the Burgers vector to the periphery of the boundary. This also determines the resultant sliding. At the same time, it is well known that the lattice dislocation, penetrating into the boundary, forms an unstable state and during some period of time t_s, referred to as the spreading time, penetrates into the system of the pure grain boundary dislocations, increases the density of the dislocations and accelerates grain boundaries sliding in accordance with the relationship (5.7). There are several models of the spreading of the lattice dislocations [10]. However, all the models estimate t_s using approximately the same procedure:

$$t_s = \frac{l^3}{G\Omega}\frac{kT}{\delta D_B}, \tag{5.8}$$

where G is the shear modulus; $l \approx 10^2\, b_B$ is the characteristic distance over which the dissociation products are distributed.

Thus, for a realistic description, the model of the stimulated grain boundary sliding must take into account the increase of the density of the grain boundary dislocations as a result of the spreading of the lattice dislocations supplied by the grains of the bicrystal.

Numerical estimates relating to the experimental results, shown in Fig. 5.1, will now be carried out and they can be used to estimate the level and value of the density of the grain boundary dislocations in the formation of the sliding pattern. Using the diffusion and lattice parameters presented in [11]: for Cd $\delta D_{0B} = 5 \cdot 10^{-14}$ m^3/s, Q = 54.4 kJ/mole, $b_L = 2.93 \cdot 10^{-10}$ m; for Zn $\delta D_{0B}=1.3 \cdot 10^{-14}$ m^3/s, Q = 60.5 kJ/mole, $b_L = 2.67 \cdot 10^{-10}$ m, according to (5.3)

$$V_{\mathrm{Cd}} = 1.4 \cdot 10^{-2} C\,[\mathrm{m/s}]; \quad V_{\mathrm{Zn}} = 9.2 \cdot 10^{-6} C\,[\mathrm{m/s}], \tag{5.9}$$

where C is a constant defined by one of the relationships (5.4)–(5.6). Taking into account that in the case of both Cd and Zn there is an tilted boundary with basic slip planes in the grains misoriented by $\pi/2$, it should be assumed that the purely geometrical (by nature) dimensionless constants C_{Cd} and C_{Zn} are the quantities of the same order of magnitude. Consequently, the rate of grain boundary sliding in Cd is more than three orders of magnitude higher than that in Zn. This is not surprising because the activation energy in Cd is 6 kJ/mole lower than in Zn and the test temperature is 100 K higher. At the same time, the initial rate of grain boundary sliding in Cd ($\dot{S}_0 = 1.6 \cdot 10^{-8}$ m/s) is only three times higher than the initial rate of grain boundary sliding in Zn ($\dot{S}_0 = 5.1 \cdot 10^{-9}$ m/s). However, since the Burgers vectors in Cd and Zn are approximately identical, then it should be accepted, in accordance with the relationship (5.7), that in the case of Zn (low grain boundary diffusion rate) the dislocation density of the boundary is very high, whereas in the case of Cd (high grain boundary diffusion rate) some other mechanism operates during grain boundary sliding and removes the difference in the rates of grain boundary diffusion and ensures approximately the same level of the sliding rate.

Since the sliding rate reaches a constant level independent of time (approximately equal to the rate of pure grain boundary sliding) after interrupting intragranular dislocation activity, it should be assumed that the grain boundary diffusion rate in the test should be independent of time. Consequently, of the three co-multipliers in the right-hand part of equation (5.7), only ρ_B can depend on time. In other words, not only the initial level of the rate of grain boundary sliding is determined by the density of grain boundary dislocations but also the dependence $\dot{S}(t)$ is determined by the time dependence of the density of grain boundary dislocations.

Thus, the problem can be formulated as follows. To construct a model of stimulated grain boundary sliding it is necessary to describe the dynamics of the density of grain boundary dislocations, including the mechanism of suppressing density at high rates of grain boundary dislocations [12].

5.2. Formulation of the model of stimulated grain boundary sliding

Construction of the model is based on using the principle of minimum sufficiency. We postulate the simplest variant of the processes taking place during grain boundary sliding. The processes are described by the appropriate mathematical physics equations. The equations are solved and it is checked whether the solutions of these equations can reproduce the experimental dependences. If this is not so, it is necessary to explain which special features of the experiment are not reproduced and which processes should be added to the scenario in order to insure the required behaviour of the solution. A process is included in the equation, the equation is solved and the procedure is repeated until all the significant special features of the experiment are reproduced. This method of permanent correction of the experiment prevents us from moving towards scholastic theoretical considerations and speculative claims. The model does not contain any optional elements (which, on the basis of the general considerations, could be includes in the model), and the need for the elements which should be included is dictated by the experiments.

Taking into account the analysis results, it is assumed that in stimulated grain boundary sliding the dislocations propagating on the basic sliding planes reach the boundary plane. During the time equal to t_s each dislocation penetrates into the system of grain boundary dislocations, forming a specific number of sessile (generally speaking, changing the misorientation at the grain boundary) and sliding dislocations. These numbers are given by the expansion coefficient of the lattice Burgers vector with respect to the basis of the displacement shift completed (DSC) lattice, i.e., it is determined by the specific geometry of the bicrystal. In the simplest variant we do not take into account the changes of misorientation and do not consider sessile dislocations. Without specifying the geometry of the bicrystal, it is assumed that the sliding dislocations, formed from the lattice dislocations of different grains, have opposite signs, i.e., move in the plane of the boundary against each other. Finally, it is assumed that these two sets of the dislocations can be studied irrespective of each other (as indicated later, up to a certain moment).

In the proposed scheme, the source of grain boundary dislocations is distributed along the entire boundary plane and a sink is formed, depending on the sign of grain boundary dislocations, either through the left or right edge of the boundary. The evolution of the density of grain boundary dislocations in this coordinate system forms the spatially heterogeneous distribution $\rho(x, t)$ which can be described only by the differential equation in the partial derivatives. Considering, for determinacy, the dislocations with the same sign moving under the effect of applied stress with constant speed V_B in the positive direction of the x axis (the direction of sliding in the plane of the boundary), the following equation is obtained

$$\frac{\partial\rho(x,t)}{\partial t}+V_B\frac{\partial\rho(x,t)}{\partial x}=I(x,t) \quad (0\le t; 0\le x\le L), \tag{5.10}$$

where L is the length of the boundary in the sliding direction, $I(x, t)$ is the source, i.e., the number of grain boundary dislocations formed in unit time in the unit length of the boundary.

In this scheme, the source is formed by lattice dislocations. In HCP metals one basic slip system operates in each grain, i.e., lattice dislocations move on parallel planes and laminar plastic yielding (according to Cottrell's terminology [13]) takes place. Under these conditions it may be assumed that the distances between the slip planes are defined by the expression

$$h=\frac{Gb_L}{8\pi(1-v)\sigma_L},$$

where σ_L is the shear stress acting in the basic plane, v is the Poisson coefficient.

In this case, $1/(h \sin\varphi)$ basic slip planes reach the unit length of the boundary (here φ is the dihedral angle between the plane of the boundary and the basic slip plane). During loading the slip system becomes 'harder' and the number of actively acting planes decreases. The condition $\sigma/G \sim 10^{-4}$, i.e., the first stage of plastic flow, was realised in the experiments [1, 2]. The factors which determine this hardening are well known: a) locking of the Frank–Read sources, acting in the slip planes, by the reversed stress from the emitted dislocations; b) multipolar hardening, formed during the interaction of dislocation elements with the opposite signs and located in the adjacent slip plane; c) formation of low-angle boundaries as a result of polygonisation by slip. However, there is no dynamic model of the first stage of the plastic flow [14, 15]. We restrict our considerations to the following. The number of planes working at the initial moment of time in a grain is $N_0 = L/(h \sin\varphi)$. Let the plane with the number $(1 \le i \le N)$ closes at time t_i as a result of hardening. It is also assumed that t_i is a random quantity having the values of the half-line $[0, \infty)$. Finally, let all the values t_i be independent and distributed in the same manner in accordance with some probability density $W(t)$. The assumption on this independence is obviously an approximate. For example, in the case of multipolar hardening the active planes are knocked out from action by pairs, and the values t_i and $t_i + 1$ are strictly correlated. Nevertheless, accepting this scheme, we can immediately write the number of active slip planes at the moment of time t. It is clear that at this moment the planes whose working time is shorter than t are closed, and the fraction of these planes in the total number N_0 is $\int_0^t W(t')dt'$. Thus, the number of working planes is

$$N(t) = N_0 - N_0 \int_0^t W(t')dt' = N_0 \int_t^\infty W(t')dt'.$$

Assuming that the lattice dislocation penetrates into the boundary during time t_s and this is accompanied by the formation of n mobile grain boundary dislocations, we can write the number of grain boundary dislocations appearing in unit time in the unit length of the boundary: $nN(t)/(Lt_s)$. Introducing the notation

$$\rho_0 = n \frac{8\pi(1-\nu))}{b_L} \frac{\sigma_L}{G} \sin \varphi \tag{5.11}$$

we determine the required source for the equation (5.10)

$$I(x,t) = \frac{\rho_0}{t_s} \int_t^\infty W(t')dt'. \tag{5.12}$$

The formal solution of equation (5.10) with the source (5.20) and the initial condition $\rho(x, 0) = 0$ qualitatively reproduces the shape of the time dependence of the extent of sliding in Cd. However, this solution does not reproduce the results in Zn regardless of the parametrisation used. If we specify the condition of accurate parametrisation of the initial sliding rate, we obtain the unrealistically small total value of sliding at the moment of completion of the stage of stimulated grain boundary sliding. However, if the total value of sliding is parametrised, we obtain unrealistically high initial rates. Returning to the experimental results, it should be noted that in the case of Zn the sliding rate is almost completely constant in the first 600 seconds, i.e., the density of grain boundary dislocations is also constant. The model (5.10), (5.12) does not contain the mechanism of stabilisation of density and, therefore, does not work in situations in which the rate does not decrease monotonically from the initial value. The results were obtained assuming that

$$W(t) = \tau^{-1} \exp(-t/\tau),$$

where τ is the mean time of closure of a single basic slip plane. If we use formally the gamma distribution with an integer exponent

$$W(t) = \frac{1}{\tau m!}\left(\frac{t}{\tau}\right)^m \exp(-t/\tau),$$

as $W(t)$, then we may obtain a small section with constant density. However, it should be assumed in this case that the intragranular processes in Cd and Zn take place by completely different mechanisms and this appears to be completely unrealistic (laminar flow in the first stage). Thus, it is necessary to find another mechanism of stabilising the density of grain boundary dislocations.

During the formation of the high density of grain boundary dislocations the boundary acquires additional elastic fields which prevent the entry of lattice dislocations into the boundary. In particular, this circumstance must be taken into account in this stage. The following expression [16] is used as the approximate estimate of the spatial distribution of the stress field from the dislocations uniformly distributed in the plane of the boundary:

$$\sigma_{xy} \sim \frac{2\pi Gby}{(1-\nu)D^2}\exp\left(-\frac{2\pi y}{D}\right),$$

where y is the distance from the boundary plane; D is the distance between the dislocations. Consequently, the equation of movement of the lattice dislocation in the effective stress field $\sigma_L - \sigma_{xy}$ (taking into account only the retarding exponential tail) has the following form

$$\dot{y}(t) = -\left[1 - \frac{\rho}{\rho_0}\exp(-2\pi\rho y)\right]V_L,$$

where V_L is the speed of the lattice dislocations at a relatively large distance from the boundary. Solving this equation, we obtain the estimate of the additional time in comparison with the case of free propagation. This time is required by the lattice dislocation to penetrate into the boundary

$$\Delta t_s = -\frac{1}{V_L \rho} \ln\left(1 - \rho / \rho_0\right).$$

Thus, in the expression for the source (5. 12) the spreading time should be replaced by some effective time

$$t_{s\,ef} = t_s + \Delta t_s,$$

which transforms equation (5.10) into a non-linear equation. Although the equation of this type can be easily solved (at the characteristics this equation transforms to the conventional differential equation of the first order with separated variables), the resultant solution contains an unknown number of artefacts caused by the extremely approximate description of both the field from the boundary (for more details see [16]), and of the process of propagation of the lattice dislocation in the field. Therefore, only the qualitative relationships resulting from this analysis will be used. Thus, if the actual density ρ tends to ρ_0, quantity t_s tends to infinity and the effect of the source is interrupted ($I\ (x,t) \to 0$). However, since the movement of the grain boundary dislocation in the boundary plane continues and the dislocations leave the plane, the density of the dislocations decreases, the source is activated and density again reaches the value ρ_0. This process activates the mechanism of stabilisation of density and, consequently, ensures the constant rate of grain boundary sliding. This behaviour of the solution can be achieved within the framework of the linear equation, assuming that

$$I(x,t) = I_0(t)\left[1 - \frac{\rho(x,t)}{\rho_0}\right], \tag{5.13}$$

where the following notation is introduced to shorten the equations

$$I_0(t) = \frac{\rho_0}{t_{s\,ef}} \int_t^{\infty} W(t')dt'. \tag{5.14}$$

In this case

$$t_{s\,ef} = t_s + \frac{1}{V_L \rho_0}, \tag{5.15}$$

where t_s is determined by equation (5.8) and the second term in the right-hand part has the meaning of time during which the lattice source forms a single loop [17].

The solution of equation (5.10) with the source (5.30) describes efficiently the shape of the curve $S(t)$ in both Zn and Cd. However, the absolute values of the grain boundary sliding rate in Cd are very high because the density $\rho(x,t)$ reaches almost the same value ρ_0 as in the case of Zn. However, according to preliminary estimates, the density of grain boundary dislocations in Cd should be orders of magnitude lower than the density in Zn. Thus, the model (5.10), (5.13) contains the mechanism of stabilisation of density and does not have the mechanism for suppressing density stabilisation at high speeds of grain boundary dislocations.

In accordance with the results, the only mechanism which produces a sink of grain boundary dislocations and which depends on the speed of the dislocations, is the annihilation of the dislocations of the opposite signs supplied from different grains. If the density of dislocations of the same sign (for example, the dislocations moving in the positive direction of the axis x with the speed V_1) reaches $\rho_1(t)$, and that of the opposite side (i.e., moving towards these dislocations with the speed V_2) is $\rho_2\ (x,t)$, then the number of dislocations of both signs knocked out from the system during unit time from the unit length of the boundary will be $(V_1 + V_2)\ \rho_1\rho_2$. The evolution of the dislocation density is described by the system of differential equations

$$\frac{\partial\rho_1}{\partial t}+V_1\frac{\partial\rho_1}{\partial x}=I_{01}(t)\left(1-\frac{\rho_1}{\rho_{01}}\right)-(V_1+V_2)\rho_1\rho_2; \tag{5.16a}$$

$$\frac{\partial\rho_2}{\partial t}-V_2\frac{\partial\rho_2}{\partial x}=I_{02}(t)\left(1-\frac{\rho_2}{\rho_{02}}\right)-(V_1+V_2)\rho_1\rho_2. \tag{5.16b}$$

The solution of this equation is associated with considerable difficulties and will be used only for the evaluation of the limiting density of the grain boundary dislocations achieved at the boundary. Since the dislocations of the systems 1 and 2 in the accepted model differ only in the sign, it is not justify to assume that the speeds V_1 and V_2 differ. For qualitative evaluation it is assumed that ρ_{01} and ρ_{02} and also $I_{01}(t)$ and $I_{02}(t)$ are approximately the same. Consequently, substitution $\rho_2(x,t) = \rho_1(L{-}x,t)$ reduces the system to a single non-linear equation in the partial derivatives with the displaced argument:

$$\frac{\partial\rho_1(x,t)}{\partial t}+V\frac{\partial\rho_1(x,t)}{\partial x}=I_0(t)\left[1-\frac{\rho_1(x,t)}{\rho_0}\right]-2V\rho_1(x,t)\rho_1(L-x,t). \qquad (5.17)$$

At low densities $\rho_1(x,t)$ (i.e., in the stage of increase of density from zero initial value and in the stage of decrease to 0 after interrupting intergranular dislocation activity) the non-linear quadratic terms can be ignored. The role of the non-linear term becomes important in the range of the density maximum and, consequently, this term determines the maximum itself. It is estimated as follows. Linear transformation transfers the range of variation of the spatial variable $x\in[0, L]$ to the interval $\mu\in[-1,1]$. Density $\rho_1(x,t)$ changes to $\tilde{\rho}(\mu,t)$. The dependence on μ is reproduced by a series using the complete set of the functions in the interval $[-1,1]$. This set is represented by the Legendre polynomials

$$\tilde{\rho}(\mu,t)=\sum_{n=0}^{\infty}\frac{2n+1}{2}P_n(\mu)q_n(t), \qquad (5.18)$$

Substituting (5.18) into (5.17), multiplying the resultant equation by $P_l(\mu)$ and integrating with respect to μ in the interval $[-1,1]$, using the composition theorem for the Legendre polynomials we obtain an infinite non-linear system (l = 0, 1, 2,...) the conventional differential equations. Splitting this system, we obtain the following equation for the expansion coefficient $q_0(t)$ in the rough approximation $q_{l\geq 1} = 0$:

$$\dot{q}_0(t)=I_0(t)\left(1-\frac{q_0}{\rho_0}\right)-Vq_0^2 \qquad (5.19)$$

When the maximum of function q_0 is reached (this function is proportional to the density of the grain boundary dislocations, averaged out along the boundary), the derivative q_0 with respect to t converts to zero. The value of q_{max} is estimated by equating the right-hand part of equation (5.19) to 0. As indicated by the estimates, the time required to reach the maximum is approximately equal to L/V, which is considerably shorter than the length of the interval of stimulated grain boundary sliding. Therefore, in estimating q_{max} we can use the condition $\int_{t_{max}}^{\infty}W(t')dt'\approx 1$. Thus, from equation (5.19) we obtain

$$q_0\leq q_{max}=\frac{\sqrt{1+4Vt_{s\,ef}\rho_0}-1}{2Vt_{s\,ef}},$$

which leads to the final form of the model. In the process of grain boundary sliding the density of the grain boundary dislocation $\rho(x,t)$ is described by the differential equation

$$\rho_t + V_B \rho_x = \frac{\rho_0}{t_{s\,ef}} \left(1 - \frac{\rho}{\rho^*}\right) \int_t^\infty W(t')dt', \tag{5.20}$$

where ρ_0 is given by the relationship (5.11), and the effective spreading time $t_{s\,ef}$ is determined by spreading time t_s (equation (5.8)) and by the mean time of formation of a single active dislocation loop (equation (5.15)); ρ^* has the meaning of the limiting density of the grain boundary dislocations which can be obtained under the given loading conditions and is determined by the equation

$$\rho^* = \frac{\sqrt{1 + 4Vt_{s\,ef}\rho_0} - 1}{2Vt_{s\,ef}},$$

The factor

$$\alpha(t) = \int_t^\infty W(t')dt' \tag{5.21}$$

defined as the dynamics of hardening in intergranular slip. The limiting density ρ^* at low speeds V_B tends to 0, and at high speeds to $\sqrt{\rho_0 / \left(V_B t_{s\,ef}\right)} < \rho_0$.

5.3. Formal solution and its analysis

The characteristics of equation (5.20) have the form of the straight lines $x = x_0 + V_B t$. The equation at the characteristics is the conventional linear differential equation. Solving this equation (taking into account the initial condition $\rho(x, 0) = 0$) and returning to the variables x, t, gives

$$\frac{\rho(x,t)}{\rho^*} = 1 - \exp\left[-\frac{\rho_0}{t_{s\,ef}\rho^*} \int_{\max(0,\, t - x/V_B)}^{t} \alpha(t')dt' \right]. \tag{5.22}$$

For direct comparison with the experiment we determine the mean value of density along the length of the boundary:

$$\overline{\rho}(t) = \frac{1}{L} \int_0^L \rho(x,t)dx.$$

Integrating with respect to the spatial variable, we obtain the following results. For the time range $0 \le t \le L/V_B$, the mean value of density is described by the equation

$$\frac{\rho_1(t)}{\rho^*} = 1 - (1 - \frac{V_B t}{L}) \exp\left[-\frac{\rho_0}{t_{sef}\rho^*} \int_0^t \alpha(t')dt' \right] - \frac{V_B}{L} \int_0^t \exp\left[-\frac{\rho_0}{t_{sef}\rho^*} \int_{t'}^t \alpha(t'')dt'' \right] dt'. \tag{5.23}$$

and for the interval $t > L/V_B$ by the equation

$$\frac{\rho_2(t)}{\rho^*} = 1 - \frac{V_B}{L} \int_{t-L/V_B}^{t} \exp\left[-\frac{\rho_0}{t_{sef}\rho^*} \int_{t'}^t \alpha(t'')dt'' \right] dt'. \tag{5.24}$$

Differentiating (5.23) and (5.24) with respect to t, it may easily be established that the function $\rho_1(t)$ has a maximum in the definition range, and $\rho_2(t)$ is a non-increasing function of time. If t^* is the point at which the extreme of the function $\rho_1(t)$ is reached, the maximum speed of the grain boundary dislocations is given by the relationship

$$\dot{S}_0 = 2b_B V_B \rho_1(t^*). \tag{5.25}$$

The total value of the sliding rate is determined by integrating the speed with respect to time:

$$S_{tot} = 2b_B V_B \int_0^\infty \overline{\rho}(t)dt. \tag{5.26}$$

We determine the time during which the stimulated grain boundary sliding takes place as the interval characterised by 90% of thc total value of the grain boundary sliding. Consequently, the required time t might is determined by solving the equation

$$2b_B V_B \int_0^{t_m} \overline{\rho}(t)\,dt = 0.9 S_{tot} \tag{5.27}$$

The values $\dot{S}_0$, S_{tot} and t_m, like the time dependence of the extent of sliding,

$$S(t) = 2b_B V_B \int_t^\infty \overline{\rho}(t)dt, \tag{5.28}$$

is recorded directly in the experiments so that direct comparison can be carried out.

In a general case, the values defined by the equations (5.25)–(5.28) can be determined only numerically. However, approximate estimates, permitting qualitative examination, can be obtained for exponential density $W(t)$.

We calculate the integral in the index of the exponent of the expression (5.24):

$$\frac{\rho_2(t)}{\rho^*} = 1 - \int_0^1 \exp\left[-\frac{\rho_0\tau}{\rho^* t_{s\,ef}} \exp\left(-\frac{t}{\tau}\right)\left(\exp\frac{Lx}{V_B\tau} - 1\right)\right] dx.$$

Although the remaining interval is expressed through the integral exponential function, we obtain an approximate equation which is however more suitable for analysis. Under the condition $L/(V_B\tau) \ll 1$ which is undoubtedly fulfilled for Cd, the following estimate is valid

$$\exp\frac{Lx}{V_B\tau} - 1 \approx \frac{Lx}{V_B\tau}.$$

Consequently

$$\frac{\rho_2(t)}{\rho^*} = 1 - \frac{1 - \exp(-A(t))}{A(t)}, \tag{5.29}$$

where

$$A(t) = A_0 \exp\left(-\frac{t}{\tau}\right); \quad A_0 = \frac{\rho_0 L}{\rho^* V_B t_{s\,ef}}.$$

Since $A_0 \gg 1$, at a low t (t < t) we have $\rho_2(t) \approx \rho^*$. $A(t)$ decreases monotonically with time. When it becomes lower than unity, which takes place if the inequality

$$t > \tau \ln \frac{\rho_0 L}{\rho^* V_B t_{s\,ef}},$$

is fulfilled, the equation (5.9) is reduced to the expression

$$\frac{\rho_2(t)}{\rho^*} \cong \frac{\rho_0 L}{2\rho^* V_B t_{s\,ef}} \exp\left(-\frac{t}{\tau}\right),$$

i.e., describes the exponential attenuation of density.

After determining the qualitative special features of the behaviour of function (5.29), it is quite easy to obtain the following approximate estimates of the experimentally recorded parameters (because of the approximations, these estimates are quite accurate for high speeds of grain boundary dislocations; however, in a general case, it is necessary to use the exact solutions of the equations (5.25)–(5.27)):

$$\dot{S}_0 = 2b_B V_B \rho^*; \quad S_{tot} = 2b_B V_B \rho^* \tau \ln\frac{A_0}{A^*}; \quad t_m = \tau \ln\frac{A_0}{2\varepsilon \ln A_0 / A^*}, \tag{5.30}$$

where A^* is a constant which determines the inflection point of the time dependence (5.29) and is approximately equal to 1.793, ε is the experimental error of the determination of the total extent of sliding.

The dependence of the rate of stimulated grain boundary sliding on stress and temperature is determined by the expression:

$$\dot{S}_0 \cong b_B \frac{\sqrt{1 + 4V_B t_{s\,ef} \rho_0} - 1}{t_{s\,ef}}, \tag{5.31}$$

which defines a far more general functional form than (5.1). If the inequality $V_B t_{sef} \rho_0 \ll 1$ is fulfilled, equation (5.31) determines the dependence of the type

$$\dot{S}_0 \sim \sigma^2 \exp(-E_B / kT).$$

If the reversed inequality is fulfilled, equation (5.31) has the form: $\dot{S}_0 \cong 2b_B \sqrt{V_B \rho_0 / t_{sef}}$ and, depending on the ratio of the quantities t_s and Δt_s, can lead to different forms of the dependence on σ and T. If the process of grain boundary sliding is controlled by the spreading of the lattice dislocations, i.e., at a high value of t_s, we have

$$\dot{S}_0 \sim \sigma \exp(-E_B / kT).$$

However, if the intragranular processes are limiting, i.e., at high values of t_s, then

$$\dot{S}_0 \sim \sigma^2 \exp\left(-\frac{E_L + E_B}{2kT}\right).$$

Thus, even the approximate analytical expression (5.31) reproduces qualitatively the main experimental relationships expressed by (5.1).

We calculate the time dependence $S(t)$ of the extent of sliding using (5.28). Parametrisation is based on the relationships (5.30) which can be used to determine τ, $b_B V_B \rho^*$ and A_0 at the known (from the experiments) $\dot{S}_0$, S_{tot} and t_m. For Cd

$$\tau = 107 \text{ s}; \quad b_B V_B \rho^* = 0.112 \cdot 10^{-1} \text{mm/s}; \quad \frac{\rho_0 L}{\rho^* V_B t_{s\,ef}} = 10.$$

For Zn (curve corresponding to σ = 1.2 MPa)

$$\tau = 82 \text{ s}; \quad b_B V_B \rho^* = 0.332 \cdot 10^{-2} \text{mm/s}; \quad \frac{\rho_0 L}{\rho^* V_B t_{s\,ef}} = 10^5$$

and (σ = 0.9 MPa)

$$\tau = 68 \text{ s}; \quad b_B V_B \rho^* = 0.258 \cdot 10^{-2} \text{mm/s}; \quad \frac{\rho_0 L}{\rho^* V_B t_{s\,ef}} = 2 \cdot 10^6.$$

The dependences $S(t)$, calculated from equation (5.28) (taking into account the correction for the contribution of pure grain boundary sliding, $\dot{S}(t)$ for the determined values of the parameters are presented in Fig. 5.1. They reproduce quite efficiently the special features of the experimental curves, including the section with the constant speed of grain boundary sliding for Zn and its monotonic decrease in Cd.

5.4. Special features of pure grain boundary sliding

When analysing the mechanisms of pure grain boundary sliding it is initially necessary to answer the following question: are there sources of grain boundaries dislocations at the boundary or, as assumed by Gates [8], sliding is the movement of structural grain boundary dislocations? The first argument in favour of the existence of sources (absence of any large change of misorientation during sliding) was discussed previously. The second argument follows from the following estimate. In pure grain boundary sliding in Cd the total extent of sliding S reaches 5 μm. At $b_B \sim b_L/10 = 3 \cdot 10^{-4}$ μm it is necessary for this purpose that $N = S/b_B \sim 10^4$ grain boundary dislocations. At the length of the boundary $L \sim 7 \cdot 10^3$ μm the distance between the dislocations is $L/N \sim 10^4\, b_B$, i.e., they are determined with sufficient accuracy, and the total extent of sliding can be explained completely by the movement of structural grain boundary dislocations present initially in the boundary, if the speed of the grain boundary sliding would not be constant over the entire examination period. If the initial number of the dislocations were $\sim 10^4$, then at the moment of selecting the extent of sliding as 5 μm its speed would be zero ($\rho \to 0$), whereas in Cd the speed remains almost completely constant. It may be assumed that the remote dislocations form a small part of initial density which can be increased by 2–3 orders of magnitude without any loss of the accuracy of resolution of the GBD (no overlapping of the cores). However, here we obtain the accurate rate of grain boundary sliding at the end of the measurement period and a much higher rate at the start of the period. The latter can be easily confirmed as follows. If the pure grain boundary sliding is the movement of the structural grain boundary dislocations, leaving the plane of the boundary only at exit to its periphery, the dislocation density satisfies the equation

$$\frac{d\rho}{dt}=-\frac{V}{L}\rho,$$

which leads to the following dependence of the grain boundary sliding rate on time:

$$\dot{S}=b_B V\rho_0\exp(-Vt/L), \tag{5.32}$$

where ρ_0 is the initial dislocation density. Thus, regardless of initial density, the grain boundary sliding rate decreases *e* times during the period L/V, which is a very small value even for Zn (and even smaller for Cd). As shown by experimental results, the grain boundary sliding rate remains constant over a considerably longer period of time which, in our opinion, indicate unambiguously the existence of the sources of grain boundary dislocations in the boundary.

Regardless of the nature of the sources, the set of the sources is characterised by two parameters: density ρ_s (the number of sources per unit length of the boundary in the sliding direction) and time τ_0 of the generation of a single dislocation which will be regarded as the same for all the sources. In this case, if ρ_1 and ρ_2 are the densities of the segments of the dislocation loops with the opposite signs and moving in the opposite direction along the sliding direction, the dynamics of dislocation density is described by the previously formulated hyperbolic system (5.16). In fact, it would be strange if the evolution of the dislocation ensembles, differing only in the origin (in the case of stimulated grain boundary sliding these are the products of the spreading of the lattice dislocations, and in the case of pure grain boundary sliding they are the segments of dislocation loops, formed in the plane of the boundary), would be described by the completely different equations because in this case the movement of the grain boundary dislocations is also accompanied by both the process of annihilation (the last term in the right-hand part of (5.16)) and by a decrease of the productivity of the sources by the reversed stress of the emitted dislocations (the second co-multipliers of the first term in the right-hand part of (5.16)). The main difference in comparison with the case of stimulated grain boundary sliding is the re-determination of the quantity (5.14) defining the productivity of the system of sources. Firstly, the time dependence which in the case of stimulated grain boundary sliding is determined by intragranular hardening, is no longer there. Secondly, ρ_0 has the meaning of the density of grain boundary sources ρ_s, and $t_{s\,ef}$ is replaced by the time τ_0 of generation of a single loop. The dependence of the parameters ρ_s and τ_0 on the loading conditions is determined by the nature of the sources. This problem has not as yet been completely solved, although one special feature of the sources can be mentioned: they should form a specific family of dislocations with Burgers vectors belonging to the boundary plane. The latter circumstance follows directly from the fact that pure grain boundary sliding is not accompanied by the migration of the boundary, whereas in the case of stimulated grain boundary sliding sliding and migration act in parallel. An example of calculating τ_0 for the lattice dislocations was published by Barton [17]. Qualitative evaluation can be carried out using the approximation $\tau_0 \sim \lambda/V_B$, where λ is the initial length of the growing segment of the dislocation loops. If it is assumed that ρ_s is a purely geometrical parameter proportional to the number of growth segments in the boundary plane, then the application of equation (5.31) for estimating the speed of pure grain boundary sliding leads directly to the dependence of type (5.2). The lower rate of pure grain boundary sliding in comparison with stimulated grain boundary sliding is explained by the low productivity of the two-dimensional source in comparison with the productivity of the three-dimensional one.

The following pattern of the formation of the processes taking place during stimulated grain boundary sliding in bicrystals with the HCP lattice can be proposed on the basis of these considerations. The lattice dislocations, propagating on the basic slip planes of two grains, reach the boundary of the bicrystal and as a result of dissociation (during some characteristic period of time) they penetrate into the grain boundary, forming two moving sets of the dislocations with the opposite Burgers vectors. At low speeds of grain boundary dislocations (either low temperatures or high activation energy of grain boundary diffusion), the dislocations of both sets move slowly against each other (ensuring local sliding from which the resultant sliding forms). Over a long period of time, essential for meeting and annihilation with a dislocation of the opposite sign, additional lattice dislocations can penetrate into the boundary and form a high density of mobile grain boundary dislocations. A certain level of density, determined by the ratio of the rate of inflow and annihilation, is established. A high level of the density of grain boundary dislocations forms in the vicinity of the boundary additional elastic fields complicating the penetration of the lattice dislocations into the boundary and weakening the efficiency of the source. At high speeds of grain boundary dislocations (or high temperatures or low activation energy of grain boundary diffusion), the lifetime of the dislocations in the boundary becomes very short (as a result of a high rate of annihilation) and the lattice source does not manage to generate high densities. The time dependence of the density of grain boundary dislocations actually tracks the rate of hardening in intragranular slip. Thus, the boundary of the bicrystal demonstrates to some extent the special features of behaviour of the systems with a negative feedback. The formalisation of this scenario makes it possible to describe efficiently the results of experimental investigations of grain boundary sliding in Cd and Zn.

The results contain the dependence on the dimensional factor L (the length of the boundary in the sliding direction) so that it is possible to carry out additional verification by means of appropriate experiments. The formal dependence on L may cause temptation to extrapolate the results for describing grain boundary sliding in polycrystals with small grains. However, these attempts would not be successful because an additional factor operates in the polycrystal – the non-free sink of grain boundary dislocations through the edge of the boundary (triple junction) resulting in the dynamics differing completely from that investigated previously.

The effective verification of the model may be carried out in the experiments in which the factor responsible for the negative feedback, i.e., annihilation, is not considered. For this purpose, the bicrystal should be loaded to ensure that the shear stresses on the basic slip planes in one of the grains are equal to zero and in the other grain intergranular slip would be possible. Since this case is characterised by the formation of the grain boundary dislocations of mostly the same sign in the boundary, annihilation is almost completely suppressed and high dislocation density is made to form in the boundary even at a high dislocation speed. In this case, the grain boundary sliding rate should be higher than that in the case of operation of two grains at the same temperatures and identical other conditions. With the increase of temperature, i.e., at an even higher speed of grain boundary dislocations, the dislocation density reaches the maximum value which complicates the entry of the lattice dislocations into the boundary. The non-realised lattice dislocations, remaining in the grain, increase the degree of cold working of the grain and this is directly recorded by the change of the microhardness across the bicrystal boundary.

The microhardness of the non-working grain should not change. The degree of cold working should correlate strictly with the length of the section of the constant (time-independent) density of the grain boundary dislocations. This section is found in the experimental dependence obtained for Zn in [5]. The study also notes (according to the results of microhardness measurements) the higher degree of cold working of the near-boundary region. The indirect confirmation of the pattern observed here is the established fact of the higher speed of grain boundary sliding at the interface boundaries in comparison with intergranular boundaries.

Detailed analysis of the behaviour of $S(t)$ at low values of t and low speeds of grain boundary dislocations may detect the special features associated with the increase of density from zero to its maximum value. The duration of this range should be determined by the length of the boundary.

Analysis of the experimental results obtained for the bicrystals with the cubic lattice shows that the main special features are associated with the mechanisms of hardening by intergranular slip with a large variety of the lattice dislocations supplied to the boundary as a result of the effect of a large number of slip systems. If general formalism is unchanged, this may be taken into account by re-determining the values (5.11) and (5.12). For the parametrisation of this type it is necessary to carry out a more detailed examination of the dynamics of hardening of the first stage of the plastic flow of the cubic lattice metals [18].

The proposed formalism may be used efficiently for solving numerous problems: on the basis of the experimentally recorded dependence $S(t)$ it is quite easy to restore the parameters of intergranular slip and the speed of grain boundary dislocations. Combined processing of the of investigations of pure and stimulated grain boundary sliding increases the accuracy of determination of the parameters.

5.5. Local migration of the grain boundary as the mechanism of reorganisation of the triple junction. Weak migration approximation

The model of grain boundary sliding discussed in the previous section, cannot be used to describe the mutual shift of two conjugate grains in the polycrystal. Generally speaking, the only obstacle to this is the given condition of the free sink of grain boundary dislocations,leaving the boundary plane. In the polycrystal conditions, the boundary does not travel to the free surface and makes contact with the boundary of another pair of grains along the so-called triple junction line. The direct transition of the dislocation segment from one plane of the grain boundary to another plane which is in contact with it is forbidden in a general case because of crystallographic considerations. The propagating dislocations form a pile-up with the head dislocation, situated in the triple junction or in the immediate vicinity of the junction. The pile-ups of the grain boundary dislocations formed in the initial stage of plastic deformation of fine-grained materials, determined various important physical processes such as the development of cavitation in the creep conditions, the formation of a crack at high loading rates [19], the rearrangement of the configuration of the triple junction at stresses characteristic of the superplastic state [20], inelastic relaxation of the material after unloading [21, 22]. Depending on the power of the resultant pile-up and the stress field determined by the pile-up, further evolution may take place by different mechanisms. The heterogeneous stress fields activates vacancy flows and this may result in the formation of microporosity [23, 24]. At a high rate of formation of the pile-up and at low temperatures the head dislocations can merge together and a microcrack nucleus can form [25]. The stress from the pile-up may prove to be sufficient for activating slip in the contacting grains. In particular, this process is the basis of accommodation of grain boundary sliding by intergranular slip which has been accepted as the model of superplastic deformation. The answer to the question as what takes place in the conditions characteristic of superplastic deformation was obtained as a result of experimental investigations.

When investigating the distribution of the values of the dihedral angles in the triple junctions of a polycrystal it was established [26] that the distribution in the undeformed material is, as expected, relatively 'sharp', with the mean value of $2\pi/3$ characteristic of the equilibrium configuration of the junction. Superplastic deformation is followed by a rapid increase of the fraction of the angles equal to π, i.e. the triple junctions are 'straightened' as a result of local migration (swelling) of one of the boundaries (with the maximum shear stresses acting in the the boundary plane) to the rearrangement of the configuration of the triple junction ensuring compatibility of shear in two out of three boundaries forming the given junction. Consequently, the evolution of the pile-up can be investigated in the 'frozen' configuration of the triple junction, as was usually the case. As a result of the absence of the free sink the pile-up + grain boundary set is regarded as an independent dynamic object. The possibility of reorganisation, leading to changes of the dihedral angle in the triple junction, was also noted by Todd [22].

In a general case, any deviation of the configuration of the boundary from the flat boundary increases its energy by the amount proportional

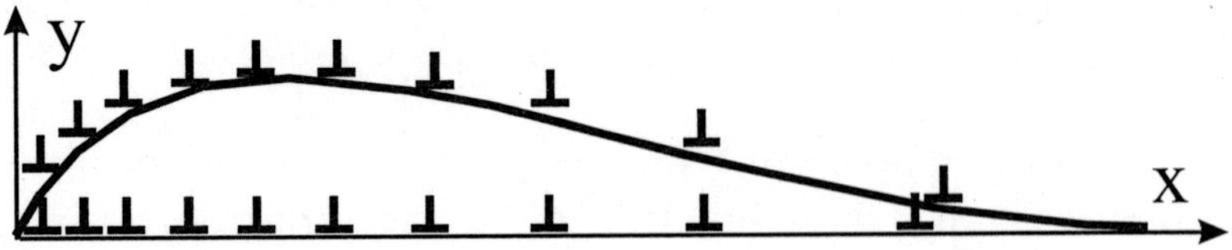

Fig. 5.2. Local migration of a boundary.

to the increase of the area. On the other hand, the dislocation pile-up compressed by the acting stress in the triple junction can be distributed over a large length and reduce intrinsic energy (Fig. 5.2). On the whole, the situation is very similar to the classic Euler problem of the loss of stability of an elastic bar under coaxial loading. In complete analogy if some critical value of the capacity of the produced pile-up is exceeded (the analogue of elastic coaxial loading) the boundary tries to leave the flat configuration and release the cold working energy.

The condition of the preferred non-planar configuration will now be estimated on the basis of energy considerations [27]. The density of straight dislocations in a pile-up, determined as the number of dislocations per unit length of the boundary, satisfies the well-known equation

$$\int_0^{x_s} \frac{\rho(x')dx'}{x'-x} = -\frac{\tau}{\mu_0 b}, \tag{5.33}$$

where $\mu_0 = \mu/2\pi(1-\nu)$; μ is the grain boundary shear modulus (generally speaking, it is lower than the lattice modulus [28]); ν is the Poisson coefficient; b is the modulus of the Burgers vector of the grain boundary dislocations; τ is the shear stress in the boundary plane in the slip direction; x_s is the length of the interval in which the pile-up is distributed. If the pile-up is formed by N dislocations, the normalisation of density is given by the expression

$$\int_0^{x_s} \rho(x)dx = N. \tag{5.34}$$

Solution of (5.33), (5.34) has the following form [29]

$$\rho(x) = \frac{2N}{\pi x_s}\sqrt{\frac{x_s - x}{x}}; \quad x_s = \mu_0 b \frac{2N}{\tau}. \tag{5.35}$$

The number of dislocations in a pile-up is determined by the mechanism of formation of the dislocations in the boundary. If the boundary has its own source with the threshold stress τ_c and the coordinate $x = L/2$ (where L is the length of the boundary in the sliding direction), then after emitting N dislocations the reversed field from the pile-up blocks this source. The blocking condition has the form

$$\int_0^{x_s} \frac{\rho(x')dx'}{x' - L/2} = -\frac{\tau - \tau_c}{\mu_0 b} \tag{5.36}$$

and taking into account (5.35) determines the number of dislocations in the pile-up:

$$N = \frac{L}{4b}\frac{\tau}{\mu_0}\left(1 - \frac{\tau_c^2}{\tau^2}\right). \tag{5.37}$$

Consequently

$$x_s = \frac{L}{2}\left(1 - \frac{\tau_c^2}{\tau^2}\right). \tag{5.38}$$

The energy of the 'pile-up + boundary' system (per unit length of the boundary in the direction of the dislocation line) is given by the expression

$$\frac{E}{H} = \frac{E_0}{H} + \frac{W_0}{2}\int_0^{x_s} \rho(x) \int_0^{x_s} \rho(x') \ln \frac{R}{|x - x'|} dx' dx + \alpha x_s, \tag{5.39}$$

where α is the coefficient of surface tension of the boundary; $W_0 = \mu_0 b^2$; E_0 is the intrnsic energy of the non-interacting dislocations in the pile-up (the constant term).

As a result of migration the initially flat boundary ($y = 0$) acquires the profile given by some function $y = y(x)$, determined in the interval $0 \leq x \leq x_s$. The number of dislocations N of the pile-up is distributed in the section with the length

$$l_s = \int_0^{x_s} \sqrt{1 + [y'(x)]^2}\, dx. \tag{5.40}$$

Generally speaking, the energy of interaction between the dislocations on the migrating boundary changes not only as a result of the change of the distances between the dislocations but also as a result of mutual rotation of their Burgers vectors. Examining the case of weak migration as a preliminary estimate, we ignore this circumstance. In this case, it may be assumed that the new dislocation density satisfies as previously equation (5.33) but in this case integration should be carried out along the curve $y = y(x)$. Consequently, the expression for the density has the following form

$$\rho(l) = \frac{2N}{\pi l_s}\sqrt{\frac{l_s - l}{l}}, \tag{5.41}$$

where l_s is determined by expression (5.40), and l is a natural parameter counted along the curve $y = y(x)$ from the point of the triple junction. Expression (5.41) is normalised as previously with respect to the total number of the dislocations N but in this case l_s is an independent parameter not linked with N by the equation of type (5.37). The energy of the new configuration becomes equal to

$$\frac{E_1}{H} = \frac{E_0}{H} + \frac{W_0}{2}\int_0^{l_s}\rho(l)\int_0^{l_s}\rho(l')\ln\frac{R}{|l-l'|}dl'dl + \alpha l_s.$$

Taking into account the equality $l' = x'\frac{l_s}{x_s}$

$$\int_0^{l_s}\sqrt{\frac{l_s - l'}{l'}}\ln\frac{R}{|l-l'|}dl' = \frac{l_s}{x_s}\int_0^{x_s}\sqrt{\frac{x_s - x'}{x'}}\ln\frac{R}{|l - x'l_s/x_s|}dx',$$

the double integral is converted to the form

$$\int_0^{l_s}\rho(l)\int_0^{l_s}\rho(l')\ln\frac{R}{|l-l'|}dl'dl = \int_0^{x_s}\rho(x)\int_0^{x_s}\rho(x')\ln\frac{R}{|x-x'|}\frac{x_s}{l_s}dl'dl =$$

$$= \int_0^{x_s}\rho(x)\int_0^{x_s}\rho(x')\ln\frac{R}{|x-x'|}dx'dx + \ln\frac{x_s}{l_s}\int_0^{x_s}\rho(x)\int_0^{x_s}\rho(x')dx'dx.$$

As a result of the relationships (5.34), (5.35) we obtain

$$\frac{E_1}{H} = \frac{E}{H} + \frac{W_0 N^2}{2}\ln\frac{x_s}{l_s} + \alpha(l_s - x_s), \tag{5.42}$$

where E/H is determined by (5.40), and N by the expression (5.41). Equation (5.1), regarded as the function l_s, has the minimum value at $l_s = l_s^* = W_0 N^2/2\alpha$. The inequality $l_s^* > x_s$ defines the situations in which migration is preferred from the energy viewpoint. The explicit form of this condition is:

$$\tau > \sqrt{\frac{16\alpha\mu_0}{L} + \tau_c^2}, \quad \tau_c \approx \frac{\mu_0 b}{l_c}. \tag{5.43}$$

This condition determines the threshold stresses which depend on the length of the boundary and the surface tension coefficient. If these values are exceeded the rearrangement of the configuration of the boundary in the vicinity of the triple junction becomes possible.

The equation (5.43) holds in the case in which the capacity of the pile-up is formed only by the source of grain boundary origin. However, if the grain boundary dislocations also form as a result of the spreading of the lattice dislocations, the equation becomes more complicated and the condition of migration is formulated more efficiently in the variables 'the length of the boundary–the capacity of the pile-up'. If the head dislocation is placed at the point of a triple junction, the density of the dislocations in the pile-up with the capacity N, situated in the interval $0 \le x_L \le x \le x_R$, it is defined by the relationship

$$\rho(x) = \frac{\sqrt{(x_R - x)(x - x_L)}}{\pi\sqrt{x_L x_R} x}, \tag{5.44}$$

$$\sqrt{x_L x_R} = \frac{\mu_0 b}{\tau}; \quad \frac{x_L + x_R}{2} = (N+1)\frac{\mu_0 b}{\tau}. \tag{5.45}$$

Replacing (5.36) by density (5.44) and repeating considerations for the condition (5.44), we obtain the migration condition

$$N(N+4) > \frac{L\alpha}{\mu_0 b^2}. \tag{5.46}$$

In the case of the action of olny the grain boundary source the capacity of the pile-up (taking into account the dislocation situated at the triple junction) is

$$N + 1 = \frac{L}{4b}\frac{\tau}{\mu_0}\left(1 - \frac{\tau_c^2}{\tau^2}\right). \tag{5.47}$$

However, if the effect of the acting stresses is sufficient to activate intergranular dislocations sources then, as in the case of describing the stimulated grain boundary sliding in the bicrystals, the power of the pile-up increases by the value of the order

$$\Delta N \sim \frac{\mu_0 b}{4\mu_{0L} b_L}\left(\frac{\tau L}{\mu_0 b}\right)^2\left(1-\frac{\tau_{cL}^2}{\tau^2}\right), \tag{5.48}$$

as a result of the spreading of the lattice dislocations. Here b_L, μ_{0L}, τ_{cL} are the intergranular lattice analogues of the previously introduced grain boundary quantities.

In the case of large grains (and the large length of the grain boundary) local migration is suppressed by a process with high intensity - dislocation creep. The lower limit for dislocation creep is the well-known condition [30]

$$\frac{\tau}{\mu_0} > \frac{K b_L}{2L}, \tag{5.49}$$

where K is the dimensionless constant with the characteristic value for metallic materials of ~10–15.

Expressions (5.47)–(5.49) determine the range of the length of the grain boundaries in which local migration of the grain boundaries is observed and leads to the rearrangement of the configuration of the triple junction (the difference between the intergranular and grain boundary shear moduli is ignored here):

$$\frac{K}{2}\frac{\mu_0}{\tau} = \frac{L_D(\tau)}{b_L} \geq \frac{L}{b_L} \geq \frac{L_{LM}(\tau)}{b_L} \cong$$
$$\cong \begin{cases} \left[\chi(T)\dfrac{b_B}{b_L}\left(2\dfrac{\mu_0}{\tau}\right)^4\right]^{1/3} & \text{at} \quad \dfrac{\tau}{\mu_0} < \dfrac{\chi(T)}{8}\left(\dfrac{b_L}{2b_B}\right)^2; \\ 8\dfrac{b_B}{b_L}\dfrac{\mu_0}{\tau} & \text{at} \quad \dfrac{\tau}{\mu_0} > \dfrac{\chi(T)}{8}\left(\dfrac{b_L}{2b_B}\right)^2, \end{cases} \tag{5.50}$$

where $\chi(T) = \alpha(T)/\mu_0 b_L = \chi_0(1-T/T_m)^{2\gamma}$, $\gamma \cong 0.65$ [31].

The region determined by the inequalities (5.50) is shown schematically in Fig. 5.3. If the parameters characterising the boundary (L, b_B, α_{GB}) would have the same value for the entire material then either no triple junction would be modified (both inequalities (5.50)

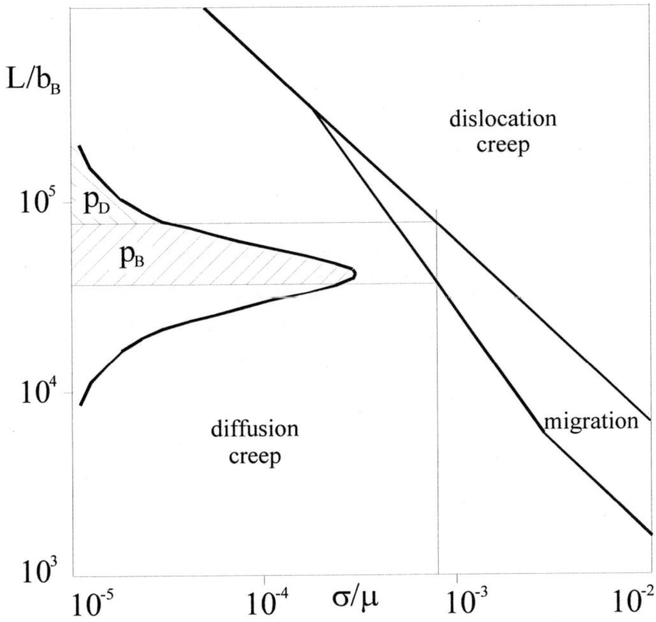

Fig. 5.3. Formation of partial contributions of deformation mechanisms for the given size distribution of grains.

are not satisfied) or all the junctions would be modified. In the latter case, deformation would be uniform not only on the macro- but also mesolevel. The experimentally detected nonuniformity of the deformation on the mesolevel is the direct reflection of the fact that the polycrystal is a system with the distributed parameters. It should be mentioned that in general considerations the problem of the distributed parameters should already arise when analysing the manifestation of the micromechanisms on the macrolevel. Thus, to count the number of modified triple junctions, it is necessary to know the combined distribution of the quantities L, b_B and α_{GB}. For an isotropic material, ignoring correlations, we can restrict our considerations to the size distribution of the grains and grain boundaries with respect to the misorientation (on the condition that the relationship between the misorientation on the given boundary and its surface tension coefficient is known). Consequently, the task of calculating the number of migrated triple junctions is reduced to the integration in space of the misorientation (axis–angle)–grain size in the region given by the inequality (5.50). However, if the two-particle distribution function of the misorientations is reduced to a single particle distribution (distribution of the orientations [32–38]), this task becomes unjustifiably complicated and, in addition to this, we do not yet know the relationship between the misorientation at the boundary and the surface tension coefficient α_{GB}.

In the approximation of the fixed surface tension coefficient at the given temperature and stress the number of the boundaries whose length in the sliding direction is in the interval $[L_{LM}, L_D]$ is determined only by the function of distribution of the length of the grain boundaries in the given material (again in the absence of a correlation between the length of the boundary and the shear stress acting in the plane of the boundary; otherwise, it is necessary to consider the combined distribution with respect to L and τ which may be the case in the materials with a distinctive texture). Using the log–normal distribution

$$W(L) = \frac{1}{\sqrt{2\pi}DL}\exp\left[-\frac{1}{2D^2}\ln^2\left(\frac{L}{\langle L\rangle}\exp\frac{D^2}{2}\right)\right],$$

where $D^2 = \ln\frac{\langle L^2\rangle}{\langle L\rangle^2}$,

the probability of one randomly selected boundary taking part in the formation of CGBS band has the form

$$p_B(\tau,\langle L\rangle) = \int_{L_{LM}(\tau)}^{L_D(\tau)} W(L)dL = \frac{1}{2}\left[\Phi\left(\frac{1}{\sqrt{2}D}\ln\frac{L_D(\tau)\exp(D^2/2)}{\langle L\rangle}\right) - \Phi\left(\frac{1}{\sqrt{2}D}\ln\frac{L_{LM}(\tau)\exp(D^2/2)}{\langle L\rangle}\right)\right], \quad (5.51)$$

where $\Phi(x)$ is the error function.

Examples of the dependence of p_B on the mean grain size and stress for two homologous temperatures and the dispersion of the grain size are shown in Fig. 5.4 constructed on the basis of equation (5.51). It is clear that in the material with a large mean grain size the conditions for local migration do not exist at any of the stresses. A reduction of the mean grain size leads to the formation of migration conditions in a specific stress range. When temperature is increased local migration takes place in wider ranges of both the grain size and stresses. The increase of dispersion results in a decrease of the number of the grain boundaries taking part in the formation of the CGBS bands. This reproduces one of the well known conditions of superplastic deformation – homogeneity of the structure.

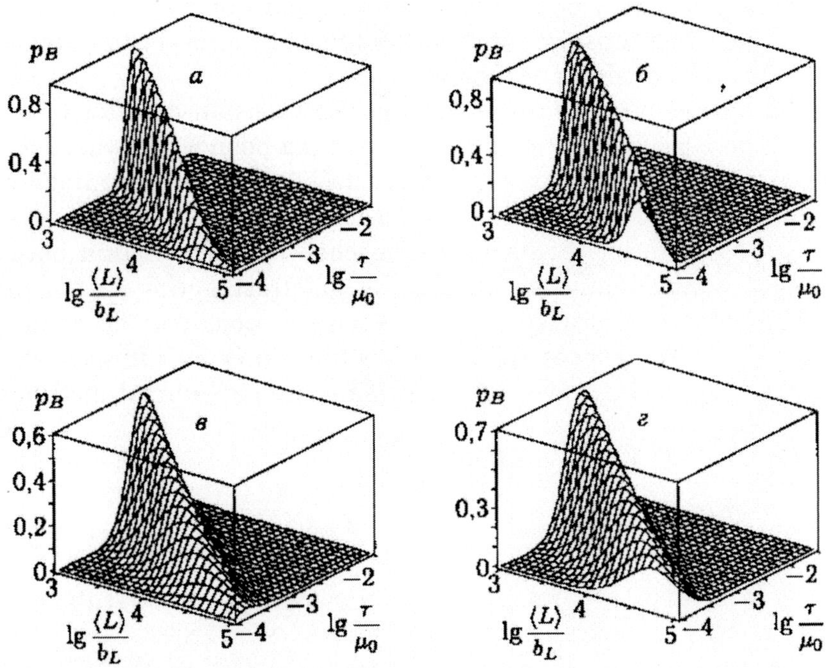

Fig. 5.4. Dependence of the probability of local migration on the mean grain size and stress: a) D^2 = 0.1, T/T_m = 0.5; b) D^2 = 0.1, T/T_m = 0.8; c) D^2 = 0.5, T/T_m = 0.5; d) D^2 = 0.5, T/T_m = 0.8.

Generally speaking, the expression (5.51) solves the given task: to determine the number of triple junctions modified under the given loading conditions. However, a large number of approximations was made when deriving this relationship, with the most restricting approximation in our view being the following: 1) the dislocation density in a pile-up is determined for a flat boundary and is used in the analysis of the energy relationships on the migrating boundary; 2) the interaction of the pile-ups formed at the conjugated boundaries of the investigated triple junction is ignored. The quality of these approximations will now be verified.

5.6. Variance formulation of the system of equations for the shape of the boundary and pile-up density

The interaction energy per unit length of two straight parallel dislocations is determined by the well-known Nabarro equation [18]:

$$\frac{W_{12}}{H} = -\frac{\mu(\mathbf{b}_1\xi)(\mathbf{b}_2\xi)}{2\pi}\ln\frac{R}{R_0} - \frac{\mu}{2\pi(1-\nu)}\left[(\mathbf{b}_1\times\xi)(\mathbf{b}_2\times\xi)\right]\ln\frac{R}{R_0} - \\ -\mu\frac{\left[(\mathbf{b}_1\times\xi)\mathbf{R}\right]\left[(\mathbf{b}_2\times\xi)\mathbf{R}\right]}{2\pi(1-\nu)R^2}.$$

For the configuration shown in Fig. 5.5, the vector defining the direction of dislocation is $\xi = (0, 0, 1)$, the Burgers vector of the i-th dislocation $\mathbf{b}_i = b(\cos\varphi_i, \sin\varphi_i, 0)$, and $\mathbf{R} = (x_2-x_1, y_2-y_1, 0)$. In the notations

$$\frac{x_2 - x_1}{\sqrt{(x_2-x_1)^2+(y_2-y_1)^2}} = \cos\Phi_{12}, \qquad \frac{y_2 - y_1}{\sqrt{(x_2-x_1)^2+(y_2-y_1)^2}} = \sin\Phi_{12}$$

we obtain

$$E(x_1,x_2) \equiv \frac{W_{12}}{HW_0} = \cos(\varphi_1-\varphi_2)\ln\frac{R_c}{R} + \frac{1}{2}\cos(\varphi_1+\varphi_2-2\Phi_{12}), \tag{5.52}$$

where $R_c = R_0 \exp(-1/2)$. Calculating the appropriate derivatives, we determine the force acting from the side of the first dislocation on the second dislocation. Its projections on the direction of the Burgers vector of the second dislocation ($F_n^{(2)}$), and also on the normal to the plane given by the Burgers vector and the direction of the dislocation ($F_n^{(2)}$) are equal to

$$\frac{F_\tau^{(2)}}{HW_0} = \frac{\cos\Phi_{12}\cos(\varphi_1-\Phi_{12})\cos\left[2(\varphi_2-\Phi_{12})\right]}{x_2-x_1}; \tag{5.53}$$

$$\frac{F_n^{(2)}}{HW_0} = -\frac{\sin(\varphi_1-\Phi_{12})+\cos(\varphi_1-\Phi_{12})\sin\left[2(\varphi_2-\Phi_{12})\right]}{x_2-x_1}\cos\Phi_{12}. \tag{5.54}$$

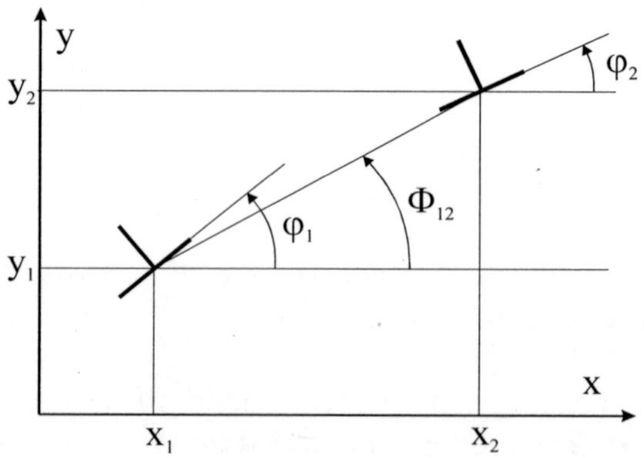

Fig. 5.5. Parameters defining the position of dislocations.

To avoid a non-physical (integrated) singularity at the origin of the coordinates taking into account the expression (5.54), the pile-up will be investigated in the discrete-continual approximation [39] separating from it the head dislocation and placing it at the origin of the coordinates. This can be carried out on the basis of the fact that we consider the boundary situated in the plane of the maximum shear stresses, and that this boundary should actually occupy the triple junction. The functional of the total energy of the pile-up + boundary system has the form

$$E_T \equiv \frac{W_T}{W_0 H} = \int_{x_L}^{x_R} E(0,x)\rho(x)\,dx + \frac{1}{2}\int_{x_L}^{x_R}\int_{x_L}^{x_R} E(x',x)\rho(x')\rho(x)\,dx'dx + \frac{\alpha}{W_0}\int_0^L \sqrt{1+\dot{y}^2(x)}\,dx, \tag{5.55}$$

where x_L and x_R are the abscissas of the left and right boundary of the pile-up, respectively; $y(x)$ is the function which defines the form of the migrating boundary; the kernel $E(x',x)$ is determined by the expression (5.52). It is assumed that the orientation of the Burgers vectors in the pile-up is determined unambiguously by the function $y(x)$: $\operatorname{tg}\varphi(x) = dy(x)/dx$. The equilibrium equation is written as the condition of the balance of tangential forces acting on each dislocation

$$\left.\frac{\cos\Phi_{12}\cos(\varphi_1-\Phi_{12})\cos[2(\varphi_2-\Phi_{12})]}{x_2-x_1}\right|_{x_1=0} + \int_{x_L}^{x_R}\rho(x_1)\frac{\cos\Phi_{12}\cos(\varphi_1-\Phi_{12})\cos[2(\varphi_2-\Phi_{12})]}{x_2-x_1}dx_1 = \frac{b\sigma}{2W_0}\sin[2(\beta+\varphi_2)] \tag{5.56}$$

where the right-hand side determines the Pitch–Keller force from the external stress σ; β is the angle between the loading axis in the initial plane of the boundary.

In this formulation, it is necessary to determine both the density of the dislocations in the pile-up and the form of the migrating boundary, associated with this pile-up. We consider a more restricted task: determine the dislocation density in a pile-up concentrated at the boundary of the given form. In this case, the kernel of the equation (5.56) is determined unambiguously. Taking dimensionless variables

$$x = \frac{x_R + x_L}{2} + \frac{x_R - x_L}{2} t; \;\; \varphi(x) = \varphi\left(\frac{x_R + x_L}{2} + \frac{x_R - x_L}{2} t \right) = \psi(t);$$

$$\frac{x_R - x_L}{2} \rho\left(\frac{x_R + x_L}{2} + \frac{x_R - x_L}{2} t \right) = \tilde{\rho}(t); \;\; \Delta = \frac{x_R + x_L}{x_R - x_L} > 1,$$

which retains the normalisation of the density with respect to the number of dislocations in a pile-up, the following equation is obtained

$$\frac{F(-\Delta, t_2)}{t_2 + \Delta} + \int_{-1}^{1} \frac{F(t_1, t_2)\tilde{\rho}(t_1)}{t_2 - t_1} dt_1 = F_0(t_2),$$

where

$$F_0(t_2) = b\sigma \frac{x_R - x_L}{4W_0} \sin\{2[\beta + \psi(t_2)]\};$$

$$F(t_1, t_2) = \cos\psi_{12} \cos(\psi_1 - \Psi_{12}) \cos[2(\psi_2 - \Psi_{12})].$$

After regularisation of the integral term, it can be determined the form of a singular integral equation with the Cauchy kernel:

$$\int_{-1}^{l} \frac{\tilde{\rho}(t_1) dt_1}{t_1 - t_2} = \omega(t_2), \tag{5.57}$$

where

$$\omega(t_2) = \omega_0(t_2) + \omega_1(t_2);$$

$$\omega_0(t_2) = \left[F_0(t_2) - \frac{F(-\Delta, t_2)}{t_2 + \Delta} \right] \frac{1}{F(t_2, t_2)};$$

$$\omega_1(t_2) = \frac{1}{F(t_2, t_2)} \int_{-1}^{1} \frac{F(t_1, t_2) - F(t_2, t_2)}{(t_2 - t_1)} \tilde{\rho}(t_1) dt_1.$$

It should be noted that $F(t, t) \equiv \cos\psi(t)$. Using the standard methods [40], we find out for the first node of the equation (5.57) ($t_2 = -1$) λ_1 is equal to either 0 or 1, and for its second node ($t_2 = 1$) λ_2 is equal to either –1 or 0 (i.e., both nodes are non-singular). Fixing the class of the solution by selecting $\lambda_1 = 1$, $\lambda_2 = 0$ (the index of the class $\chi = -1$), the solution and the condition of its existence can be written in the following form

$$\tilde{\rho}(t) = \frac{\sqrt{1 - t^2}}{\pi^2} \int_{-1}^{1} \frac{\omega_0(t') + \omega_1(t')}{t' - t} \frac{dt'}{\sqrt{1 - t'^2}}; \int_{-1}^{1} \frac{\omega_0(t') + \omega_1(t')}{\sqrt{1 - t'^2}} dt' = 0. \tag{5.58}$$

The resultant expression is the Fredholm equation with the regular kernel:

$$\tilde{\rho}(t)=\rho_0(t)+\frac{\sqrt{1-t^2}}{\pi}\int_{-1}^{1}\Re^{(0)}(t,t')\tilde{\rho}(t')dt',$$

where

$$\rho_0(t)=\frac{\sqrt{1-t^2}}{\pi}\int_{-1}^{1}\frac{\omega_0(t')dt'}{(t'-t)\sqrt{1-t'^2}};$$

$$\Re^{(0)}(t,t_1)=\frac{1}{\pi}\int_{-1}^{1}\frac{f(t_1,t_2)}{(t_1-t_2)(t_2-t)}\frac{dt_2}{\sqrt{1-t_2^2}};$$

$$f(t_1,t_2)=F(t_1,t_2)/F(t_2,t_2)-1.$$

The series of the perturbation theory through the integrated kernels has the following form:

$$\tilde{\rho}(t)=\rho_0(t)+\frac{\sqrt{1-t^2}}{\pi}\sum_{t=0}^{\infty}\int_{-1}^{1}\Re^{(i)}(t,t_1)\rho_0(t_1)dt_1, \tag{5.59}$$

where

$$\Re^{(i)}(t,t_1)=\int_{-1}^{1}\ldots\int_{-1}^{1}\Re^{(0)}(t,t_{i+1})\frac{\sqrt{1-t_{i+1}^2}}{\pi}\Re^{(0)}(t_{i+1},t_i)\ldots$$

$$\ldots\Re^{(0)}(t_3,t_2)\frac{\sqrt{1-t_2^2}}{\pi}\Re^{(0)}(t_2,t_1)dt_2\ldots dt_{i+1}.$$

The regular function $f(t_1,t_2)/(t_1-t_2)$ is represented by a series with respect to Chebysheff's polynomials

$$\frac{f(t_1,t_2)}{t_1-t_2}=\sum_{k,n=0}^{\infty}f_{kn}U_k(t_1)T_n(t_2)$$

with the coefficients

$$f_{kn}=\frac{1}{1+\delta_{n0}}\frac{4}{\pi^2}\int_{-1}^{1}\int_{-1}^{1}\frac{f(t_1,t_2)}{t_1-t_2}U_k(t_1)T_n(t_2)\frac{\sqrt{1-t_1^2}}{\sqrt{1-t_2^2}}dt_1dt_2$$

and we introduce the matrix of the coefficients displaced by the column

$$\gamma_{kn}=f_{k,n+1}\qquad(k=0,1,2,\ldots;\ n=-1,0,1,2,\ldots).$$

In these notations, the integrated kernels have the following form

$$\Re^{(i)}(t,t_1)=\frac{1}{2^i}\sum_{k,n=0}^{\infty}U_k(t_1)\left[\tilde{\gamma}^{i+1}\right]_{kn}U_n(t)$$

and are expressed through the arbitrary degree of the matrix $\tilde{\gamma}$. It may easily be seen that the equation (5.58) does not contain any term linear with respect to the migration amplitude, and expansion begins with the quadratic term. Thus, in the linear approximation with respect to the migration amplitude the density of the dislocations at the migrating boundary is in complete agreement with the density at the flat boundary. In this case, the equation for the shape of the boundary can be derived varying the functional (5.59) with respect to the function $y\ (x)$ assuming that density ρ is independent of the function y:

$$\left[D(x,0)+\int_{x_L}^{x_R}D(x,x')\rho(x')dx'\right]\rho(x)-$$
$$-\frac{d}{dx}\left\{\rho(x)\left[D^*(x,0)+\int_{x_L}^{x_R}D^*(x,x')\rho(x')dx'\right]\right\}= \quad (5.60)$$
$$=\frac{\alpha_0}{W_0}\frac{\ddot{y}(x)}{(1+\dot{y}^2)^{3/2}}.$$

The accurate forms of the kernels, obtained in formal variation, are

$$D(x,x')=\frac{2\delta(\delta-\dot{y})(\delta-\dot{y}')/(1+\delta^2)-\delta(1+\dot{y}\dot{y}')-(2\delta-\dot{y}-\dot{y}')}{(x-x')(1+\delta^2)\sqrt{1+\dot{y}^2}\sqrt{1+\dot{y}'^2}};$$

$$D^*(x,x')=\frac{(\delta-\dot{y}')(1+\delta\dot{y})/(1+\delta^2)-(\dot{y}-\dot{y}')ln\left[R_0/\left(|x-x'|\sqrt{1+\delta^2}\right)\right]}{(1+\dot{y}^2)^{3/2}\sqrt{(1+\dot{y}'^2)}},$$

where

$$\dot{y}=\frac{dy(x)}{dx};\qquad \dot{y}'=\frac{dy(x')}{dx'};\quad \delta=\frac{y(x)-y(x')}{x-x'}.$$

As regards the method of derivation, equation (5.15) has any meaning only as a linear equation with respect to y. Consequently

$$D(x,x')=\frac{\dot{y}+\dot{y}'}{x-x'}-3\frac{y-y'}{(x-x')^2};$$

$$D^*(x,x')=-(\dot{y}-\dot{y}')\ln\frac{R_0}{|x-x'|}+\frac{y-y'}{x-x'}-\dot{y}'.$$

Thus, we obtain an integro-differential equation with the Cauchy kernel (the logarithmic kernel is removed applying integration by parts, and the second degree in the denominator D is regularised). The resultant equation can not be solved and the form of the boundary can not be restored even in the approximation of weak migration. As the migration condition we use the qualitative estimate which does not make it possible to restore the form of the boundary but can be used to determine the states preferred from the viewpoint of energy.

5.7. The power of pile-ups of grain boundary dislocations

The effect of interaction of the pile-ups formed on the conjugate boundaries of the investigated triple junction will now be estimated.

Let the region $|x|\le x_0$ be an obstacle to the movement of the dislocations and this region be regarded as the kernel of a triple junction with the crystal structure differing from the structure of the boundaries. The equations (5.35)–(5.38) are generalised and the dislocations are prevented from occupying the region in the immediate vicinity of the triple junction. In this case, the density of the dislocations in the isolated one-sided pile-up, distributed in the interval $x_0 \le x \le x_s$, is given by the equation

$$\int_{x_0}^{x_s} \frac{\rho_0(x')dx'}{x'-x} = -\frac{\tau}{\mu_0 b}. \tag{5.61}$$

If the dislocations are emitted by a source situated at the point with the coordinate $x = L/2$ and with the threshold stress $\tau_c \sim \mu b/l$, the length of the pile-up is determined by the condition of blocking the source by the inverse field of the emitted dislocations:

$$\tau + \mu_0 b \int_{x_0}^{x_s} \frac{\rho_0(x')dx'}{x'-L/2} = \tau_c. \tag{5.62}$$

Consequently

$$\begin{aligned} \rho_0(x) &= \frac{\tau}{\pi\mu_0 b}\sqrt{\frac{x_s - x}{x - x_0}}; \\ x_s - x_0 &= \left(\frac{L}{2} - x_0\right)\left(1 - \frac{\tau_c^2}{\tau^2}\right). \end{aligned} \tag{5.63}$$

Consequently, the number of the dislocations in the pile-up is

$$N_0 = \int_{x_0}^{x_s} \rho_0(x)\,dx = \frac{\tau(x_s - x_0)}{2b\mu_0}. \tag{5.64}$$

For a triple junction in the general position the pile-ups of grain boundary dislocations can form at all three boundaries and the interaction of the dislocations greatly modifies the derived equations. The problem for an arbitrary triple junction is reduced to a complicated system of three integral equation. As a preliminary estimate of the effect of interaction of the pile-up we investigate two modelling examples, permitting an exact analytical solution.

Let us also assume that in addition to the already mentioned pile-up in the vicinity of the triple junction ($x = 0$) there is also a second pile-up of dislocations with the opposite sign distributed in the interval $-x_s \leq x \leq -x_0$, formed by the source situated at the point $x = -L/2$ (Fig. 5.6). Since according to the symmetry considerations the dislocation density in the two pile-ups is the same, we obtain an equation for the density:

$$\int_{x_0}^{x_s} \frac{\rho_1(x')dx'}{x' - x} + \int_{x_0}^{x_s} \frac{\rho_1(x')dx'}{x' + x} = -\frac{\tau}{\mu_0 b}, \tag{5.65}$$

and the condition of blocking of the sources

$$\tau + \mu_0 b \int_{x_0}^{x_s} \frac{\rho_1(x')dx'}{x' - L/2} + \int_{x_0}^{x_s} \frac{\rho_1(x')dx'}{x' + L/2} = \tau_c. \tag{5.66}$$

Replacing the independent variable $x^2 = y$ [39], equation (5.65) is reduced to the equation of type (5.61) and has the solution

$$\rho_1(x) = \frac{\tau}{\pi\mu_0 b}\sqrt{\frac{x_s^2 - x^2}{x^2 - x_0^2}}. \tag{5.67}$$

Subsequently, from (5.66) we obtain

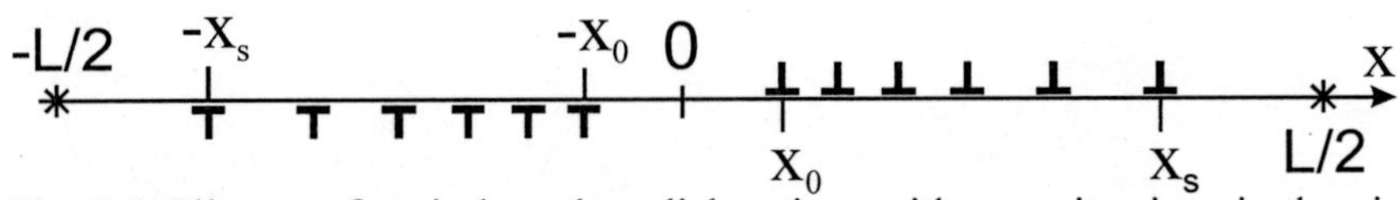

Fig. 5.6. Pile-ups of grain boundary dislocations with opposite signs in the vicinity of a pseudotriple function.

$$x_s^2 - x_0^2 = \left(\frac{L^2}{4} - x_0^2\right)\left(1 - \frac{\tau_c^2}{\tau^2}\right). \tag{5.68}$$

The number of dislocations in the pile-up is equal to

$$N_1 = \frac{\tau x_s}{\pi \mu_0 b}\left[\mathrm{K}(q) - \mathrm{E}(q)\right], \tag{5.69}$$

where x_s is determined by the relationship (5.68); $\sqrt{x_s^2 - x_0^2} / x_s$; K and E are the total elliptical integrals of the first and second kind [41].

If L is the length of the grain boundary, then at the point $x = L$, like at the point $x = 0$, there is a triple junction which is an obstacle to the propagation of the dislocations. The source, situated at the point $x = L/2$, forms not only a dislocation pile-up in the interval $x_0 \leq x \leq x_s$ but also a dislocation pile-up with the opposite sign in the interval $L - x_s \leq x \leq L - x_0$. The interaction of these pile-ups can both weaken and intensify the effect of attraction of the pile-ups investigated in the first example. The interaction changes the magnitude of the stress concentration in the region of the triple junction ($|x| \leq x_0$). The second modelling example will be investigated to evaluate the effect.

Let there be obstacles to the movement of dislocations along the straight line at the points $x_i = \pm nL$ (where $n = 0, 1, 2, \ldots$) and at the points $x_m = \pm mL/2$ (where $m = 1, 3, 5, \ldots$) sources with the same threshold stress. Consequently, the dislocation density in any pile-up satisfies the equation

$$\sum_{n=-\infty}^{\infty} \int_{x_0}^{x_s} \frac{\rho_2(x')dx'}{x' + x - nL} + \sum_{n=-\infty}^{\infty} \int_{x_0}^{x_s} \frac{\rho_2(x')dx'}{x' - x - nL} = -\frac{\tau}{\mu_0 b},$$

which, taking into account the equality [42]

$$\sum_{n=-\infty}^{\infty} \frac{1}{\xi - n\alpha} = \frac{\pi}{\alpha} \operatorname{ctg} \frac{\pi \xi}{\alpha}$$

has the following form

$$\frac{2\pi}{L} \int_{x_0}^{x_s} \frac{\rho_2(x') \sin(2\pi x' / L)}{\cos(2\pi x' / L) - \cos(2\pi x' / L)} dx' = -\frac{\tau}{\mu_0 b} . \tag{5.70}$$

The condition of blocking of the source is determined by the equation

$$\tau - \frac{2\pi\mu_0 b}{L}\int_{x_0}^{x_s} \frac{\rho_2(x')\sin(2\pi x'/L)}{1+\cos(2\pi x'/L)}dx' = \tau_c. \tag{5.71}$$

As in the first example, when replacing the independent variable cos $(2\pi x/L) = y$ equation (5.70) is reduced to the equation of type (5.61) with the solution

$$\rho_2(x) = \frac{\tau}{\pi\mu_0 b}\sqrt{\frac{\cos(2\pi x/L) - \cos(2\pi x_s/L)}{\cos(2\pi x_0/L) - \cos(2\pi x/L)}}. \tag{5.72}$$

From (5.71)

$$\cos\frac{\pi x_s}{L} = \frac{\tau_c}{\tau}\cos\frac{\pi x_0}{L}. \tag{5.73}$$

Finally, the number of dislocations in the pile-up is

$$N_2 = \frac{\tau L}{\pi^2 \mu_0 b}\sqrt{\frac{1-v_s}{1+v_0}}\left[\mathrm{K}(t) - \frac{1-v_0}{1-v_s}\Pi(\pi/2, n, t)\right], \tag{5.74}$$

where

$$v_0 = \cos(2\pi x_0/L), \quad v_s = \cos(2\pi x_s/L),$$

$$n = \frac{v_0 - v_s}{1-v_s}, \quad t = \sqrt{\frac{2(v_0 - v_s)}{(1-v_s)(1+v_0)}},$$

and Π is the elliptical integral of the third kind.

The dependence of the capacity of the isolated one-sided pile-up of the grain boundary dislocations on the length of the boundary L as a result of the condition $L \gg x_0$ ($x_0 \sim 5b$) is almost linear. The linear form of the dependence of N_0 on is disrupted only in the vicinity of the threshold stress τ_c. In Fig. 5.7 the dependences N_1, N_2 are presented in the unit of the power N_0 of the isolated pile-up. At low stresses, the effect of interaction of the two pile-ups is especially strong at small L. With increase of both L and τ the role of interaction of the pile-ups becomes less distinctive. However, even in this case the interacting pile-ups contain 5–6 times more dislocations than the isolated ones. The screening effect of the chain of pile-ups is also most evident at low stresses: in comparison with the two-sided pile-up, the power is

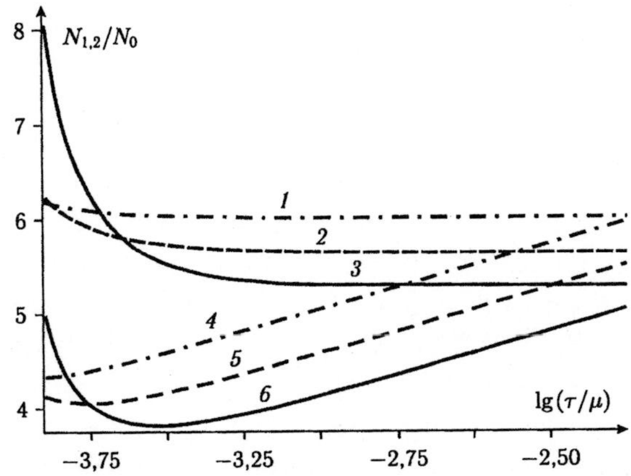

Fig. 5.7. Capacity of interacting pile-ups of grain boundary dislocations in units of the power of the isolated pile-up: N_1 (1–3), N_2 (4–6); lg (L/b) = 4.5 (1, 4), lg (L/b) = 4.25 (2, 5), lg (L/b) = 4 (3, 6).

approximately halved, although its level is still four times higher than the power of the isolated pile-up. The increase of stress balances the effect of screening; the capacities of the chain of pile-ups and of the two-sided pile-up become quite similar.

The equations (5.67) and (5.72) leads to a considerably stronger stress concentration at the point of the triple junction in comparison with the one-sided pile-up (see (5.63)). High strain rates are characterised by especially suitable conditions for crack formation.

Simple and relatively accurate estimates of the equations (5.67) and (5.72) are given by the main terms of the expansions with respect to the small parameter $x_0/L \ll 1$:

$$N_1 \cong \frac{\tau L}{2\pi\mu_0 b}\sqrt{1-\frac{\tau_c^2}{\tau^2}}\ln\left(\frac{2L}{x_0}\sqrt{1-\frac{\tau_c^2}{\tau^2}}\right);$$

$$N_2 \cong \frac{\tau L}{\pi^2\mu_0 b}\sqrt{1-\frac{\tau_c^2}{\tau^2}}\ln\left(\frac{4L}{\pi x_0}\sqrt{\frac{\tau^2}{\tau_c^2}-1}\right).$$

Thus, the interaction of the pile-ups of the grain boundary dislocations, formed on the conjugate boundaries of the triple junction, can increase several times the capacity of the pile-up at the active boundary and, at the same time, improve the conditions for local migration of the grain boundaries. However, it must be taken into account that the investigated modelling examples give the upper estimate of the power; for the triple junction with a configuration close to the equilibrium of the interaction of the pile-ups is weaker because of the rotation of the slip planes. On the other hand, in superplastic deformation, the powers of the pile-ups are determined not only by the effect of the grain boundary sources. Because of the activity of the grains they can also be regarded as independent quantities. This weakens the role of the factor of interaction of the pile-ups in the formation of their capacity. Finally, it should be mentioned that the accuracy of the theoretical investigations in the conditions of high indeterminacy of the values of the physical parameters should not be overestimated. For example, when determining the condition (5.50) we already ignore the difference between the intragranular and grain boundary shear moduli because of the indeterminacy of the latter. The diffusion parameters of the boundaries, surface tension coefficients, and their dependence on misorientation are also undetermined to the same degree.

References

1. Fukutomi H., et al., Trans. Japan Inst. Metals, 1982, vol. 23, No. 10, 579–584.
2. Valiev R.Z., et al., Izv. VUZ, Fizika, 1991, No. 3, 93–103.
3. Kokava H., et al., Phil. Mag. A, 1981, vol. 44, No. 6, 1239–1254.
4. Fukutomi H. and Kamijo T., Proc. of ICSMA-7, 1987, 677–682.
5. Takahashi T. and Horiuchi R., Proc. of JIMIS-4, 1986, 707–714.
6. Ando H., et al., Journal of Materials Science Letters, 1990, vol. 9, 314–315.
7. Perevezentsev V.N., et al., Acta Metall. Mater. 1992, vol. 40, No. 5, 887–894.
8. Gates R.S., Acta Metallurgica, 1973, vol. 21, 855–864.
9. Reading K. and Smith D.A., Phil. Mag. A, 1985, vol. 51, No. 1, 71–78.
10. Garbacz H., et al., W., Journal of Materials Science Letters, 1987, vol. 6, 1325–1328.
11. Frost H.J. and Ashby M.F., Deformation–mechanism maps, Pergamon Press, 1982.
12. Pshenichnyuk A.I., et al., Phil. Mag. A, 1998, vol. 77, No. 4, 1093–1106.
13. Cottrell A.H., Dislocations and plastic flow in crystals [Russian translation], Moscow, 1958.
14. Friedel G., Dislocations, Mir, MOscow, 1967.
15. Shtremel' M.A., Strength of alloys, II. Deformation, MISIS, Moscow, 1997.
16. Kurzydlowski K.J. and Garbacz A., Phil. Mag. A, 1985, vol. 52, No. 5, 689–697.
17. Barton B., Phil. Mag. A, 1982, vol. 45, No. 4, 657–675.
18. Nabarro F.R.N., Proc. of the 7th International Conference on the Strength of Metals and Alloys, Canada, Pergamon Press. 1985.
19. Wu M.S., and Niu J., Phil. Mag. A., 1995, vol. 71, No. 4, 831–854.
20. Astanin V. V., et al., Scripta Met. et Mater., 1997, vol. 36, No. 1, 117–122.
21. Schneibel J.H. and Hazzledine P.M., Acta Metall., 1982, vol. 30, 1223–1230.
22. Todd R.I., Acta Metall. Mater., 1984, vol. 42, No. 9, 2921–2928.
23. Evans A.G., and Rana A., Acta Metallurgica, 1980, vol. 28, 129–141.
24. Beere W., Acta Metallurgica, 1980, vol. 28, 143–150.
25. Wu M.S. and He M.D., Phil. Mag. A., 1999, vol. 79, 271–292.
26. Astanin V. V. and Kaibyshev O.A., Materials Science Forum, 1994, vol. 170–172, 23–28.
27. Astanin V. V., et al., Mater. Sci. Forum, 1997, vol. 243–245, 41–46.
28. Valiev R.Z., et al., Key Engineering Materials, 1994. vol, 97–98, 59–64.

29. Landau L.D. and Lifshits E.M., Theory of elasticity, Nauka, Moscow, 1987.
30. Gittus J.W., High-temperature deformation of two phase structure., Phil. Trans., R. Soc. Lond. A, 1978, vol. 288, 121–146.
31. Landau L.D. and Lifshits E.M., Statistical physics, vol. 1, Nauka, Moscow, 1996.
32. Gertsman V.Yu., et al., Material Science Forum, 1990, vol. 62–64, Pt. II, 669–672.
33. Gertsman V. Yu., et al., Acta Metall. Mater., 1992, vol. 40, No. 6, 1433–1439.
34. Gertsman V.Yu., et al., Material Science Forum, 1993, vol. 126–128, 277–280.
35. Zhilyev A.P., et al., Acta Metall. Mater., 1993, vol. 41, No. 9, 2657–2665.
36. Zhilyaev A.P., et al., Trans. Mat. Res. Soc. Jpn., 1994, vol. 16B, 1377–1380.
37. Zhilyaev A.P., et al., Strength of Materials, Proceedings of the ICSMA-10, Japan, 1994, 887–890.
38. Aleksandrov I.V., et al., Modelling Simul. Mater. Sci. Eng., 1995, vol. 3, 149–159.
39. Khannanov Sh.Kh., Distribution of dislocations, BNTs UrO RAN, Ufa, 1992.
40. Muskhelishvili N.I., Singular integral equations, Nauka, Moscow, 1968.
41. Gradshtein I.S. and Ryzhik I.M., Tables of integrals, sums, series and products, Nauka, Moscow, 1971.
42. Prudnikov A.P., et al., Integrals and series, vol. 1, Elementary functions, Nauka, Moscow, 1981.

6

PERCOLATION MECHANISM OF DEFORMATION PROCESSES IN ULTRAFINE-GRAINED POLYCRYSTALS

6.1. Percolation mechanism of the formation of a band of cooperative grain boundary sliding

Experiments show [1–3] that when the stable flow stage is reached (the indication of establishment of the superplastic deformation regime [4]) the independent shear sections merge together and deformation takes place by matched shear along the surfaces passing through the entire cross-section of the specimen and oriented very close to the planes with the maximum shear stresses. These surfaces are the bands of cooperative grain boundary sliding (CGBS). The transition from the independent processes, taking place in different sections of the deformed specimen, to the coherent processes on the scale of the entire specimen is identical to the tendency to the infinity of the correlation radius in the phase transitions of the second kind which can be described adequately by percolation models. The tasks, investigated in the previous chapter, make it possible to estimate the number of re-organised triple junction representing independent flow sections. The determination of the conditions of formation of the bands and the evaluation of the number of the bands under the given loading conditions are based on the assumption (to which the investigations of the structure combined with the analysis of special features of the stage of increase of the loading curves lead) according to which the band forms by the percolation transition mechanism. Determining the total number of the grain boundaries taking part in the formation of the CGBS bands, it is quite easy to calculate the number of bands which can be constructed from the given boundaries. To solve the problem, it is necessary to determine the active band. The band will be regarded as some non-smooth surface consisting of

n grain boundaries. If half of these boundaries locally migrate, then each boundary will 'bring' one adjacent boundary to deformation and the entire surface starts to operate as a coherent shear band. However, this is valid only in the case in which the migrating boundaries are distributed with the maximum uniformity along the band, i.e., when two migrating grain boundaries are separated by a single boundary included in cooperation. Otherwise, half of the migrating boundaries may prove to be insufficient for the formation of the active band. On the other hand, the experimental data indicate that the dislocation clusters, formed at some boundary, is capable of causing cooperation of more than one adjacent boundary [5, 6]. In this case, the active band can form when less than half of the boundaries n, forming the band, migrate. Taking these considerations into account, it is assumed that if the band consists of n grain boundaries and part of these bands, equal to $p_c n$, locally migrate, then the active band of the CBGS is activated. The percolation threshold p_c ($0 < p_c < 1$) is regarded as a free parameter.

The maximum possible number of the bands in the specimen is equal to $M_0 = l/\langle d \rangle$, where l is the length of the specimen. Each band in the sliding direction consists of $n = 2^{1/2}h / \langle L \rangle$ grain boundaries, where h is the thickness of the specimen in the sliding direction. The total number of the grain boundaries, situated in all the bands, is nM_0. The probability of local migration for each boundary is equal to the known value p_B. Since $nM_0 >> 1$, the total number of the grain boundaries, undergoing local migration, can be regarded as differing only slightly from the mean value $p_B nM_0$. The problem is reduced to the distribution of this number of grain boundaries in M_0 bands and calculation of the number of bands in which the number of migrated boundaries is higher than $n_c = p_c n$. In the case of the uniform distribution of the migrated boundaries in the volume of the specimen (spatially uniform structures) the probability of i-th band ($i = 1, 2, \ldots, M_0$) containing n_i aligned boundaries is determined by the multidimensional hypergeometrical distribution:

$$w(n_1, n_2 \ldots, n_{M_0}) =$$

$$= \binom{n}{n_1}\binom{n}{n_2}\ldots\binom{n}{n_{M_0}}\delta\left(n_1 + \ldots + n_{M_0} - p_B nM_0\right)\binom{nM_0}{p_B nM_0}^{-1}.$$

Consequently, the probability of the number of migrated boundaries in M bands exceeding n_c is

$$W(M)=\binom{M_0}{M}\sum_{n_1=n_c}^{n}\dots\sum_{n_M=n_c}^{n}\sum_{n_{M+1}=0}^{n_c-1}\dots\sum_{n_{M_0}=0}^{n_c-1}w(n_1,n_2,\dots,n_{M_0}). \tag{6.1}$$

Using the representation

$$\delta(p)=\frac{1}{2\pi}\int_{-\infty}^{\infty}\exp(ipx)\,dx,$$

summation is performed and each sum is calculated independently of the others. In the notations

$$s_1(x)=\sum_{k=0}^{n_c-1}\binom{n}{k}\exp(ikx),\quad s_2(x)=\sum_{k=n_c}^{n}\binom{n}{k}\exp(ikx),$$

$$s_1(x)+s_2(x)=(1+\exp(ix))^n$$

we obtain

$$W(M)=\binom{M_0}{M}\binom{nM_0}{p_BnM_0}^{-1}\frac{1}{2\pi}\int_{-\infty}^{\infty}\exp(-ip_BnM_0x)s_1^{M_0-M}s_2^M\,dx. \tag{6.2}$$

$W(M)$ is the density of a discrete ($M = 0, 1, 2,\dots,M_0$) random quantity: its positive determinacy is evident from the equation (6.1), and normalisation is confirmed by the following equalities:

$$\sum_{M=0}^{M_0}W(M)=\frac{1}{2\pi}\binom{nM_0}{p_BnM_0}^{-1}\int_{-\infty}^{\infty}(s_1+s_2)^{M_0}\exp(-ip_BnM_0x)\,dx=$$

$$=\frac{1}{2\pi}\binom{nM_0}{p_BnM_0}^{-1}\int_{-\infty}^{\infty}(1+\exp(ix))^{nM_0}\exp(-ip_BnM_0x)\,dx=$$

$$=\frac{1}{2\pi}\binom{nM_0}{p_BnM_0}^{-1}\int_{-\infty}^{\infty}\sum_{k=0}^{nM_0}\binom{nM_0}{k}\exp(ix(k-p_BnM_0))\,dx=$$

$$=\binom{nM_0}{p_BnM_0}^{-1}\sum_{k=0}^{nM_0}\binom{nM_0}{k}\delta(k-p_BnM_0)\equiv 1.$$

Substituting $W(M)$ into the expression

$$\langle M\rangle = \sum_{M=0}^{M_0} M W(M)$$

for the mean number of active bands and using the equality

$$\sum_{M=0}^{M} M \binom{M_0}{M} s_1^{M_0-M} s_2^{M} = M_0 s_2 (1+\exp(ix))^{n(M_0-1)},$$

we obtain

$$\frac{\langle M\rangle}{M_0} = \frac{1}{2\pi}\binom{nM_0}{p_B nM_0}^{-1} \times$$
$$\times \sum_{k=n_c}^{n}\binom{n}{k}\int_{-\infty}^{\infty}\exp\big(i(k-p_B nM_0)x\big)\big(1+\exp(ix)\big)^{n(M_0-1)}\,dx =$$
$$= \binom{nM_0}{p_B nM_0}^{-1}\sum_{k=n_c}^{n}\binom{n}{k}\sum_{j=0}^{n(M_0-1)}\binom{n(M_0-1)}{j}\delta\big(k+j-p_B nM_0\big) =$$
$$= \binom{nM_0}{p_B nM_0}^{-1}\sum_{j=\max\{0,(p_B M_0-1)n\}}^{\min\{n(M_0-1),\,p_B nM_0-n_c\}}\binom{n(M_0-1)}{j}\binom{n}{p_B nM_0-j},$$

Since $M_0 >>1$, all the physically significant cases are exhausted by the condition

$$\frac{1}{M_0} < p_B < 1-\frac{1-p_c}{M_0}$$

and the resultant expression is simplified to the following form:

$$\frac{\langle M\rangle}{M_0} = \binom{nM_0}{p_B nM_0}^{-1}\sum_{j=p_B nM_0-n}^{p_B nM_0-n_c}\binom{n(M_0-1)}{j}\binom{n}{p_B nM_0-j} = \binom{nM_0}{p_B nM_0}^{-1}\sum_{k=n_c}^{n}\binom{n}{k}\binom{n(M_0-1)}{p_B nM_0-k}. \tag{6.3}$$

A more symmetric formula can be obtained for other ordering of the factorials

$$\frac{\langle M\rangle}{M_0} = \binom{nM_0}{n}^{-1}\sum_{k=n_c}^{n}\binom{p_B nM_0}{k}\binom{(1-p_B)nM_0}{n-k}. \tag{6.4}$$

Using the Stirling equation for the non-zero factorials in the equation (6.3), we obtain a simple estimate:

$$\frac{\langle M\rangle}{M_0} \approx \sum_{k=n_c}^{n} \binom{n}{k} p_B^k (1-p_B)^{n-k}. \tag{6.5}$$

The 'tail' of the binomial distribution in the conditions of applicability of Gaussian approximation $np_B\ (1 - p_B) \geq 9$ can be easily estimated; we obtain the following ratio for the number of active bands:

$$\frac{\langle M\rangle}{M_0} \cong \frac{1}{2}\left[1-\Phi\left(\frac{\sqrt{n}(p_c-p_B)}{\sqrt{2p_B(1-p_B)}}\right)\right]. \tag{6.6}$$

Regardless of the large number of the assumptions made, the quality of the approximation (6.6) is relatively high. Figure 6.1 compares calculations carried out using the equations (6.3) and (6.6) for the values of the parameters $M_0 = 10$, $n = 10$ and $p_c = 1/2$. The stepped form of the dependence, given by the equation (6.3), is associated with the fact that the binomial coefficients are determined only for the integer values $n_B = p_B n$ and $n_c = p_c n$. In the case of higher values of M_0 and n the quality of approximation (6.6) becomes obviously even higher: at M_0 ~100 the stepped curve coincides with the smoothed curve.

Examples of the dependence of the ratio M/M_0 on the mean grain

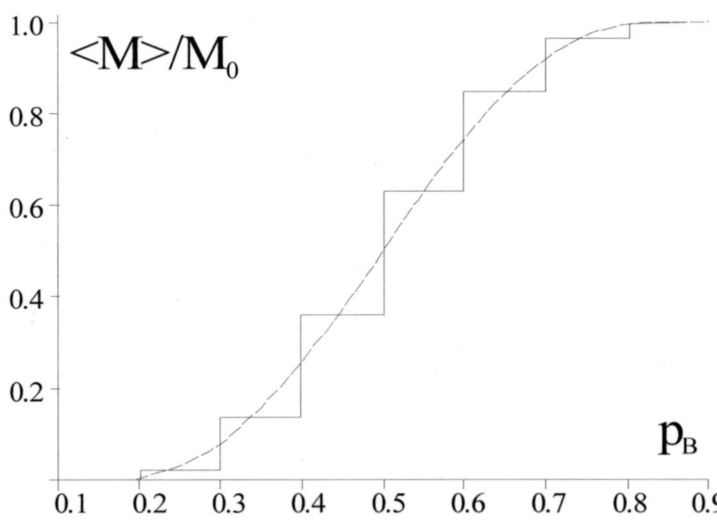

Fig. 6.1. Comparison of calculations carried out using equation (6.3) (solid curve) and (6.6) (dotted curve) for the values of the parameters $M_0 = 10$, $n = 10$ and $p_c = 1/2$.

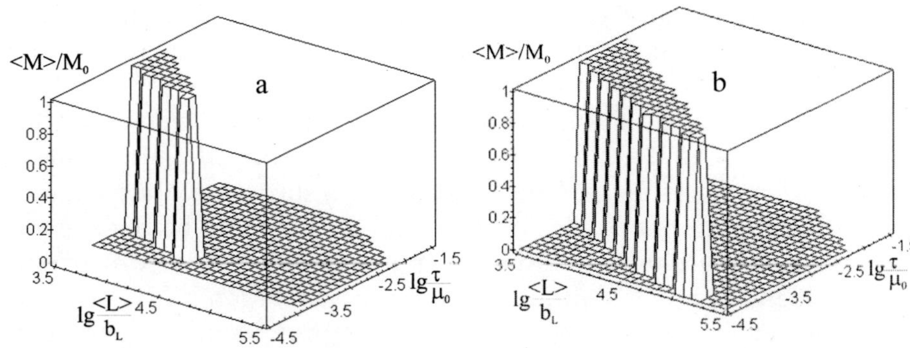

Fig. 6.2. Dependence of the mean number of the CGBS bands on the mean grain size and stress for two homologous temperatures: a) $T/T_m = 0.5$; b) $T/T_m = 0.9$ ($D^2 = 0.5$; $p_c = 0.2$).

size and stress for two homologous temperatures are shown in Fig. 6.2. This result determines the range of the stresses (for the given mean grain size) in which active CGBS bands form, i.e., in fact the range of superplastic deformation. Thus, equation (6.6) solves the given task.

6.2. Conditions of formation of CGBS bands as the condition of realisation of the superplastic deformation regime

The conditions of the formation of the CGBS bands will be determined as the conditions of realisation of the superplastic deformation regime. Factor M/M_0 determines the range of superplastic deformation in the space of the parameters of the material and the loading conditions. Analysis of this interval gave the following results [7, 8].

Since in the equation (6.6) $n >>1$, the essential condition for $\langle M\rangle$ differing from zero is reduced to the inequality

$$\max_{\tau,\langle L\rangle}\{p_B(\tau,\langle L\rangle)\} > p_c \tag{6.7}$$

which, using approximate representation $\Phi(x) \approx \sqrt{1-\exp(-4x^2/\pi)}$ (x > 0) has the following form

$$D^2 < \ln^2\left(\frac{K}{16}\frac{b_L}{b_B}\right)\left\{2\pi\left|\ln\left(1-p_c^2\right)\right|\right\}^{-1}, \tag{6.8}$$

It is well known that superplastic deformation results in the formation of a uniform equiaxed structure in the material. As indicated by the inequality (6.8), the uniformity condition is the necessary condition for the formation of the CGBS bands. The absence of bands in the case of a wide grain size distribution is confirmed by experiments [9]. The upper limit of dispersion is determined by the percolation threshold p_c, parameter K, controlling the transition to the dislocation creep regime, and by the ratio of the Burgers vectors of the lattice of grain boundary dislocations, i.e., in fact, by the presence of a sufficiently large number of arbitrary grain boundaries in the material. It should be remembered that the initial grain distribution of the grains may have a dispersion greater than the upper limit, dictated by the condition (6.8). However, the stage of transfer to the stable flow regime should occur in the conditions of active and dynamic recrystallisation ensuring structural processing of the material. However, when the stable flow stage is reached, the final grain size distribution should be characterised by a small dispersion because this is essential for the formation of bands.

At the given test temperature any material has the maximum permissible value of the mean grain size (see expression (6.6) and Fig. 6.2). The materials with the large grain size do not show the formation of the CGBS bands and, consequently, the superplastic deformation regime cannot be established in these materials. The equations (6.5) and (6.6) gives the explicit condition for the mean grain size

$$\frac{\langle d\rangle}{b_L} < \frac{\langle d\rangle_M}{b_L} = \frac{2}{\chi(T)}\left(\frac{K}{4}\right)^4 \frac{b_L}{b_B}\exp\left[\frac{D^2}{2} - 7D\sqrt{\frac{\pi}{2}\left|\ln(1-\mathrm{p}_c^2)\right|}\right]. \tag{6.9}$$

It is well known that the increase of temperature enables the superplastic deformation regime to be established in materials with larger grains. Reproducing this tendency, equation (6.9) can also be used to formulate a more convincing claim: the temperature dependence of the maximum permissible mean grain size is determined by the temperature dependence of the surface tension of the grain boundaries. The value $\langle d\rangle_M$ under similar conditions is controlled by the values of dispersion and percolation threshold, i.e., the physical parameters which are not taken into account by the models of superplastic deformation. The dependence of $\langle d\rangle_M$ on D and p_c is shown in Fig. 6.3, constructed for the following values of the parameters: χ_0 = 0.02; $K = b_L/b_B$ = 10; T/T_m = 0.75.

When the conditions (6.8) and (6.9) are fulfilled, it is quite easy to determine the range of stresses in which the CGBS bands form:

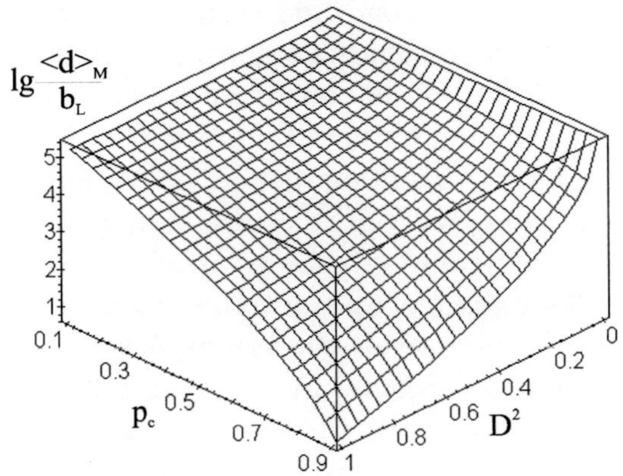

Fig. 6.3. Dependence of the maximum attainable mean grain size on the values of the percolation threshold and the dispersion of the size distribution of the grains.

$$\tau_{th}^{(1)} < \tau < \tau_h, \tag{6.10}$$

where

$$\frac{\tau_{th}^{(1)}}{\mu_0} = 2\chi(t)\frac{b_B}{b_L}\left(\frac{4}{K}\right)^3 \exp\left[\frac{3}{4}R + A\right]; \tag{6.11}$$

$$\frac{\tau_h}{\mu_0} = 2\chi(t)\frac{b_B}{b_L}\left(\frac{4}{K}\right)^3 \exp\left[R - \frac{3}{4}A\right]. \tag{6.12}$$

To shorten the explanation, the following notations will be introduced here:

$$R^2 = \ln^2\left[\frac{b_L}{b_B}\frac{2b_L}{\langle d\rangle}\left(\frac{K}{4}\right)^4 \frac{\exp(D^2/2)}{\chi(T)}\right] + 14D^2\ln\frac{4}{3};$$

$$A = \frac{D}{\sqrt{2}}\left(\sqrt{\pi\left|\ln(1-p_c^2)\right|} - \frac{2\ln 2}{\sqrt{\pi\left|\ln(1-p_c^2)\right|}}\right).$$

Quantity $\tau_{th}^{(1)}$ is the threshold stress. If these stresses are exceeded, active CGBS bands appear in the material. The right-hand boundary of the interval (6.10) τ_h determines the transition from the superplastic deformation condition to the dislocation creep regime. It may easily be seen that the expressions (6.11) and (6.0) undergo the reference limiting transition $D \to 0$ and degenerate in this case to the relationships (5.49) and (5.50) derived in chapter 5. In this case, τ_h ceases to depend on temperature, is proportional to constant K and inversely proportional to the mean grain size. The threshold stresses at $D \to 0$ depend on temperature as $(1 - T/T_m)^{\nu/2}$ and on the grain size as $\langle d\rangle^{-3/4}$. In a general case $(D \to 0)$ the temperature dependence of $\tau_{th}^{(1)}$ may intensify to $\propto(1 - T/T_m)^{\nu/2}$; the dependence on the mean grain size weakens to complete disappearance at relatively high values of D. The analysis of the temperature dependence of the experimental values of the threshold stresses is usually carried out assuming that $\sigma_{th} \propto \exp(E/kT)$. This is often justified and leads to an accurate value of activation energy E [10]. However, it may be seen that the experimental results for high-rate superplastic deformation [11] are approximated more accurately by the linear ($\sigma_{th} \propto a - bT$) and not exponential dependence. The resultant indeterminacy of interpretation may be caused by the following circumstance. The strain rate in the range (6.10) is determined by the value of the shear rate V along the band. The mechanism, controlling the shear rate, may have the intrinsic threshold stress $\tau_{th}^{(2)}$, differing from $\tau_{th}^{(1)}$. If $\tau_{th}^{(2)} < \tau_{th}^{(1)}$, the range of superplastic deformation is determined by the condition (6.8). However, if the reversed relationship is fulfilled, i.e. $\tau_{th}^{(2)} < \tau_{th}^{(1)}$, the range of superplastic deformation becomes smaller and the role of threshold stresses is transferred to $\tau_{th}^{(2)}$ (2). Thus, the actual form of the temperature dependence of the threshold stresses is determined by the temperature dependence of the maximum values of $\tau_{th}^{(1)}$ and $\tau_{th}^{(2)}$. The threshold stresses $\tau_{th}^{(1)}$, responsible for the formation of the CGBS bands depend, in accordance with (6.11), exponentially on temperature, and the stresses $\tau_{th}^{(2)}$ associated with accommodation processes causing shear along the CGBS band are thermoactivated. Thereby, the considerations regarding superplastic deformation introduce the concept of threshold stresses of the first and second kind: threshold $\tau_{th}^{(1)}$ determines the transition to the formation of the CGBS bands, and the threshold $\tau_{th}^{(2)}$ specifies the conditions of activation of the mechanism controlling the shear rate along the band.

6.3. Shear rate along the CGBS band

The final characteristic of the CGBS band, determined on the mesolevel, is the shear rate along the band at the given stress level. Assuming the accommodation nature of slip in the grains, the Orowan relationship is used for preliminary estimate: $V = b_B \langle\rho\rangle V_B$. The speed of grain boundary dislocations is given by the relationship $V_B = b_B \sigma_B \dfrac{\delta D_B}{kT} C$.

The mean density of the grain boundary dislocations in the band is determined by the number of the dislocations formed in all the boundaries of the given band. At the boundary of the given length the number of the dislocations is characterised by the relationships (4.7) and (4.8) (depending on the contribution of intragranular activity). Transferring from the number of dislocations to the density ($\rho = (N+1)/L$) and averaging the realisations (4.7) and (4.8) with respect to random quantity L (using, as previously, the log–normal distribution of L) gives

$$b_B\langle\rho(\tau)\rangle \approx \frac{\tau}{4\mu_0} p_B(\tau,\langle L\rangle) + \frac{\langle L\rangle}{4b_L}\left(\frac{\tau}{\mu_0}\right)^2 \exp(D^2) p_D(\tau,\langle L\rangle), \tag{6.13}$$

where p_D describes the partial contribution from intragranular dislocation slip:

$$p_D(\tau,\langle L\rangle) = \frac{1}{2}\left[1-\Phi\left(\frac{1}{\sqrt{2}D}\ln\frac{L_D(\tau)\exp(D^2/2)}{\langle L\rangle}\right)\right] \tag{6.14}$$

and the scheme of formation of this quantity of the given grain size distribution is shown in Fig. 5.3.

It should be mentioned that when the intragranular dislocation slip becomes the main mechanism ($p_D \to 1$), the speed in the band is determined by the cube of applied stress (two are given by the number of dislocations, and one by their speed). However, if $p_D \to 1$ (strictly speaking, this is possible only at zero dispersion of the grain size distribution), the power of the stresses decreases to the second power. The actual dependence of the strain rate on the applied stress is determined not only by the speed in the CGBS band but also by the number of these bands which form the range of the permissible values of p_B through the percolation threshold – the quantities which also depend on the applied stress.

All the characteristics of the CGBS bands, formed with participation of the processes on the mesolevel and required for investigating these problems, have now been determined. We now transfer to the properties of a large-scale flow pattern. The construction of the dependence of the strain rate on stress in the stable flow stage is the content of the standard (and probably the only one) tests of any physical model of superplastic deformation. In experiments, this dependence is restored from the results of macroexperiments, whereas calculations are always reduced to investigating the mechanism controlling the rate of transfer of strain through some triple junction – limiting local characteristic. The empirically determined nonuniformity of deformation on the mesolevel requires development of a more substantiated formalism which would make it possible to link the characteristics of the macrolevel with the characteristics of the CGBS bands.

6.4. Kinetics of deformation in CGBS bands

Let the region (specimen) be given in the coordinates (x_1, x_2) by the inequalities $|x_1| \leq r_0$, $|x_2| < l_0$, at the moment of time $t = 0$ (here r_0 is the half width of the specimen; l_0 is the half length of the specimen). Movements (deformation) in the system of the centre of masses will be investigated in the conditions where for any $t > 0$ the coordinate of the centre of masses is $x_1 = x_2 = 0$. Deformation takes place by shear along two systems of the CGBS bands. The first system is defined by the equations $x_2 = z_n - x_1$, where the index n determines the number of the band, and z_n is the coordinate of intersection of the band with the axis x_2. It is convenient to assume that the number of the band in the system is $M^{(-)} + 1$, where $M^{(-)}$ is an even number. Consequently $n = \{-M^{(-)}/2, -M^{(-)}/2+1, ..., -1, 0, 1, ..., M^{(-)}/2-1, M^{(-)}/2\}$. The second system of the bands is given by the equation $x_2 = z_j + x_1$, where $j = \{M^{(+)}/2, ..., M^{(+)}/2\}$. The conditions of formation of the bands were determined previously. However, the duration of formation of the band and the magnitude of shear up to blocking the given band are difficult to determine in experiments and by theoretical considerations. Therefore, it is necessary to postulate the individual considerations regarding the dynamics of switching over of the bands. If at some moment of time shear takes place along a band with the number n of the system 1, then in this case all the bands of the system 2 intersecting this band are blocked, at least for some period of time Δt after which the band of the system 1 becomes blocked and the bands of the transverse system start to operate. Consequently, the active bands of the systems 1 and 2 at the given moment of time are distributed in space, i.e., in some section there are bands of one or the other system. However, other courses of the process are possible in which only the system of the bands $M^{(-)}$ operates during some period of time Δt, and this is followed by switching to the system $M^{(+)}$, and so on. The variant of distribution of the bands in time (not in space) is simpler to formalise and we shall refer to it.

The field of the speeds during the action of the first system of the bands is described by the following equation:

$$\mathbf{V}(x_1,x_2)=\frac{V_B}{\sqrt{2}}\begin{pmatrix}-1\\1\end{pmatrix}\sum_{n=-M^{(-)}/2}^{M^{(-)}/2}\left[2\theta\left(x_2+x_1-z_n\right)-1\right], \tag{6.15}$$

where V_B is the shear rate along the band (assuming that it is the same for all bands); $\theta(x)$ is the Heaviside function. Let us estimate the time during which shear takes place along the system 1. Since the grips cannot move in direction x_1, deformation is accompanied by the change of the angle between the direction of shear along the band and the loading axis. This reduces the value of the Schmidt factor in the acting system of the bands and they are blocked. According to the relationship (6.15), the speeds of the upper and lower grips are equal to respectively

$$\frac{V_B}{\sqrt{2}}\begin{pmatrix}-1\\1\end{pmatrix}\left(M^{(-)}+1\right);\ -\frac{V_B}{\sqrt{2}}\begin{pmatrix}-1\\1\end{pmatrix}\left(M^{(-)}+1\right).$$

Consequently, the angle between the loading axis and the shear direction along the band is $\psi = \pi/4-\alpha$, where

$$\operatorname{tg}\alpha=\frac{\left(M^{(-)}+1\right)V_B\Delta t}{\sqrt{2}l+\left(M^{(-)}+1\right)V_B\Delta t}$$

(the ratio of transverse shear to the actual base).

Since the shear stress in the band system is equal to $(\sigma/2)\sin 2\psi$, from the condition of the decrease of the stress to the threshold value σ_{th} we obtain the time during which shear takes place along the first system of bands:

$$\Delta t=\frac{l(t)}{\sqrt{2}(M^{(-)}+1)V_B}\sqrt{\frac{\tau}{\tau_{th}}-1}\left(\sqrt{\frac{\tau}{\tau_{th}}-1}+\sqrt{\frac{\tau}{\tau_{th}}+1}\right), \tag{6.16}$$

where $l\,(t)$ is the actual length. After time Δt the system of the bands $M^{(+)}$ is opened. The corresponding field of the speeds is:

$$\mathbf{V}(x_1,x_2)=\frac{V_B}{\sqrt{2}}\begin{pmatrix}1\\1\end{pmatrix}\sum_{j=-M^{(+)}/2}^{M^{(+)}/2}\left[2\theta\left(x_2-x_1-z_j\right)-1\right]. \tag{6.17}$$

On the basis of the general considerations it is assumed that $M^{(+)} = M^{(-)}$. Consequently, this system of bands acts for approximately the same period of time Δt (because of a different initial length). On the macroscopic structural level of description, the successive effect of the bands in the time period $2\Delta t$ is an elementary act of shape changes. The equation of movement (radial component) of some point of the region in the period $[t_0, t_0 + \Delta t]$ as a result of (6.15) has the following form

$$\frac{dx_1(t)}{dt} = -\frac{V_B}{\sqrt{2}}\left[2N^{(-)}(x_1, x_2) - M^{(-)} - 1\right],$$

where $N^{(-)}(x_1, x_2)$ is the number of the bands of the first system situated below the point (x_1, x_2). If this point is moved in the direction parallel to the bands, it is evident that the point will not intersect these bands and $N^{(-)}(x_1, x_2)$ does not change. In particular, $N^{(-)}(x_1, x_2) = N^{(-)}(0, x_1 + x_2)$. Consequently, we can introduce the function specifying the distribution of the bands along the specimen at time t:

$$p(z,t) = \sum_n \delta\left(z - z_n(t)\right).$$

Consequently, the equations of motion in the period $[t_0, t_0 + \Delta t]$ have the following form

$$\frac{dx_1(t)}{dt} = \frac{V_B}{\sqrt{2}}\left[2\int_{-l(t)}^{x_2(t)-x_1(t)} p(z,t)dz - \int_{-l(t)}^{l(t)} p(z,t)dz\right];$$

$$\frac{dx_2(t)}{dt} = -\frac{dx_1(t)}{dt}.$$

Similarly, for the period $[t_0 + \Delta t, t_0 + 2\Delta t]$ taking into account the equality $N^{(+)}(x_1, x_2) = N^{(+)}(0, x_2 - x_1)$ we obtain

$$\frac{dx_1(t)}{dt} = \frac{V_B}{\sqrt{2}}\left[2\int_{-l(t)}^{x_1(t)+x_1(t)} p(z,t)dz - \int_{-l(t)}^{l(t)} p(z,t)dz\right];$$

$$\frac{dx_2(t)}{dt} = -\frac{dx_1(t)}{dt}.$$

The initial point is fixed: $x_1(t_0) = x_{10}$, $x_2(t_0) = x_{20}$. Since the value Δt is small, the following solution can be written:

in the first interval

$$x_1(t) = x_1(t_0) - \frac{V_B}{\sqrt{2}}\left[2\int_{-l(t_0)}^{x_1(t_0)+x_2(t_0)} p(z,t_0)dz - \int_{-l(t_0)}^{l(t_0)} p(z,t_0)dz\right](t - t_0);$$

$$x_2(t) = x_2(t_0) + \frac{V_B}{\sqrt{2}}\left[2\int_{-l(t_0)}^{x_1(t_0)+x_2(t_0)} p(z,t_0)dz - \int_{-l(t_0)}^{l(t_0)} p(z,t_0)dz\right](t - t_0),$$

in the second interval

$$x_1(t) = x_1(t_0+\Delta t) + \frac{V_B}{\sqrt{2}}\left[2\int\limits_{-l(t_0+\Delta t)}^{x_2(t_0+\Delta t)-x_1(t_0+\Delta t)} p(z,t_0+\Delta t)dz - \right.$$

$$\left. - \int\limits_{-l(t_0+\Delta t)}^{l(t_0+\Delta t)} p(z,t_0+\Delta t)dz\right](t-t_0-\Delta t);$$

$$x_2(l) = x_2(t_0+\Delta t) + \frac{V_B}{\sqrt{2}}\left[2\int\limits_{-l(t_0+\Delta t)}^{x_2(t_0+\Delta t)-x_1(t_0+\Delta t)} p(z,t_0+\Delta t)dz - \right.$$

$$\left. - \int\limits_{-l(t_0+\Delta t)}^{l(t_0+\Delta t)} p(z,t_0+\Delta t)dz\right](t-t_0-\Delta t);$$

Assuming that the number of the bands in switching does not change and the normalised integrals are the same, we obtain

$$\frac{x_1(t_0+2\Delta t)-x_1(t_0)}{2\Delta t} = -\frac{V_B}{\sqrt{2}}\left[\int\limits_{-l(t_0)}^{x_2(t_0)+x_1(t_0)} p(z,t_0)dz - \right.$$

$$\left. - \int\limits_{-l(t_0+\Delta t)}^{x_2(t_0+\Delta t)-x_1(t_0+\Delta t)} p(z,t_0+\Delta t)dz\right];$$

$$\frac{x_2(t_0+2\Delta t)-x_2(t_0)}{2\Delta t} = \frac{V_B}{\sqrt{2}}\left[\int\limits_{-l(t_0)}^{x_2(t_0)+x_1(t_0)} p(z,t_0)dz - \frac{1}{2}\int\limits_{-l(t_0)}^{l(t_0)} p(z,t_0)dz + \right.$$

$$\left. + \int\limits_{-l(t_0+\Delta t)}^{x_2(t_0+\Delta t)-x_1(t_0+\Delta t)} p(z,t_0+\Delta t)dz - \frac{1}{2}\int\limits_{-l(t_0+\Delta t)}^{l(t_0+\Delta t)} p(z,t_0+\Delta t)dz\right].$$

Since averaging is carried out with respect to fast time, i.e., as if $M^{(+)}$ and $M^{(-)}$ were forced to operate 'simultaneously' on the slow time scale, and at the limit $\Delta t \to 0$ we have

$$\frac{dx_1(t)}{dt} = -\frac{V_B}{\sqrt{2}}\int\limits_{x_2(l)-x_1(t)}^{x_2(t)+x_1(t)} p(z,t)dz; \qquad (6.18)$$

$$\frac{dx_2(t)}{dt} = \frac{V_B}{\sqrt{2}}\left[\int\limits_{-l(t)}^{x_2(t)+x_1(t)} p(z,t)dz - \int\limits_{x_2(l)-x_1(t)}^{l(t)} p(z,t)dz\right] = \frac{V_B}{\sqrt{2}}\int\limits_{x_1(t)-x_2(t)}^{x_1(t)+x_2(t)} p(z,t)dz. \tag{6.19}$$

The last equality is obtained for the obvious condition $p(z, t) = p(-z, t)$ because of symmetry. It is interesting to note that another condition of the non-divergence of the flow (incompressibility) is also the parity $p(z, t)$ with respect to the spatial coordinates. This is directly indicated by the expressions

$$\frac{\partial V_1}{\partial x_1} = -\frac{V_B}{\sqrt{2}}\left[p\left(x_2 + x_1, t\right) + p\left(x_2 - x_1, t\right)\right];$$

$$\frac{\partial V_2}{\partial x_2} = \frac{V_B}{\sqrt{2}}\left[p\left(x_2 + x_1, t\right) + p\left(x_1 - x_2, t\right)\right].$$

Thus, the kinetics of the flow in deformation in which CGBS bands take part is described by the equations (6.18) and (6.19).

6.5. Comparison of the calculated values with the experimental results

Substitution $x_1(\mathrm{t}) \equiv 0$, $x_2(t) = l(t)$ transforms equation (6.19) into the following equation

$$\frac{dl(t)}{dt} = \frac{V_B}{\sqrt{2}} M(t), \tag{6.20}$$

where $M(t)$ is the total number of the active bands at time t. Consequently, the macroscopic strain rate $\dot{\varepsilon} = \dot{l}(t)/l(t)$ can be easily expressed through the mesolevel characteristics

$$\dot{\varepsilon} = \frac{\langle M \rangle}{M_0}\frac{V}{\sqrt{2}\langle d \rangle}, \tag{6.21}$$

where the number of the CGBS bands is given by the relationship

$$\frac{\langle M \rangle}{M_0} \cong \frac{1}{2}\left[1 - \Phi\left(\frac{\sqrt{n}(p_c - p_B)}{\sqrt{2p_B(1-p_B)}}\right)\right],$$

and the shear rate in the band is determined by the Orowan ratio for the mean density of the grain boundary dislocations on the given level of the shear stresses in the CGBS band:

$$V = b_B \langle \rho \rangle V_D,$$

$$b_B \langle \rho(\tau) \rangle \cong \frac{\tau}{4\mu_0} p_B(\tau, \langle L \rangle) + \frac{\langle L \rangle}{4 b_L} \left(\frac{\tau}{\mu_0} \right)^2 \exp(D^2) p_D(\tau, \langle L \rangle).$$

The total strain rate is additionally determined by the rate of diffusion or dislocation creep described by the following well-known expressions [12–14]:

$$\dot{\varepsilon}_{dif} = 7\dot{\varepsilon}_0 \left(\frac{b_L}{\langle L \rangle} \right)^2 \frac{\tau}{\mu_0} \left(\frac{D_L}{D_B} + \pi \frac{b_L}{\langle L \rangle} \right), \ \dot{\varepsilon}_0 = \mu b_L \frac{D_B}{kT}; \tag{6.22}$$

$$\dot{\varepsilon}_{dis} = A \mu_0 b_L \frac{D_L}{kT} \left(\frac{\tau}{\mu_0} \right)^n p_D(\tau, \langle L \rangle). \tag{6.23}$$

Diffusion creep is regarded as the superposition of the Coble and Nabarro–Herring mechanisms. The expression for dislocation creep differs from the usual form by the presence of an additional multiplier characterising the partial contribution of intragranular deformation.

In the conditions of additivity of the rates, i.e., at

$$\dot{\varepsilon}_{tot} = \dot{\varepsilon}_{dif} + \dot{\varepsilon}_{CGBS} + \dot{\varepsilon}_{dis} \tag{6.24}$$

the dependence of the strain rate on stress is determined by the following circumstances. The ranges of the lowest and highest values of the stress ($\langle M \rangle = 0$) describe the classic behaviour of the material in the condition of diffusion and dislocation creep, respectively. In a coarse-grained material, these regions merge together and the superplastic properties are not evident at any values of τ and $\dot{\varepsilon}$. When the grain size is relatively small, a new region appears between these regions and the specific conditions on the level of the threshold stress and the percolation threshold, temperature, grain size and its dispersion are fulfilled in this new region. Fulfilment of these conditions ensures the formation of the active bands of CGBS ($\langle M \rangle \neq 0$) and the superplastic properties become evident. The diagram of formation of such a dependence is shown in Fig. 6.4. Although superplastic deformation occupies an intermediate position between diffusion and dislocation creep, it is not the superposition of these mechanisms

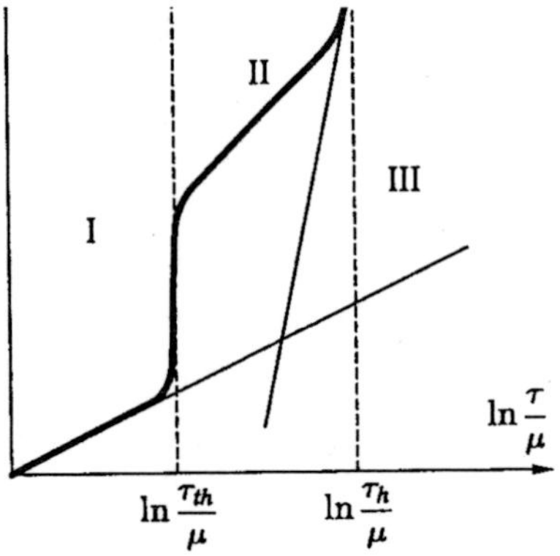

Fig. 6.4. Diagram of the formation of the sigmoidal dependence in superplastic deformation. The numbers correspond to the following ranges: I) diffusion creep ($m^{-1} = 1$, $\langle M\rangle = 0$; II) superplastic deformation ($m^{-1} \approx 2$, $\langle M\rangle \neq 0$), and III) dislocation creep ($m^{-1} > 3$, $\langle M\rangle = 0$).

and is an independent physical phenomenon realised by the specific deformation mechanism – cooperative grain boundary sliding [15, 16].

The parameters required for defining the properties of the material can be calculated using relationship (6.24). The shear modulus, with its temperature dependence taken into account (linear dependence)

$$\mu(T) = \mu_{300}\left[1 + \frac{T-300}{T_m}\frac{T_m}{\mu_0}\frac{d\mu}{dT}\right]$$

is determined by three parameters:

– μ_{300} – the shear modulus at 300 K;

– $\frac{T_m}{\mu_0}\frac{d\mu}{dT}$ – a numerical coefficient, characterising the angle of inclination of the temperature dependence;

– T_{max} – melting point of the material.

Two parameters are added by the temperature dependence

$$\alpha(T) = \alpha_0\left(1 - T/T_m\right)^{2\gamma}$$

of the surface tension coefficient of the grain boundaries:

– α_0 – the surface tension coefficient, extrapolated to zero temperature;

– 2γ – an exponent.

Four parameters determine the temperature dependences of intragranular and grain boundary diffusion coefficients:

$$D_L = D_{L0} \exp(-Q_L / RT),$$
$$D_B = \delta D_{B0} \exp(-Q_B / RT)$$

– D_{L0} – the pre-exponent of the lattice diffusion coefficient;
– Q_L – the activation energy of the lattice diffusion coefficient;
– δD_{B0} – the product of the effective width of the boundary by the pre-exponent of the grain boundary diffusion coefficient;
– Q_B – the activation energy of the grain boundary diffusion coefficient.

In addition:

– ν – the Poisson coefficient;
– b_L – the modulus of the Burgers lattice vector;
– b_B – the modulus of the grain boundary Burgers vector;
– K – a parameter controlling transition to the dislocation creep regime.

There are also two parameters defining the dislocation creep rate (6.23):

– A – the Dorn constant;
– n – the creep exponent

Finally, the following parameters should also be considered:

– d – the mean grain size;
– D – the dispersion of the grain size distribution;
– p_c – percolation threshold;
– multiplier C in the expression for the grain boundary dislocation rate (see (5.4)–(5.6)), considered in the form

$$C_{gr} = \frac{1}{10}\left(\frac{b_L}{b_B}\right)^3 \frac{\sin\varphi}{\cos^2\varphi} \approx \frac{1}{10}\left(\frac{b_L}{b_B}\right)^3.$$

Of the 19 quantities mentioned here characterising the properties of the material, the mean grain size is determined most accurately but even this parameter does not remain constant during deformation. The modulus of the Burgers vector of the grain boundary dislocations has different values at all the boundaries, like the parameters defining the temperature dependences of the surface tension coefficients and grain boundary diffusion. Since the grain size is the only distributed parameter considered in this case and the distribution of the misorientations is ignored, all the quantities relating to the boundaries are characterised by some effective values. For example, it is always assumed that $b_L/b_B = 10$. The diffusion parameters were taken from handbooks [17, 18] and are summarised in Table 6.1.

Table 6.1. Parameters of the materials used in comparison with the experimental results

Parameters	Aluminium	Lead	α-titanium	Tin
b_L	$2.86 \cdot 10^{-10}$	$3.49 \cdot 10^{-10}$	$2.95 \cdot 10^{-10}$	$4.12 \cdot 10^{-10}$
T_M, m	933	601	1933	505
μ_{300}, Pa	$2.54 \cdot 10^{10}$	$0.73 \cdot 10^{10}$	$4.36 \cdot 10^{10}$	$(1.6–1.9) \cdot 10^{10}$
$\frac{T_m}{\mu_0}\frac{d\mu}{dT}$	–0.50	–0.76	–1.20	–
D_{L0}, m^2/s	$1.7 \cdot 10^{-4}$	$1.4 \cdot 10^{-4}$	$8.6 \cdot 10^{-10}$	$(7.7–10.7) \cdot 10^{-4}$
Q_L, kJ/mol	142	109	150	107–110
ΔD_{B0}, m^3/s	$5.0 \cdot 10^{-14}$	$8.0 \cdot 10^{-14}$	$3.6 \cdot 10^{-16}$	–
Q_B, kJ/mol	84	66	97	–
N	4.4	5.0	4.3	–
A	$3.4 \cdot 10^6$	$2.5 \cdot 10^8$	$7.7 \cdot 10^4$	–

In fact, these diffusion coefficients are self-diffusion coefficients. Even they are determined with a relatively low accuracy. Various handbooks give activation energies differing by tens of kJ/mole. Comparison with the experimental results is also complicated by the fact that the materials, used in the experiments, are never completely pure: they contain a large number of alloying additions and also dispersed particles of the second phase, simply two-phase eutectics. In similar conditions, the indeterminacy of all (not only diffusion) parameters, characterising the material, becomes even greater. Thus, verification of the adequacy of the model can be quite convincing if a set of dependences can be reproduced, a single curve in the worst case but not numbers. We begin comparison with a Pb–Sn eutectic which, together with classic Zn–Al, is used widely in the physical investigations of superplasticity (low temperature, stable grains). Figure 6.5 shows the appropriate experimental dependence of strain rate on stress [19] obtained in a wide strain rate range so that it is possible to record the section with the rate sensitivity parameter $m = 1$ (the value typical of the diffusion creep) at low stresses. Since the material consists of two phases, it is clear that neither the parameters of lead nor the parameters of tin describe the properties of the eutectic. It is hoped that some effective parameters can be defined taking into account the fact that, in our view, superplasticity is determined by the processes in the grain boundaries and interface boundaries where the

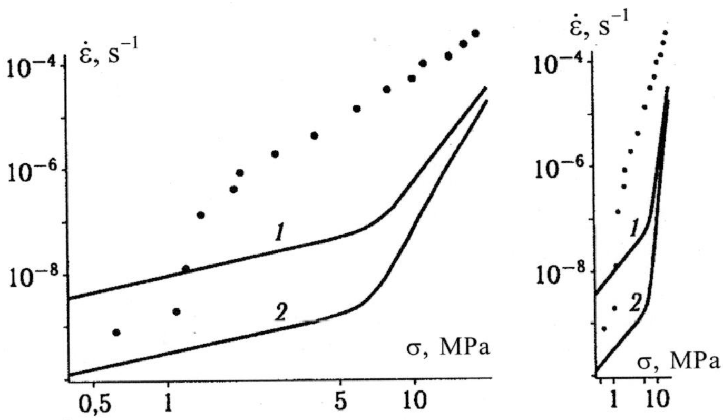

Fig. 6.5. Experimental dependence (points) of strain rate on stress in Pb–Sn [19] at d = 3.6 μm (a) and the same dependence with the same scale on the axes (b). Solid curves show the sum of the rates of diffusion and dislocation creep in pure Pb (1) and pure Sn (2).

role of the volume characteristics is not so important. The same graph shows the dependence of the sum of the diffusion and dislocation creep rates in pure Pb and Sn (the term specifying the contribution from the CGBS bands is not taken into account). In the stress range up to 1 MPa where the experimental results indicate the effect of diffusion creep only, the curve for the softer Pb restrict the experimental points at the top and for the harder Sn at the bottom. Since the dependences in pure materials are majorant for the experimental points shows that the effective parameters between the values determined for Pb and Sn can indeed be found. It is interesting to note that the experimental values of the strain rate in this stress range can be determined as mean geometrical values (but not mean arithmetic values) of the strain rate in pure Pb and Sn.

Comparison of the total strain rate, determined by the relationship (6.24), with the experimental results is shown in Fig. 6.6. All the parameters used in calculating the three theoretical curves, are presented in Table 6.2. The percolation threshold, ensuring good agreement with the experiments, is relatively low and equal to 0.2. In similar conditions, a small number of boundaries is quite sufficient for the formation of bands. It is natural to expect that these are more compliant boundaries with the parameters close to the parameters of pure lead. As seen from the plot, the threshold stresses detected at the point of the rapid increase of the rate from the level corresponding to the dislocation creep rate (the characteristic inflection point on the curve), for the hypothetical material with the parameters of pure lead, the material

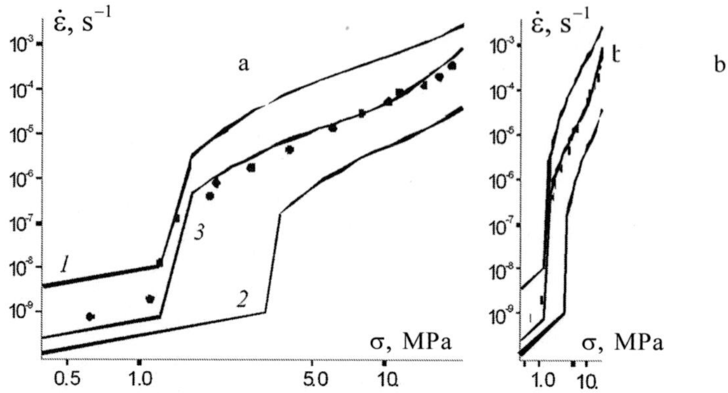

Fig. 6.6. Experimental dependence (points) of the strain rate on stress for the Pb–Sn eutectic [19] and the calculation of the dependence for hypothetical materials with the parameters of pure Pb (1), pure Sn (2) and effective parameters (3) at $d = 3.6$ μm (a). The same dependence with the same scale with respect to the axis (b).

with the effective parameters and also the experimental dependence are very similar and greatly differ from the threshold stress for the material with the pure tin parameter (this is especially clear in Fig. 6.6b with the same scale on the axes). This indicates the internal matching of the results (or at least the fact that they do not contradict each other). It should again be mentioned that the experimental values of the strain rate are very similar to the mean geometrical strain rate in pure Pb and Sn not only in the diffusion creep range but also in the superplasticity range (in the range where all three curves were plotted, i.e., above the threshold stress for tin). This observation requires confirmation because a similar fact may prove to be both a random circumstance or the manifestation of some important relationship. It is necessary to stress the ambiguity of parametrisation carried out in Table 6.2 for the Pb–Sn eutectic. For example, the value of the activation energy of grain boundary diffusion can be changed and this variation can be compensated by redetermining the values of the surface tension coefficient, shear modulus, percolation threshold, etc. A smaller number of degrees of freedom will be considered in the attempt to parametrise the family of curves.

The results obtained for the measurement of the strain rate in a material of a different type will be discussed. Figure 6.7 shows the experimental data for $Ti_{10}Co_4Al$ alloy in the ultrafine-grained condition (the grains of α-Ti with the size of 0.5 μm and approximately 23% (volume) of the dispersed particles of Ti_2Co intermetallic compound with the size of 0.2 μm) [20]. The parameters of the theoretical curves were taken from Table 6.2. Instead of the dimensional quantity α_0, the table gives the dimensionless quantities $\chi_0 = \alpha_0/(b_L\mu)$. Thus, using the model, the family of the curves can also be described with sufficient high accuracy. It should be mentioned that the exponent, which determines the temperature dependence of the surface tension coefficient, has different values for different temperatures. This may indicate that the dependence is more complicated than the exponential one. Activation energy values have not been determined. For example,

Table 6.2. Effective parameters of the materials and the model used in numerical estimates

	Pb	Sn	Pb–Sn	$Ti_{10}Co_4Al$			MA7075 + 10SiC		
μ, Pa	$0.73 \cdot 10^{10}$	$1.9 \cdot 10^{10}$	$0.73 \cdot 10^{10}$	$4.36 . 10^{10}$			$2.54 \cdot 10^{10}$		
ν	0.45	0.33	0.33	0.33			0.33		
$T_m/\mu_0/d\mu/dT$	–0.76	–0.76	–0.76	–1.2			–0.5		
b_L, m	$3.49 \cdot 10^{-10}$	$4.12 \cdot 10^{-10}$	$3.5 \cdot 10^{-10}$	$2.95.10^{-10}$			$2.86 \cdot 10^{-10}$		
T_m, K	601	505	550	1933			933		
χ_0	0.012	0.012	0.012	0.03	0.012	0.005	0.012		
2γ	1.3	1.3	1.3	1.3	2.5	3.0	1.35		
T, K	300	300	300	923	973	998	733	763	793
K	30	30	30	30			40		
D_2	0.4	0.4	0.4	0.5			0.5		
P_c	0.2	0.2	0.2	0.25	0.16	0.10	0.82	0.63	0.28
D_{L0}, m²/s	$1.4 \cdot 10^{-4}$	$90 \cdot 10^{-4}$	$10 \cdot 10^{-4}$	$8.6 \cdot 10^{-10}$			$1.7 \cdot 10^{-4}$		
Q_L, kJ/m	$109 \cdot 10^3$	$110 \cdot 10^3$	$109 \cdot 10^3$	$225 \cdot 10^3$			$234 \cdot 10^3$	$229 \cdot 10^3$	$220 \cdot 10^3$
δD_{B0}, m³/s	$8 \cdot 10^{-14}$	$360 \cdot 10^{-14}$	$80 \cdot 10^{-14}$	$3.6 \cdot 10^{-16}$			$5 \cdot 10^{-14}$		
Q_B, kJ/m	$66 \cdot 10^3$	$86 \cdot 10^3$	$79 \cdot 10^3$	$200 \cdot 10^3$	$195 \cdot 10^3$	$190 \cdot 10^3$	$130 \cdot 10^3$	$127 \cdot 10^3$	$122 \cdot 10^3$
N	5	5	5	4.3			4.4		
A	$2.5 \cdot 10^8$	$2.5 \cdot 10^8$	$2.5 \cdot 10^8$	$7.7 \cdot 10^4$			$3.4 \cdot 10^6$		

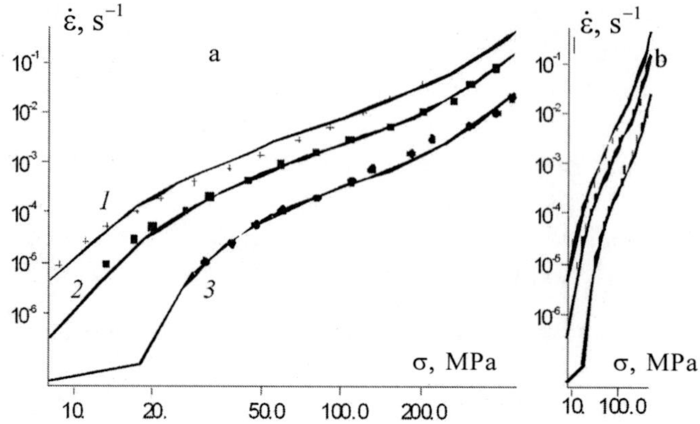

Fig. 6.7. Strain rate in the $Ti_{10}Co_4Al$ alloy (α-Ti/Ti_2Co; d = 0.5 μm). The points are the experimental results obtained at T = 998 K (1), 973 K (2) and 923 (3) [20]; the curves show the results of theoretical calculations with the parameters presented in Table 6.2 (a). The same dependence with the same scale on the axes (b).

in [17] the activation energy of self-diffusion in the α-Ti lattice is assumed to be equal to 150 kJ/mole, and in [20] the value of 190 kJ/mole is given for the same parameters. The activation energy for the strain rate is estimated in [20] as 220±10 kJ/mole. Calculations are carried out using the values Q_L = 225 kJ/mole and Q_B from 190 to 200 kJ/mole, depending on temperature. Thus, in comparison with pure α-Ti both the experiments and theoretical estimates lead to considerably higher activation energy values. Naturally, this is associated with the dispersed particles of the second phase. However, it is difficult to evaluate the significance of these differences since 150 kJ/mol reported by Frost and Ashby, and 190 kJ/mole published by Frommayer et al, should be regarded as the same.

The other material – an aluminium alloy, dispersion hardened with silicon carbide particles, shows high-rate superplasticity. Comparison of the calculations with the experimental data for the alloy is shown in Fig. 6.8 [21]. It is important to note that the percolation threshold values of this alloy are higher than those of the previously examined variants. The decrease of the percolation threshold with increasing temperature is typical of both the aluminium and titanium alloys. The activation energy of grain boundary diffusion is 1.5 times higher than the appropriate value for pure aluminium and is in agreement with the estimates of the activation energy of the strain rate made in [21] and, as in the case of the titanium alloys, is determined by the presence of the dispersed particles of the second phase.

Since the model can be used to describe with satisfactory accuracy

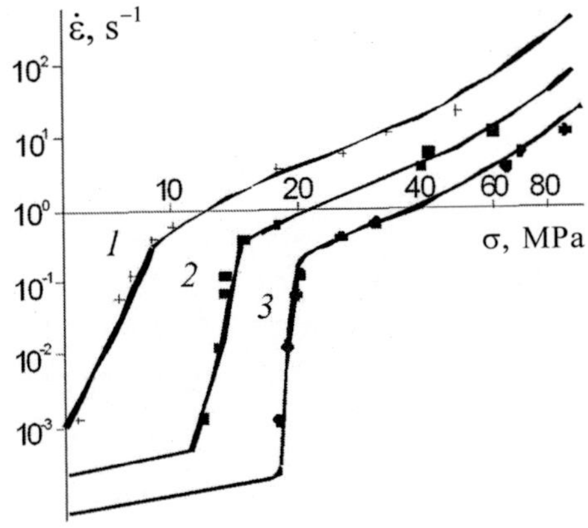

Fig. 6.8. Dependence of the strain rate on stress for the dispersion-hardened composite (MA 7075+10 SiC) at T = 793 K (1), 763 K (2) and 733 K (3) (d = 0.68 μm) [21]. The curves show the calculated results.

the relatively large and greatly differing set of the data, it may be assumed that the scheme of formation of the $\dot{\varepsilon} - \sigma$ dependence is fully realistic. This leads to important conclusions on the threshold of superplastic deformation and the method used for its experimental evaluation. This method is based on the very well known dependence:

$$\frac{\dot{\varepsilon}}{\dot{\varepsilon}_0} = \left(\frac{\sigma - \sigma_{th}}{\mu}\right)^{1/m}. \tag{6.25}$$

Using the relationship (6.25) and selecting the correct stress range, we can 'straighten' the experimental dependences and by extrapolating the resultant straight line to find the threshold stress σ_{th}. However, in accordance with (6.24) and Fig. 6.4 (the validity of these relationships is confirmed by the previously described experimental results) the superplasticity is 'activated in a jump' in accordance with the law

$$\frac{\dot{\varepsilon}}{\dot{\varepsilon}_0} \sim \theta(\sigma - \sigma_{th}) \tag{6.26}$$

(where θ is the Heaviside function) and not gradually in accordance with the law (6.25). This fact reflects the circumstance that the CGBS band either forms or does not form. If the band has not yet formed, the strain rate is determined by the diffusion creep rate; if the band has formed the rate rapidly increases. The strain rate inside the superplasticity range and the value of the threshold σ_{th} are independent quantities. No processing of the first of these quantities makes it possible to estimate the independent values of the second quantity. Thus, the considerations regarding the percolation mechanism of the formation of the CGBS bands reproduce with high accuracy the set of the $\dot{\varepsilon} - \sigma$ dependences in a wide strain rate range.

Attention will also be given to a slightly different aspect of the formation of the connected clusters, formed from the special grain boundaries and influencing grain boundary diffusion in ultrafine-grained materials.

References

1. Kaibyshev O.A., Astanin V. V. and Faizova S.N, Advanced Materials '93, III / B: Composites, Grain Boundaries and Nanophase Materials, ed. M. Sakai, Trans. Mat. Res. Soc. Jpn., 1994, 1473–1476.
2. Astanin V.V., et al., Mat. Sci. Technol., 1996, vol. 12. 489–494.
3. Astanin V.V. and Kaibyshev O.A., Materials Science Forum, ed. T. Langdon, Trans Tech Publication, Switzerland, 1994, vol. 170–172. 23–28.
4. Edington J.W., Melton K.N. and Cutler C.P., Progress in Mater. Sci., 1976, vol. 21, 61–170.
5. Astanin V.V., Kaibyshev O.A. and Faizova S.N., Scripta Met. et Mater., 1991, vol. 25, No. 12, 2663–2668.
6. Astanin V.V., Kaibyshev O.A. and Faizova S.N., Acta Metall., 1994, vol. 42, No. 8, 2617–2622.
7. Pshenichnyuk A.I., Kaibyshev O.A. and Astanin V.V., FTT, 1997, vol. 39, No. 12, 2179–2185.
8. Pshenichnyuk A.I., Kaibyshev O.A. and Astanin V.V., Phil. Mag. A, 1999, vol. 79, No. 2, 329–338.
9. Malek P., Mat. Sci. Eng. A, 1999, vol. 268, 132–140.
10. Mohamed F.A., J. Mater. Sci., 1983, vol. 18, No. 2, 582–592.
11. Kum D., Mater. Sci. Forum, 1997, vol. 243–245, 287–296.
12. Nabarro F.R.N., Report of a Conference on Strength of Solids, 78. The Physical Society, London, 1948, Herring C.J., Appl. Phys., 1950, vol. 21. 437.
13. Coble R.L., J. Appl. Phys., 1963, vol. 34, vol. 6, 1679–1682.
14. Bird J.E., Mukherjee A.K., and Dorn J.E., Quantitative Relation Between Properties and Microstructure, edited by D. G. Brandon and A. Rosen, Universities Press, 1969, 255.
15. Astanin V.V., Kaubyshev O.A. and Pshenichnyuk A.I., FMM, 1997, vol. 84, No. 6, 5–16.
16. Kaibyshev O.A., Pshenichnyuk A.I. and Astanin V.V., Acta Mater., 1998, vol. 46, No. 14. 4911–4916.
17. Frost H.J. and Ashby M.F., Deformation–mechanism maps, Pergamon Press, 1982.
18. Babichev A.P., et al., Physical quantities, ed. by Grigor'eva I.S., et al., Energoatomizdat, Moscow, 1991.
19. Geckinli A.F. and Barrett G.R., Scripta Met., 1974, vol. 8, 115.
20. Frommaer G., Hofmann H. and Herzog W., Material Science Forum, 1994, vol. 170–172, 483–488.
21. Matsuki K., Tokizawa M. and Murakami Y., Advanced Materials '93, III/B: Composites. Grain Boundaries and Nanophase Materials, ed. M. Sakai, Trans. Mat. Res. Soc. Jpn., 1994, 933–936.

7

PERCOLATION PROCESSES IN A NETWORK OF GRAIN BOUNDARIES IN ULTRAFINE-GRAINED MATERIALS

7.1. Effect of grain boundaries on oxidation and diffusion processes in polycrystalline oxide films

Zirconium and its alloys represent the main materials in nuclear power engineering. For example, Zr–1%Nb, Zr–2%Nb and Zr–2.5%Nb alloys are used for the manufacture of pipes of nuclear reactors. Consequently, it is important to investigate the mechanical and corrosion properties of these materials. The latter are determined by the presence of a zirconium oxide film (ZrO_2) on the surface which ensures efficient protection against further oxidation and also against the diffusion of hydrogen and its isotopes from the working environment of the reactor. It is generally well known that the properties of the ZrO_2 protective layer (and also of other polycrystalline materials) depend on their microstructure.

The high corrosion resistance of the zirconium oxide has been the subject of a large number of investigations and discussions regarding its nature. If the diffusion mechanism (or oxidation mechanism if the diffusion of oxygen is considered) is explained, it would be possible to prepare recommendations for improving the corrosion properties and/or predicting the safe service life of components produced from materials of this type. The latter is especially important for applications such as nuclear power engineering and aerospace technology. Attention will be given to the zirconium oxide film formed on the surface of the Zr–2.5% Nb alloy. However, the results can also be applied to other zirconium alloys used in the world.

As shown by early investigations [1–3], the microstructure of the Zr–2.5%Nb alloy consists of elongated α-grains of Zr, surrounded by the β-phase. After thermomechanical treatment required in the production of pipes (rolling and drawing), the material has a strong texture [4]: the majority of the α-grains of Zr are oriented in such a manner that the directions $\langle 1\bar{1}00 \rangle$ are situated along the axis of the pipe, and the directions $\langle 11\bar{2}0 \rangle$ along the radius of the pipe. Preliminary annealing of completed pipes leads to the formation of a protective zirconium oxide layer on the surface with the thickness of approximately 1 μm. The microstructure of the oxide consists mostly of grains with the monoclinic lattice, described in detail in section 1.1.2. These grains are elongated in the direction normal to the surface, with the diameter of approximately 50 nm [5, 6]. In addition, there is also a certain amount of the ZrO_2 tetragonal phase. This structure of the oxide film is also characteristic of other typical zirconium alloys.

X-ray diffraction and electron microscopic studies show that the oxide film has a strong texture consisting of two main components: the so-called grain growth texture and a weaker axial component. As shown by transmission electron microscopy, these two components form in different areas of the oxide film, i.e., they are separated in space. The first component forms mainly on the α-grains of zirconium. There is a correlation between the orientation of the substrate and the oxide film formed on the substrate. The accurate type of correlation has been the subject of long discussions in the literature. Various investigators have published different relationships between the crystallographic directions of the substrate and the film. For example, Lin [5], on the basis of the results of investigations of the interface by transmission electron microscopy, proposed the following relationship:

$$[100]_m \parallel \left[4\bar{5}10\right]_\alpha \quad \text{and} \quad (100)_m \parallel (0001)_\alpha \tag{7.1}$$

The latter relationship means that the colonies of the oxide grains are oriented mainly along the radius of the cylinder. Consequently, the direction $[001]_m$ in the oxide is parallel to the direction [1120] in the α-grains of the zirconium. This relationship will be denoted by:

$$(010)[100]_m \parallel (0001)\left[11\bar{2}0\right]_\alpha \tag{7.2}$$

It should be mentioned that other relationships between the orientations of the oxide film and the pure zirconium substrate [7–9] have also been found. However, we shall refer only to (7.1) or (7.2) since they correspond to the texture detected in the investigated material.

The axial component of the texture, investigated in [5], is described as a normal to the plane $(10\bar{2})_m$. Roy and David described the axial texture with the axis close to the normal to the plane $(10\bar{4})_m$ [10].

We believe that in this case there is no separate axis and some of the crystallographic orientations are distributed along the normal to the substrate surface. In addition, if we consider the axis *c* of the monoclinic cell, whose $[001]_m$ direction is the axis for the grain growth texture (see (3.2)), then this direction is also close to the normal to the $(10\bar{6})_m$ plane. Similar reflections could not be detected in the experiments (even if they were stronger) because they are situated outside the range of experimental investigations.

From the viewpoint of the information content, high-rcsolution electron microscopy may provide an answer to many questions in cases in which the conventional experimental methods cannot. In this specific case we used high-resolution electron microscopy for obtaining information on the crystallographic orientation of the nanocrystalline grains of zirconium oxide [1].

A thin foil of the oxide film produced by ion thinning, was examined in Philips CM12 (120 kV) and JEOL 2010 (200 kV) microscopes. The brightfield image of a typical section of the microstructure and the appropriate microdiffraction pattern are shown in Fig. 7.1. Analysis of the possible coincident-site lattices in the monoclinic lattice (section 2.2) shows [12] that the zirconium oxide and evidently other ceramics [13, 14] contain preferential twin orientations. Figure 7.1 shows the characteristic features of the twin boundaries – straight sections of the grain boundaries (indicated by the arrows). Similar conclusions were also drawn in [15] in which twins in oxide films, grown on Zircalloy-4 alloy, were studied. As an example, Fig. 7.2 shows a twin boundary. Analysis of the image and the diffraction pattern makes it possible to identify the planes parallel to the boundary as (011), and the plane intersecting these planes under the angle of almost 95° as $(1\bar{1}1)$. In the experimental error range, this value coincides with the value of the angle between the given planes (93.3°) at the following lattice parameters [16]: a = 5.1490 Å, b = 5.2133 Å, c = 5.3162 Å, β = 99.228°. The interplanar spacing ($d_{(011)}/d_{(111)}$ = 1.30), calculated from the values of these parameters, is very close to the experimentally measured value (1.22). Diffraction in the converging beam (Fig. 7.2b and c) shows that the axis of the zone for both adjacent grains is the same $\langle 21\bar{1}\rangle$, but the images are mirror reflections of each other. The mirror reflection can be achieved either by reflection in the plane of symmetry (grain boundary) or by rotating through 180° around the axis, normal to this plane. The misorientation parameters of such a grain boundary can be written as 90.8° [100] = 180° [6, 35, 36]. The latter direction is the direction of the normal to the (011) plane. Following the formal theory of CSL for the monoclinic lattice [39], this twin orientation is

Fig. 7.1. Microstructure of the ZrO_2 film, in the region with the axial texture.

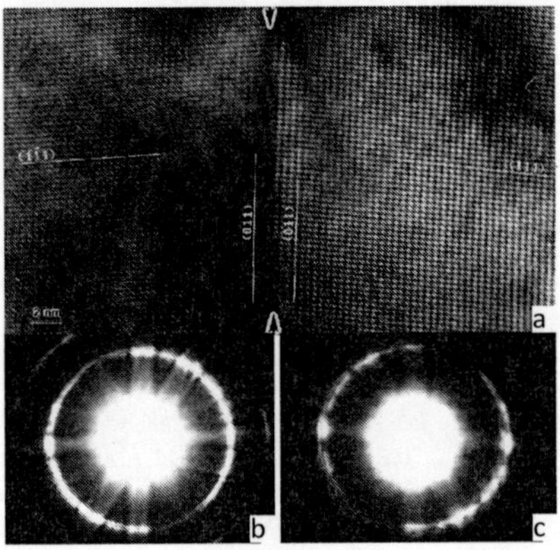

Fig. 7.2. Twin boundary in the region with the axial texture: a) the image obtained by high-resolution electron microscopy; b and c) the appropriate diffraction patterns in the converging beams.

described by the inverse density of the coincident sites $\Sigma = 71$, or more accurately $\Sigma = 71a$ [16], in accordance with the approximation of the almost coincident sites which in fact is equivalent to the theory of the 'restricted' CSL developed by King [17]. In this approach, the following approximation is used for the monoclinic lattice of zirconium oxide:

$$a = b = c, \;\; \cos\beta = -(1/6) \tag{7.3}$$

(i.e., $\beta = 99.594°$).

In this case, the matrix of the twin misorientation has the form

$$R_{71a} = \frac{1}{71}\begin{pmatrix} 71 & 12 & -12 \\ 0 & -1 & -70 \\ 0 & 72 & -1 \end{pmatrix}. \tag{7.4}$$

It may easily be verified that $R_{71a} \times [21\bar{1}] = [\bar{2}11]$.

This example has been discussed in such a detail in order to stress the following facts: it is not necessary that the twin grain boundary (and therefore any special grain boundary) should be described by the three-dimensional coincident site lattice with the lowest value of Σ. This conclusion is in agreement with the assumption made by King [17]. In this specific case we have the complete coincidence of the sites of the adjacent lattices in the grain boundary plane but the spacing of the CSL increases in the direction normal to the grain boundary plane.

7.2. High-resolution electron microscopy of zirconium oxide: grain clusters, surrounded only by special boundaries

As already mentioned in the beginning of the chapter, regions with different texture (growth and axial textures) in the microstructure of ZrO_0 are separated in space. The main component is the growth texture ($[001]_m$), which forms on the grains of α-zirconium. Two examples of such a region are shown in Fig. 7.3. In this case, the microstructure greatly differs from that shown in Fig. 7.1 in the same manner as the diffraction pattern which represents in this case ordered series of reflections, with some of these reflections split (see below). It is interesting to note that the orientations of the crystal planes in the entire region of the microstructure have the same appearance. At the same time, the boundaries between the grains are clearly visible. The atom columns in the direction normal to the examination plane coincide with the vector [001]. Therefore, the planes perpendicular to this direction are the planes (100) and (010) with similar interplanar spacings of 5.080 and 5.213 Å, respectively. Therefore, even high-resolution microscopy does not make it possible to separate visually crystallographic directions in the adjacent grains. The presence of the intercrystalline boundaries helps to describe the crystallographic orientations of the adjacent grains on the micrograph shown in Fig. 7.3. The rotation of the monoclinic lattice through the angle of 180° around the [001] axis leads to the twin misorientation Σ3b 180° [001] with the coherent boundary in the (100) plane for both adjacent grains [39]. Consequently, the sets of the planes (100) and (010) in the adjacent grains are parallel to each other. Another twin misorientation around the [001] axis is the grain boundary Σ71b 90.8° [001] = 180° [36, 35, 6]. The last direction ([36, 35, 6]) is normal to the (110) plane. These twins (Σ71b), situated under the angle of 45° to the edge of the photograph, are clearly visible in Fig. 7.3a. For the two types of twin boundaries of the grains (Σ3b and 71b) we analyse the type of the grain boundary,

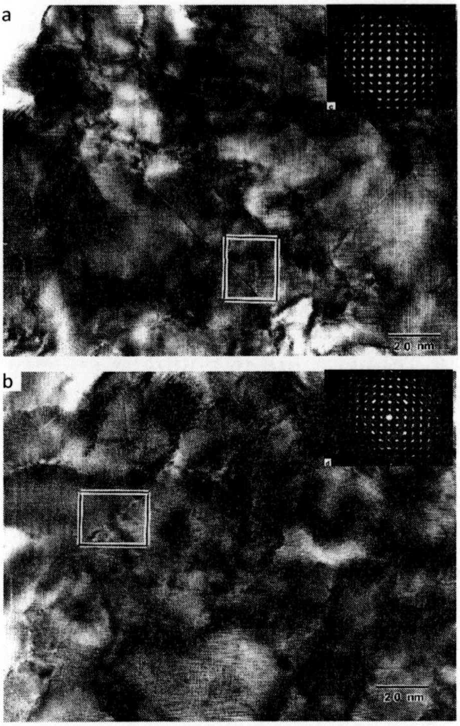

Fig. 7.3. High-resolution electron images of the microstructure of the ZrO_2 film in the region of the growth texture [001] and the appropriate diffraction patterns.

produced in the triple junction at contact of these two boundaries. The misorientation of the twin grain boundaries is written in the form

$$R_{3b} = \frac{1}{3}\begin{pmatrix} -3 & 0 & 0 \\ 0 & -3 & 0 \\ -1 & 0 & 3 \end{pmatrix}; \quad R_{71b} = \frac{1}{71}\begin{pmatrix} -1 & -72 & 0 \\ 70 & -1 & 0 \\ -12 & -12 & 71 \end{pmatrix}. \tag{7.5}$$

Consequently, the interaction between the misorientations Σ3b and Σ71b leads to the new misorientation described by the following matrix:

$$R_{3b}\,R_{71b} = \frac{1}{213}\begin{pmatrix} 3 & 216 & 0 \\ -210 & 3 & 0 \\ -35 & 36 & 213 \end{pmatrix}, 89.2^{\circ}, [001]. \tag{7.6}$$

This misorientation differs from the misorientation Σ71b, 90.8°, [001] only by $\Delta\theta = 1.6°$. The interaction between the two boundaries of the type Σ71b leads to the misorientation:

$$R_{71b}\,R_{71b}\,\frac{1}{5041}\begin{pmatrix} -5039 & 144 & 0 \\ -140 & -5039 & 0 \\ -1680 & 24 & 5041 \end{pmatrix}, 178.8^{\circ}, [001], \tag{7.7}$$

which again differs by only Δθ = 1.6° from the misorientation 3b, 180° [001]. In both cases, the deviation in the misorientations is only twice the deviation of the misorientation angle of the grain boundary Σ71b 90.8° [100] from 90°. Thus, there is a unique situation in the ZrO_2 film with the monoclinic lattice: two twin boundaries, intersecting in the triple junction, form a third twin grain boundary; as a result, the entire ensemble of the grain boundaries in these section of the microstructure consists only of twins. Several possible geometrical configurations of a similar microstructure are shown in Fig. 7.4. It is important to note the surprising agreement of the configurations shown in Fig. 7.3 and 7.4. For example, the facets (100) for Σ3b and {110} for Σ71b are the coherent boundaries, and the facets (010) for Σ3b and (100) || (010) for Σ71b are non-coherent twins. It should be mentioned that similar microstructures were found in corrosion films formed on monoclinic particles in single crystals of tetragonal zirconium, stabilised with yttrium [18], and also in particles of monoclinic ZrO_2, precipitated in the rhombohedral phase of the ZrO_2–ZrN system [19]. Several special features of the microstructure, observed by high-resolution electron microscopy, will be discussed. Firstly, it is a deficit in the misorientation θ = 1.6° formed when two boundaries Σ71b or the boundaries Σ71b and Σ3b meet at a triple junction. This deviation from the total angle of 360° can be described using the model of a joint disclination with a power of 1.6°. This is in agreement with the assumption on the joint disclinations in $YBa_2Cu_3O_{7-}$ ceramics, made in [17]. One can propose a careful hypothesis according to which the distortion of the lattice on the high-resolution electron micrograph is caused by joint disclinations. However, it should be remembered that this distortion is probably an artefact, and distortions are consequences of the bending of the thin foil (section 4.1.2). Another interesting fact is found on the magnified image of the high-resolution electron micrograph (Fig. 7.5). A planar defect in the form of a facet along the (100) crystallographic plane rotates to the left through 90° (along the (010) plane) and breaks up in the centre of the grain. It is possible that this defect is a stacking fault. At the same time, the configuration (i.e., the facets along the (100) and (010) planes) resembles strongly the twin

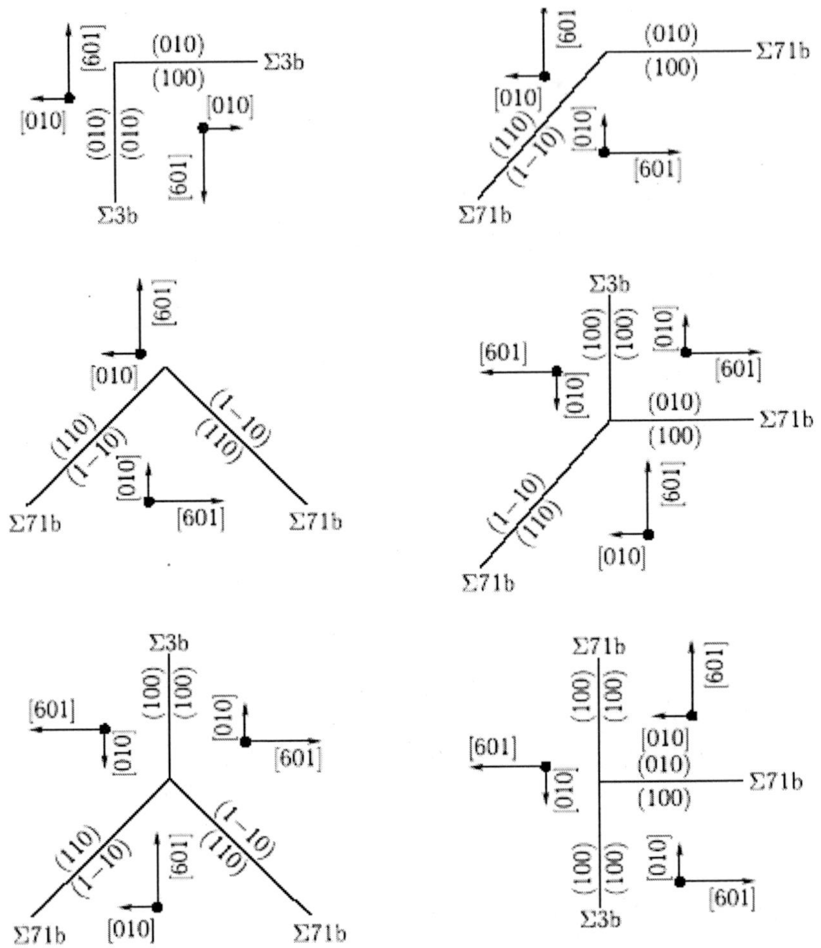

Fig. 7.4. Possible types of triple junctions and facets, formed in the ZrO_2 only by twin boundaries. The grains are oriented along the [100] direction normal to the plane of the figure.

Σ3b. The stacking fault should form as a result of grain growth and not due to the deformation of the crystal. Its energy is twice the energy of the twin boundary situated in the same crystallographic plane. Is the formation of a defect with such high energy (in comparison with twins) in the lattice possible? In addition, there was no faceting of the stacking fault, although this is possible from the crystallographic viewpoint. Another hypothesis regarding the origin of this defect is that it is in fact a disclination at the end of the twin boundary 3b, ruptured in the body of the grain. Formally, this may also be a wedge disclination with a power of either 180° or ≈19.2°, if it is assumed that the structure of the grain boundary Σ3b forms in the (010) plane by

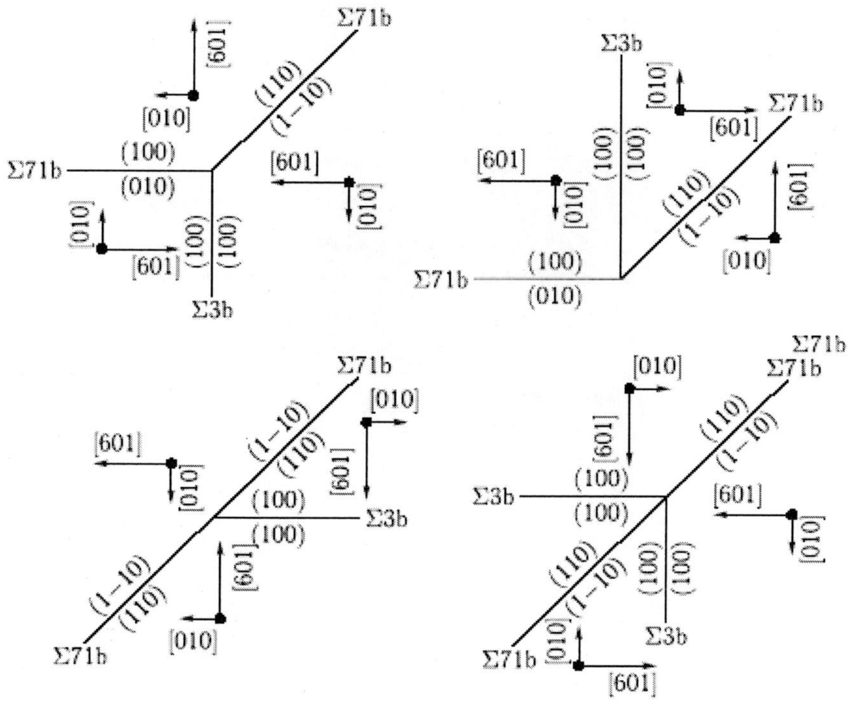

Fig. 7.4. (Continued)

shearing through the angle of 2β –180°. In the literature there are no experimental data on the existence of disclinations with such power. Another explanation may be proposed on the basis of the experimental observation of a small amount of the ZrO_2 tetragonal phase, both in x-ray diffraction studies [20] and Raman spectroscopy [5]. The monoclinic and tetragonal lattices have relatively similar parameters, and the interphase boundary can be almost coherent, especially for small grains. Similar boundaries are practically invisible under specific diffraction conditions. Appropriate observations were described in [19] where Van Tendello did not detect any sharp boundary between the grains of the orthorhombic and monoclinic phases with similar orientations. Also, there was no visible boundary between the tetragonal core of ZrO_2 and the surrounding monoclinic shell in the zirconium oxide particles [21]. It should be mentioned that these investigations were carried out in a high-resolution electron microscope. Thus, instead of the sharp interphase boundary there is a smooth transition from the lattice of one phase (monoclinic) to the lattice of another phase (tetragonal) (Fig. 7.6).

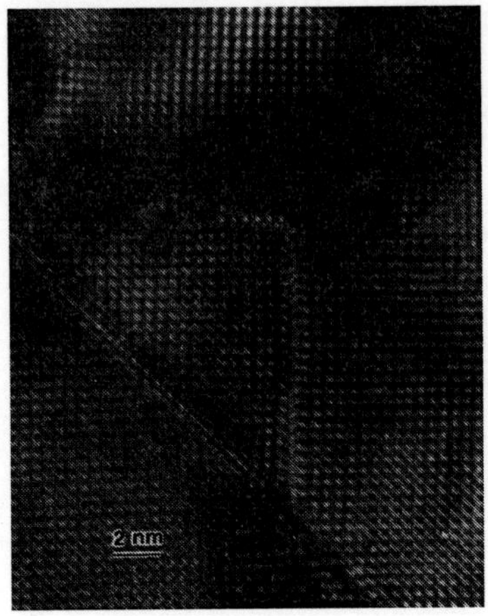

Fig. 7.5. Magnified image of the section indicated by the rectangle in Fig. 7.3a.

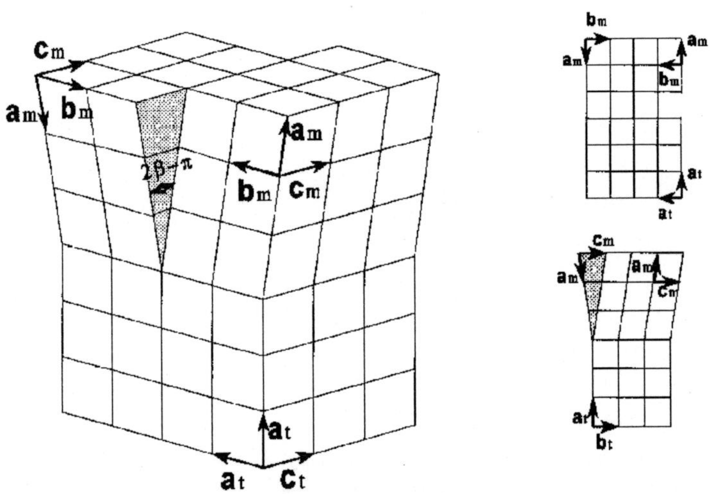

Fig. 7.6. Schematic representation of the non-coherent twin boundary Σ3b (010) broken up at the coherent interphase boundary with the tetragonal phase.

7.3. Effect of the statistics of the grain boundaries on diffusion in zirconium oxide

The crystals in the zirconium oxide film are very small (around 50 nm) so that direct methods of measuring the misorientation of the grain boundaries cannot be used. The methods described in chapters 2 and 3 helped to solve this problem by computer modelling. The information on the microstructure, presented in the previous sections, can be used for a more realistic interpretation of the spectra of the misorientation of the grain boundaries produced in this manner. In [16] similar grain boundary spectra were determined on the basis of available data on the type of texture present in the ZrO_2 film and a number of hypothetical considerations regarding the possible types of correlations which can form in the microstructure of zirconium oxide., Analysis can be carried out using the data obtained in high-resolution electron microscopy and the fact that the two types of texture components (growth texture and axial texture) in the microstructure of thc zirconium oxide film are separated in space. This greatly simplifies the given task.

As already shown, in regions in which the growth component is the main component, i.e., in the region where the oxide film is growing on the surface of α-Zr, the monoclinic lattice of the oxide film is oriented in such a manner that the direction $[001]_m$ is parallel to the normal to the substrate plane. In this case, the microstructure contains only the grain boundaries of the types Σ3b and Σ71b. Since they again form boundaries of the type Σ71b in the triple junction, the fraction of these boundaries is equal to 2/3 and the fraction of the type Σ3b boundaries is equal to 1/3. In regions in which the component of the growth texture $[100]_m$ can form, the boundaries Σ3a, Σ71a, Σ3b, Σ71b can form together with non-coherent twins Σ1′ and pseudo-twins Σ3c (all these boundary types were described in section 1.1.2). Thus, only a limited number of the grain boundary types can be present in the microstructure with the growth texture. Naturally, the previously described grain boundaries may have parameters which differ from the ideal misorientation. Similar deviations depend mainly on the sharpness of the texture of the α-grains of the substrate, i.e. zirconium. It is now necessary to analyse the types of grain boundaries existing in regions with the axial texture and in regions in the vicinity of the boundary. The misorientation distribution of the grain boundaries in the microstructure with the axial texture can be calculated using the algorithm proposed in [16]. In the present case, we use the following types of correlation: from all possible misorientations we select initially the twin misorientations and subsequently low-angle (<50°) grain boundaries. The sharpness of the axial texture, given by the half width of the Gaussian distribution, is equal to 5° (strong texture) and 15° (weak texture). It was assumed that in addition to the texture element $[001]_m$ there may also be another component, $[100]_m$, and the CGBS for the mixture of these components, taken at a ratio of 1:1, was determined. Table 7.1 shows the misorientation distribution of the grain boundaries (only the distributions with respect to Σ are given). Analysis shows the presence of a large fraction of special and low-angle grain

Table 7.1. Spectrum of the misorientation of the grain boundaries in the ZrO2 film for regions with the axial texture

Type of boundary	Type of texture			
	$[001]_m$		$[001]_m+[100]_m$	
	Sharp	Weak	Sharp	Weak
LAB	13.7	11.4	7.6	7.0
$\Sigma1'$	0.0	0.0	8.4	4.8
$\Sigma3a$	0.8	5.2	3.5	3.1
$\Sigma3b$	14.6	6.9	5.1	3.3
$\Sigma71a$	0.0	0.0	9.3	4.4
$\Sigma71b$	24.3	13.6	11.0	2.9
$\Sigma3c$	0.0	0.0	1.0	2.9
Others ($\Sigma< 30$)	1.0	2.5	0.6	1.7
HAB (arbitrary)	45.6	60.4	53.5	69.9

boundaries. This is evident in particular in the case of the 'sharp' axial texture. However, these regions do not consist exclusively of the special boundaries, as in the case of the growth texture where all the boundaries had a twin or similar orientation. As expected, the blurring of the axial texture increases the fraction of high-angle arbitrary boundaries (to 60.4% in the presence of only one texture component $[001]_m$, and up to 69.9% in the case of a mixture). All possible orientations in the boundary regions (where both the growth texture and axial texture are present) can be calculated as the misorientations between the grain boundaries relating to the growth texture, and the orientations of the grains with the axial texture. The misorientation distribution of the grain boundaries, obtained by this procedure, is shown in Table 7.2. Calculations were carried out using the same type of correlation – initially special and then low-angle grain boundaries. The results show that the main fraction of the grain boundaries for these regions is represented by the high-angle grain boundaries of the arbitrary type. Now we can estimate the total misorientation distribution of the grain boundaries for a ZrO_2 film formed on the surface of the Zr–2.5% Nb alloy. As already mentioned, the microstructure of the zirconium alloy has been studied in detail. The microstructure consists of elongated grains of α-zirconium whose dimensions are expressed approximately by the ratio 25:5:1, and the longest side is about 5 μm. X-ray diffraction studies show [20] that approximately 3/4 of all α-grains have a favourable orientation for the formation on them

Table 7.2. Spectrum of the misorientation of the grain boundaries in the ZrO_2 film for the region with the mixed texture

Type of boundary	Type of texture			
	$[001]_m$		$[001]_m + [100]_m$	
	Sharp	Weak	Sharp	Weak
LAB	7.2	2.1	3.8	1.0
Σ1′	0.1	0.0	0.4	0.0
Σ3a	0.0	0.1	0.3	0.1
Σ3b	0.8	0.0	0.2	0.0
Σ71a	0.0	0.0	0.5	0.1
Σ71b	0.0	0.0	0.3	0.0
Σ3c	1.5	0.1	0.4	0.1
Σ71c	0.0	0.0	0.8	0.1
Others (Σ< 30)	2.0	2.3	2.0	2.1
HAB (arbitrary)	88.4	95.4	91.3	96.5

of the oxide film with the growth texture. Consequently, in the total CGBS we should add up the twins Σ3b with the weight coefficient 1/4, and the grain boundaries Σ71b with the coefficient 1/2. If it is assumed that the remaining 25% of the grain boundaries are regions with the axial and mixed texture in the same proportion, we can calculate the distribution of the grain boundaries with respect to Σ in the ZrO_2 film formed on the surface of the Zr–2.5% Nb alloy during annealing. The results are presented in Table 7.3.

Now we can estimate the effect of the ensemble of the grain boundaries on the transport properties of polycrystalline materials (oxidation processes – diffusion of oxygen and hydrogen) on the example of the zirconium oxide film. The main type of diffusion at service temperatures in zirconium pipes is grain boundary diffusion. The variation of the kinetics of these processes in the ensemble of the grain boundaries with different misorientation distributions in the grain boundaries will be determined. As the model we use a two-dimensional system of grains consisting of regular hexagons (an example of such a system is shown in section 1.2.2) and use the approach described in chapter 1. For the textures present in the ZrO_2 film we determine the growth texture, the axial texture and a mixture of these textures, also the CGBS differing in the type of correlation and orientation of the adjacent grains. A similar study [16] was conducted prior to obtaining the experimental data on the microstructure of the ZrO_2 film by the methods of high-resolution electron microscopy (section 3.1.1)

Table 7.3. The total misorientation distribution of the grain boundaries in the ZrO_2 film

Type of boundary	Type of texture			
	$[001]_m$		$[001]_m+[100]_m$	
	Sharp	Weak	Sharp	Weak
LAB	3.2	2.6	1.8	1.6
Σ1′	0.0	0.0	1.9	1.1
Σ3a	0.2	1.1	13.3	13.2
Σ3b	28.2	26.5	13.6	13.2
Σ71a	0.0	0.0	27.1	26.0
Σ71b	55.3	53.0	27.4	25.6
Σ3c	0.0	0.0	0.2	0.6
Others (Σ< 30)	0.3	0.6	0.2	0.4
HAB (arbitrary)	12.7	16.2	14.6	18.3

and, therefore, the study used simulation considerations regarding the possible types of correlation in the orientation of the adjacent grains. However, the important feature of the results is the detection of the dependence of the kinetic parameters of diffusion on the type of spectrum of the grain boundaries in the polycrystal. Now, when the results obtained in high-resolution electron microscopy are available (see previous sections), we can evaluate with a high accuracy the type of ensemble of the grain boundaries in the zirconium oxide film. The distributions with respect to the inverse density of the coincident sites (Σ) are shown in Table 7.3. They greatly differ for different types of texture and the types of correlation in the orientation of the adjacent grains.

The method of modelling diffusion in a network of the grain boundaries has the form of the random walk algorithm. It can be described as follows. On the network of the hexagonal boundaries, each boundary has its own index of the low-angle boundaries (LAB), high-angle boundaries (HAB) or Σ (special boundary) in accordance with the partial fraction in the spectrum of the grain boundaries. The diffusion permeability of the individual boundary corresponds to its index. We selected the following parameters: HAB (arbitrary) has grain boundary diffusion permeability; LAB – diffusion permeability of the dislocations; for Σ–boundaries diffusion permeability is equal to lattice permeability. Since the length of all the grain boundaries is the same (ℓ), we estimate that time during which the 'test particle' travels this distance is:

$$\tau \propto \ell^2 / D, \text{ where } D = D_0 \exp\left(-\frac{Q}{RT}\right), \tag{7.8}$$

and D_0 and Q are taken for three types of diffusion (D_{0V}, Q_V is volume (bulk) diffusion; D_{0d}, Q_d is dislocation diffusion; D_{0gb}, Q_{gb} is grain boundary diffusion). We used the parameters of volume, dislocation and grain boundary diffusion of oxygen in zirconium oxide [28]. Using for temperature T = 673 K, we obtain the characteristic times for every type of grain boundary.

Subsequently, in a triple junction we select with the probability of 1/3 a grain boundary along which the 'test particle' travels. The particle travels along the entire grain boundary during the time corresponding to its type. Our system consists of 25 × 25 = 625 grains. Periodic boundary conditions were introduced: the 'test particle', leaving the system at the boundary appeared at the equivalent point of the opposite side and continued to move. The distance travelled by the particle during a relatively large number of steps characterises the kinetics of these processes.

Figure 7.7 shows the logarithmic dependence of the mean path $\langle R^2 \rangle^{1/2}$ on time (or on the number of steps) for ensembles of the grain

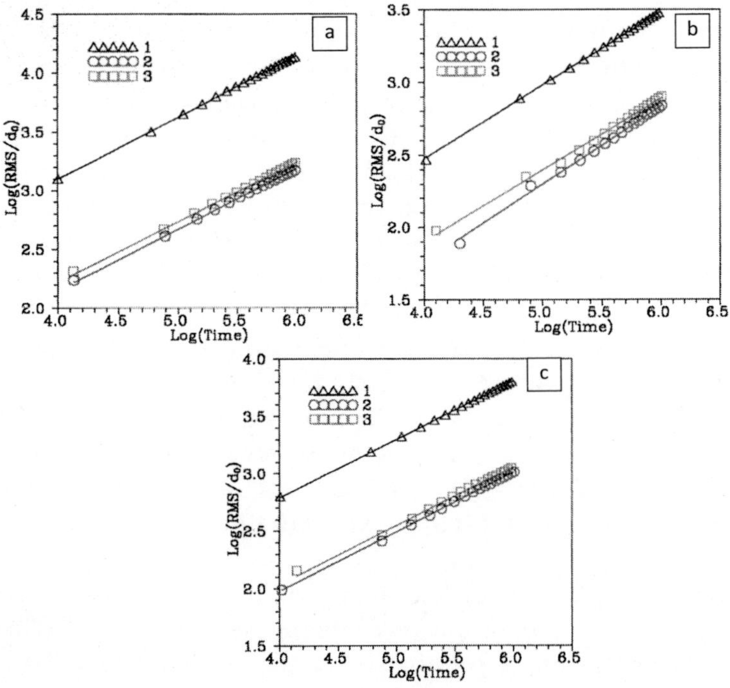

Fig. 7.7. Kinetics of oxygen diffusion through ZrO_2 film for CGBS.

boundaries with different types of correlation. It may be seen that the HAB correlation results in higher diffusion mobility of such a system (in fact, this was also expected). The difference in the oxidation rate may reach an order of magnitude. Thus, it may be predicted that the Zr–2.5% Nb alloy will be characterised by the nonuniform growth of the oxide film in boundary regions between the zones with the gross texture and the axial texture. Indeed, such 'protuberances' of the oxide film were detected in experiments at the boundaries between a columnar structure and grains with a more equiaxed structure [5, 6]. The columnar structure of the grains is a clear indication of the growth texture and the more equiaxed structure (also stretched in the growth direction) consists of the regions with the axial texture.

7.4. Special features of oxidation kinetics under the effect of stresses at the metal/oxide boundary

In the previous section it was explained that the total kinetics of oxidation of the Zr–2.5% Nb alloy is determined by the fastest of the processes – grain boundary diffusion in the regions between the microstructure with the growth texture and the axial texture. However, the oxidation kinetics on the macrolevel is also influenced by the stress formed at the metal/oxide interface during the growth of the ZrO_2 film because of differences in the lattice parameters of the substrate and the growing film and also the differences in their thermal expansion coefficient. The components of the stress tensor differ from zero in the substrate plane and the tensor is the function of the thickness of the growing film. In the approximation of the isotropic medium only its component $\sigma_{yz} = \sigma(x)$ can be investigated in the plane normal to the diffusion direction. The maximum value of this stress (compressive) measured by different methods [20, 22, 23] should reach 1 GPa. It is obvious that the function $\sigma(x)$ is a linear function of the thickness of the film and disappears at the gas/oxide boundary. The critical value of the stresses leads to the formation of pores, cracks and other three-dimensional defects in the growing layer of the oxide film. The experiments show that the ZrO_2 film consists of two parts: external (porous) and internal (in which no similar defects were detected) [20]. The internal layer protects against further oxidation in corrosive media and against hydrogen diffusion. To estimate the service life of the materials, it is very important to evaluate the kinetics of diffusion processes in these systems. Regardless of the fact that the diffusion models for zirconium alloys have been studied in a large number of investigations [24–26], to solve the task described in this book the effect of the stress, formed at the metal/oxide boundary, on the oxidation kinetics of Zr–2.5% Nb alloy in the system with two moving boundaries was studied for the first time in this work.

The mathematical model is based on the description of the following physical processes taking place in the system. The oxygen molecules breakup into atoms at the gas/oxide interface. The O^{2-} atoms diffuse through the film mostly along the grain boundaries and triple junctions. Reaching the metal/oxide interface surface, they are oxidise the metal, causing further growth of the oxide film. Growth of the film is accompanied by the formation of compressive stresses in the film reducing the diffusion rate of the oxygen atoms. To stop the stress from increasing, a percolation network of pores and cracks forms in the external part of the oxide. The effective gas/oxide interface travels into the thickness of the film.

The physical processes form, with certain assumptions, the basis of the mathematical model in which a semi-infinite specimen is separated by flat gas/oxide and oxide/metal interfaces. During oxidation, the surfaces of these boundaries move into the bulk of the metal. The compressive stresses form at the oxide/metal interface. Diffusion takes place in the field of the oxygen concentration gradient and the field of the compressive stresses gradient.

Figure 7.8 shows schematically the diffusion process described by this mathematical model. Thus, we formulate the problem of one-dimensional diffusion for the semi-infinite space from an unlimited source with two moving boundary conditions. The diffusion of oxygen in the Zr–2.5% Nb alloy is ignored because the rate of this diffusion is considerably lower than that in the oxide film. Consequently, the diffusion equation for the pore-free ZrO_2 film has the following form

$$\frac{\partial C}{\partial t} = D\left(\frac{\partial^2 C}{\partial x^2} + \frac{\Omega}{RT}\frac{\partial C}{\partial t}\frac{\partial \sigma}{\partial x}\right); \quad S_0(t) \le x \le S(t); \qquad (7.9)$$

$$C(S_0(t),t) = C_0; \qquad C(S(t),t) = C_{01},$$

where C is the concentration of the O^{2-} atoms; D is the diffusion coefficient; Ω is the molar volume; R and T are the gas constant and temperature, respectively; $S_0(t)$ and $S(t)$ are the coordinates of the moving gas/oxide and oxide/metal interfaces. The Stefan condition is satisfied for the boundary $S(t)$

$$-D\frac{\partial C}{\partial x}\Big|_{x=S(t)} = C_{01}\frac{dS}{dt}. \qquad (7.10)$$

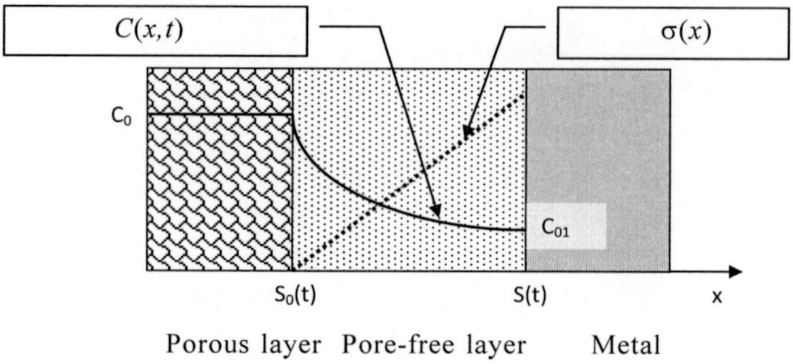

Fig. 7.8. Schematic representation of the model of diffusion of oxygen through the ZrO_2 film taking into account the effect of stress at the metal/oxide interface.

Prior to continuing investigations, it should be stressed that this model uses the 'effective' diffusion coefficient which can be calculated on the basis of assumptions on the spectrum of the grain boundaries, described in the previous section. This is a very important difference between this model and many other models found in the literature. We sutdy two conditions which greatly simplify our task and enable an analytical solution of the problem. Firstly, we select a linear relationship between the coordinates S_0 and S: $S_0(t) = \beta S(t)$. We also assume the linear dependence of the resultant stress on the thickness of the film: $\sigma(x) = \sigma_0(x-S_0)/(S-S_0)$. Transferring to the dimensionless variables we obtain the following equation

$$\frac{\partial \overline{C}}{\partial \tau} = D\left(\frac{\partial^2 \overline{C}}{\partial \overline{x}^2} + \frac{\delta}{\overline{S}-\overline{S}_0}\frac{\partial \overline{C}}{\partial \overline{x}}\right); \qquad (7.11)$$
$$\overline{S}_0(t) \le \overline{x} \le \overline{S}(\tau); \quad \overline{C}(\overline{S}_0,\tau)=1; \quad \overline{C}(\overline{S},t)=0,$$

and the Stefan condition

$$-\left.\frac{\partial \overline{C}}{\partial \overline{x}}\right|_{\overline{x}=S(\tau)} = \alpha\frac{d\overline{S}}{d\tau}, \qquad (7.12)$$

where

$$\overline{x} = x/a; \quad \tau = \frac{Dt}{a^2}; \quad \overline{S}_0 = S_0/a; \quad \overline{S} = S/a;$$

$$\overline{C} = \frac{C - C_{01}}{C_0 - C_{01}}; \quad \alpha = \frac{C_{01}}{C_0 - C_{01}}; \; \delta = \frac{\Omega \sigma_0}{RT}, \tag{7.13}$$

and a is some characteristic length. The remaining parameters are presented in Table 7.4.

Non-stationary diffusion

The solution of the equations (7.11)–(7.13) will be found in the form

$$\overline{C} = \psi(\overline{x} / \sqrt{\tau}); \; \overline{S} = \sqrt{2\gamma\tau}; \; \overline{S}_0 = \beta \overline{S}, \text{where } 0 \le \beta \le 1. \tag{7.14}$$

After simple transformations, we determine the concentration profile

$$\overline{C} = \frac{\operatorname{erf}(\sqrt{\gamma/2} + A) - \operatorname{erf}(\overline{x}/2\sqrt{\tau} + A))}{\operatorname{erf}(\sqrt{\gamma/2} + A) - \operatorname{erf}(\beta\sqrt{\gamma/2} + A)}, \text{ where } A = \frac{\delta}{\sqrt{2\gamma}(1-\beta)}, \tag{7.15}$$

and γ is the diffusion (oxidation) rate which can be determined from the transcendental equation

$$\alpha\sqrt{\frac{\pi\gamma}{2}} \exp\left[(\sqrt{\gamma/2} + A)^2\right]\left[\operatorname{erf}(\sqrt{\gamma/2} + A) - \operatorname{erf}(\beta\sqrt{\gamma/2} + A)\right] = 1. \tag{7.16}$$

Substituting the values $\delta = 0$ and $\beta = 0$ into equation (7.16) gives the well-known equation for describing the heat transfer or diffusion processes in the systems with the moving boundary conditions:

$$\alpha\sqrt{\frac{\pi\gamma}{2}} \exp(\gamma/2) \operatorname{erf}(\sqrt{\gamma/2}) = 1. \tag{7.17}$$

The following approximation can be used at high values of δ:

Table 7.4. Parameters used in equations (7.11)–(7.13)

Parameter	Decoding and values
C_0	Concentration at the gas/oxide interface (1511 kg/m) [25]
C_{01}	Concentration at the metal/oxide interface (1417 kg/m) [25]
Ω	Molar volume of ZrO_2 ($282.2 \cdot 10^{-7}$ m^3/mole)
σ_0	Maximum stress at the metal/oxide interface

$$\operatorname{erf}(x) \approx 1 - \frac{\exp(-x^2)}{x\sqrt{\pi}}.$$

Consequently, γ is determined by the equation

$$\gamma = \frac{\delta \exp(-\delta)}{\alpha(1-\beta)}.$$

Pseudo-stationary diffusion

The experiments with the measurement of the thickness of the pore-free layer show that after some time both interfaces (gas/oxide and oxide/metal) start to move in self-matched manner, i.e., the thickness of the pore-free oxide layer remains unchanged [20]. In this case, the relationship $S = S_0 + h$ can be used. Consequently, diffusion is assumed to be independent of time and takes place in the moving layer with thickness h:

$$\frac{\partial^2 \overline{C}}{\partial \overline{x}^2} + \frac{\delta}{h}\frac{\partial \overline{C}}{\partial \overline{x}} = 0; \ \overline{S}_0(\tau) \le \overline{x} \le \overline{S}(\tau); \ \overline{C}(\overline{S}_0, \tau) = 1; \ \overline{C}(\overline{S}, \tau) = 0, \tag{7.18}$$

with the solution for the concentration profile

$$\overline{C} = \frac{\exp\left[-\delta(\overline{x} - \overline{S}_0)/h\right] - \exp(-\delta)}{1 - \exp(-\delta)}. \tag{7.19}$$

The point of transition from non-stationary to stationary diffusion can be determined from the relationship:

$$\tau_{crit} = \frac{\overline{h}^2}{2\gamma(1-\beta)^2}. \tag{7.20}$$

The concentration profiles (7.15) and (7.20) should coincide.

The main task of the diffusion model is to predict the kinetic parameters, i.e., determine parameter γ and the thickness of the oxide film h. For non-stationary diffusion the dependence of the numerical solution of equation (7.16) on parameters β and δ is shown in Fig. 7.9 and 7.10. In Fig. 7.9 the diffusion rate is shown as a function of the parameter β which is referred to as the coefficient of linear response of the system and characterises the rate of removal of the stress generated during the growth of the oxide film. It is clear that the diffusion rate

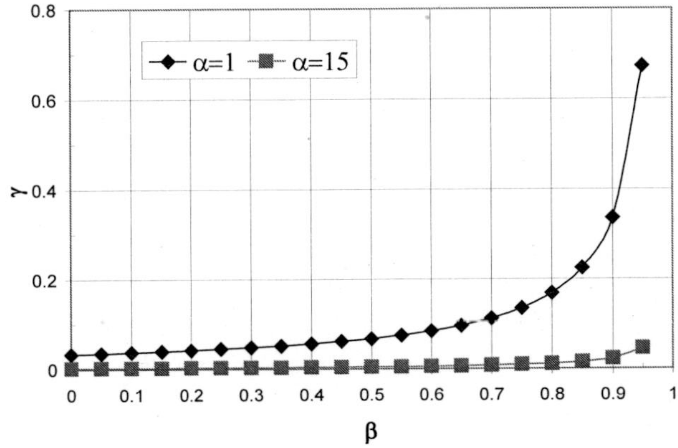

Fig. 7.9. Dependence of the diffusion rate on the 'feedback' coefficient β for two values of parameter α: 1 (1) and 15 (2).

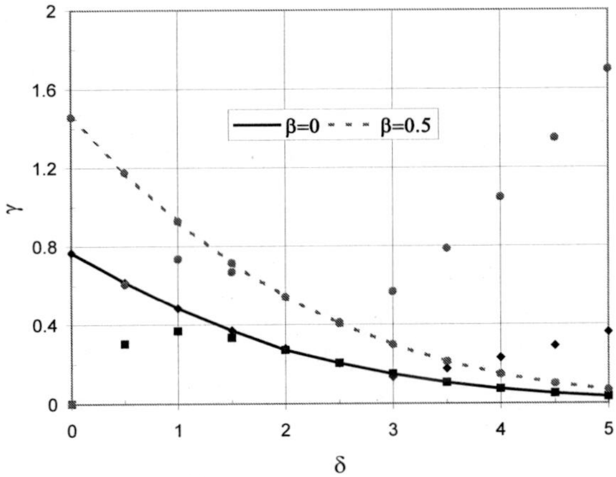

Fig. 7.10. Dependence of the diffusion rate on stress at the metal/oxide interface for two values of the parameter β: 0 (1) and 0.5 (2) (α = 15).

should decrease with increasing stress (low values of β). Figure 7.10 shows the effect of the value of the maximum stress at the oxide/metal boundary on the kinetic parameters of oxygen diffusion. The transcendental equation (7.16) was solved numerically for $\delta < 3$ using the expression for asymptotics (7.18). The diffusion rate decreases exponentially with increasing maximum stress.

The following numerical estimates were obtained on the basis of the results. Using the parameters from Table 7.4, it can be seen that $\alpha \sim 15$ and for $\sigma_0 = 1$ GPa $\delta \sim 5$. Parameter β can be determined using the experimental data for stationary diffusion as $\beta = 1 - h/w \sim 0.733$, where h and w is the thickness of the pore-free layer and the total thickness of the oxide film [27]. Using all

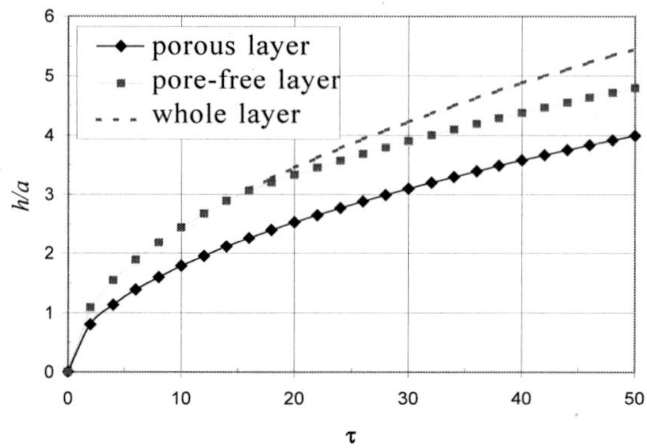

Fig. 7.11. Oxidation kinetics of zirconium for porous, pore-free and whole oxide film layers.

the data, the following estimate is obtained for the diffusion rate: $\gamma \sim 0.297$, and from equation (7.21) we determine the point of transition from non-stationary to pseudo-stationary diffusion: $\tau_{crit} \sim 15$. Using the numerical values for the oxygen diffusion coefficient [28] we can plot the dependence of the total thickness of the oxide film and of its pore-free part as a function of dimensionless time (Fig. 7.11). The equations are used for the times shorter than critical are $S(\tau) = (2\gamma\tau)^{1/2}$ and $S_0(\tau) = \beta S(\tau)$. In the case of stationary diffusion the oxidation kinetics for both interfaces is the same. Regardless of the apparently simple form of the resultant solution this solution is of considerable importance. In the experimental investigation of the oxidation kinetics of the zirconium alloys many investigators found in the thickness of the oxide layer the transition from 'quadratic' (or normal) ($h \sim t^{1/2}$) to 'cubic' dependence ($h \sim t^{1/3}$). The large numbers of attempts to explain this behaviour have not been successful. If Fig. 7.11 is plotted in the logarithmic ordinates and the so-called cubic law of oxidation kinetics is applied on the same scale (Fig. 7.12) it may be seen that for the times longer than critical the cubic law is almost completely identical with the normal oxidation kinetics. Thus, switching the type of diffusion from nonstationary to pseudo-stationary may be used as a rational explanation of the observed experimental result.

7.5. Texture and spectrum of misorientation of the grain boundaries in a NiO film on (100) and (111) substrates: modelling and experiments

Another system: nickel–nickel oxide will now be investigated. Nickel is used in the manufacture of magnetic information carriers and is

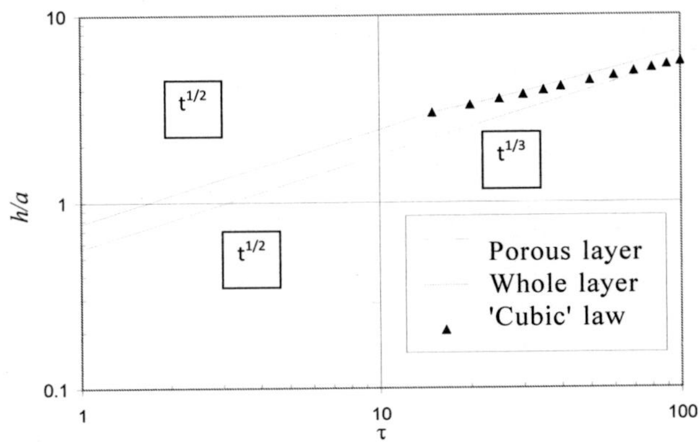

Fig. 7.12. The kinetics of oxidation of zirconium, constructed in the logarithmic coordinates, for porous and whole oxide film layers. The 'cubic' law of oxidation kinetics is shown here $(h \sim t^{1/3})$.

the main component in the majority of creep-resisting alloys used in aerospace technology [29]. In addition, at present some of the elements of the blades for gas turbine engines are produced from single crystals of nickel alloys. Therefore, the investigation of the diffusion processes and, consequently, oxidation of nickel single crystals is of considerable importance. What is the role played by the texture and grain boundaries in these processes? The point is that the oxidation kinetics is determined by the diffusion of nickel through the oxide film to the gas/oxide boundary. As in the case of oxidation of zirconium considered in the previous section, the nickel oxide film is nanocrystalline (even on the single crystal substrate) but in this case the kinetics of diffusion of the metal (nickel) is determined by the microstructure, texture and ensemble of the grain boundaries of the NiO film. The results of experimental investigations [30–32] show that the difference in the oxidation rate of the nickel single crystals (100) and (111) is more than an order of magnitude. Examination of the microstructure shows that the mean grain size of the oxide films (100)–NiO and (111)–NiO is approximately the same. The existing difference cannot increase the volume fraction of the grain boundaries to the extent sufficient for explaining this difference in the kinetics. We attempted to use the previously developed approach relating to the effect of the misorientation distribution on the diffusion properties of the investigated material. Details of these investigations are presented below.

The grain boundaries in the nickel oxide polycrystals are characterised by higher diffusion permeability for nickel at elevated temperatures[1]. Thus, they determine the appropriate corrosion properties [33]. In particular, it was found the diffusion of nickel is capable of increasing the vacancy concentration and, consequently, the volume diffusion coefficient. A similar mechanism may also operate at the grain boundaries (as shown by Atkinson and Taylor [34, 35]). Consequently, the control of the properties of the ensemble of the grain boundaries determines the shielding properties of nickel-based materials. The size of the crystals (grains) in the resultant nickel oxide film (and also in the majority of oxide films) is very small and it is not always possible to use the currently available experimental devices for direct measurements. The method of restoration of the misorientation distribution of the grain boundaries with respect to the texture data can be used in this case. Regardless of the fact that there is no unambiguous relationship between the orientation distribution function in the spectrum of the grain boundaries, it is possible to use some of its special features for determining the nature of correlation in the orientation of the adjacent grains and at the same time obtain statistically reliable data on the CGBS. Processes of formation of the NiO film on nickel single crystals with the orientation (100) and (111) will be discussed.

The experiments show that the textures of the oxide films, formed on single crystals with different orientation, greatly differ from each other. It is logical to assume that the appropriate grain boundary spectra also differ. Consequently, the NiO film on the (100) substrate also consists of the grain boundaries with no preferential orientation, whereas the oxidation of the (111) single crystal results in the formation of the ensemble of the grain boundaries which consists mainly of special grain boundaries.

The texture of the oxide was measured in a Siemens diffractometer in $Mo_{K\alpha}$ radiation. Incomplete pole figures (PF) were obtained with the tilt angle of up to 80° at 5° steps for the polar and azimuthal angles. Intensity was corrected for the absorption and defocusing conditions on the basis of the measurement of a PF standard – a textureless specimen produced from NiO powder. Figure 7.13 shows incomplete pole figures for the nickel oxide film grown on the (100) and (111) substrates.

The quantitative characteristic of the texture, ODF, can be determined by the Bunge method [17]. Figure 7.14 shows the grain orientation distribution functions for the NiO films on the (100) and (111) nickel single crystals. Analysis of the ODF shows the strong effect of the

---

[1]Discussing the elevated temperatures (~1000°C) it should be remembered that they are still lower than 0.5 T_{max} for NiO and diffusion takes place mainly along the nickel oxide grain boundaries.

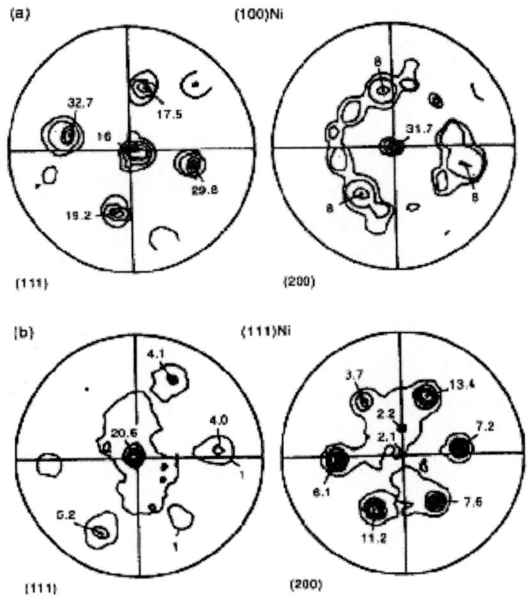

Zone	I	II	III	IV	V	VI	VII	VIII
(100)	30.9	1.8	2.1	10.7	33.4	14.0	1.6	5.5
(111)	2.5	3.4	2.2	15.6	40.0	7.4	6.9	24.1

Fig. 7.13. Grain orientation distribution function for (100)-NiO (a) and (111)-NiO (b).

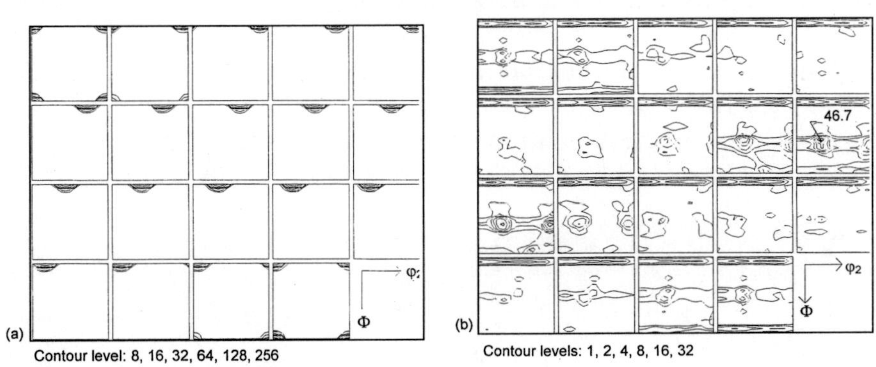

a b

Fig. 7.14. Incomplete pole figures (111) and (200) for the nickel oxide film grown on single crystal substrate: a) (100); b) (111).

substrate orientation on the produced texture of the film. For example, NiO, grown on the (100) substrate, is characterised by a strong texture component (100) [001]. The formation of this texture is determined by the epitaxial growth of the oxide film. Comparison of these ODFs with the modelling distribution function, described in section 3.1, shows that it is related with the so-called recrystallisation texture of copper. The orientation distribution function of the nickel oxide, formed on the (111) single crystal, is more complicated and is described by numerous maxima. Here it is necessary to mention the strong maximum in the section $\varphi_2 = 45°$ corresponding to the twin orientation. The presence of the maximum already indicates the existence of suitable conditions for the formation of twin boundaries (Σ3) in the system.

The resultant ODFs were used to calculate the misorientation distribution of the grain boundaries for the NiO films, grown on nickel single crystals with the orientation (100) and (111). Figure 7.15 shows the appropriate distribution function of the misorientation angles and axes. The distribution with respect to Σ is shown in Table 7.5. Here, we can divide all the boundaries into three main classes: low-angle, high-angle arbitrary and special grain boundaries. Both the misorientation distribution of the grain boundaries (Fig. 7.15) and Table 7.5 show that the ensembles of the grain boundaries in NiO strongly depend on the type of substrate. The nickel oxide film, grown on the (100) single crystal, is characterised by a high fraction of low-angle grain boundaries with the misorientation axes close to the direction of the normal to the (100) plane. These boundaries form approximately 40% of all grain boundaries. The fraction of the special grain boundaries which includes only the grain boundaries with Σ = 3, is equal to 3.1. The nickel oxide on the (111) substrate is characterised by a high fraction of the Σ3 boundaries (27.6%) and a small fraction of the low-angle grain boundaries. At approximately the same content of the high-angle arbitrary grain boundaries (56.6% for the (100) film, and 66% for the (111)) film of the oxidation kinetics is determined mainly by the difference in the fractions of the special and low-angle boundaries it

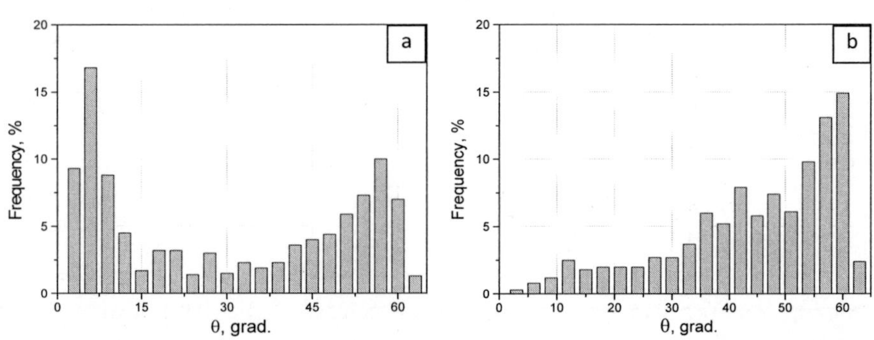

Fig. 7.15. Misorientation spectrum of the grain boundaries in the (100)–NiO (a) and (111)–NiO films (b).

Table 7.5. Distribution of grain boundaries with respect to Σ in (100)–NiO and (111)–NiO films

Substrate	Σ3	LAB	HAB (arbitrary)
(100)	3.1	40.4	56.6
(111)	27.6	6.4	66.0

has been proposed to use these special features of the misorientation distributions of the grain boundaries of the NiO films in modelling of the oxidation kinetics.

In modelling the oxidation kinetics of the nickel single crystals with different orientation we can use the random walk method modified for the actual microstructure [36]. It is assumed that an infinite source of oxygen atoms is located on the surface of the oxide film. The microstructure of the film is formed by the Monte Carlo method and its size is 4000 × 224 grains (Fig. 7.16a). The magnified fragment of this microstructure is shown in Fig. 7.16b. The grain size is selected at random in accordance with the size distribution function of the grains. In this case we use the Gaussian distribution (Fig. 7.16c). All the dimensions are presented in the units of the size of the elementary cell used in the calculations.

In accordance with the geometrical criterion the probability of a diffusing atom being in the region of the grain boundaries is given by the expression

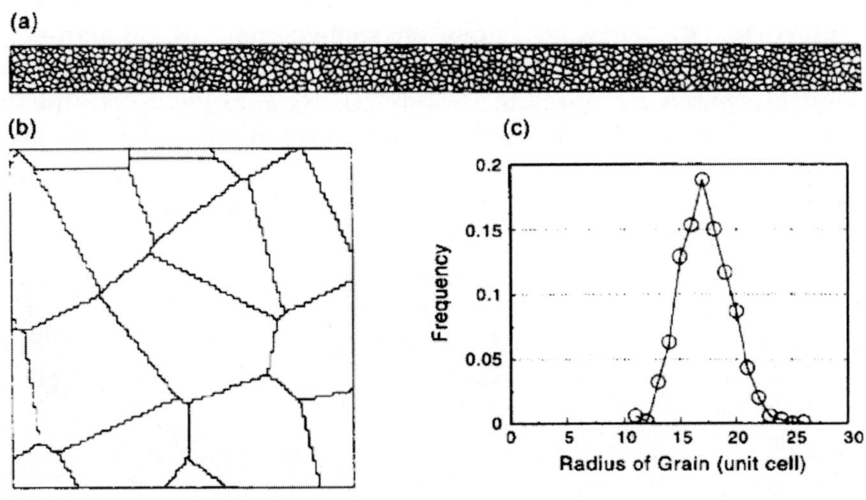

Fig. 7.16. Microstructure used in modelling the oxidation process: a) general view, b) large plan; c) the size distribution of the grains.

$$p \sim \delta / d, \tag{7.21}$$

where δ is the width of the boundary; d is the grain size. The diffusing atom is allowed to travel to the new site in one of the four possible directions. The distance over which the atom can move is determined by the equation

$$\Delta d = \sqrt{4D(x, y)\Delta t_{\min}},$$

where $D(x, y)$ is the diffusion coefficient which depends on the position of the diffusing atom; $\Delta t_{\min}$ is the minimum time period. In turn, the minimal period is calculated from the equation

$$\Delta t_{\min} = \frac{L^2}{4D_{\max}}, \tag{7.22}$$

where L is the size of the calculation cell which depend on the mean grain size; $D_{\max}$ is the maximum diffusion coefficient along the investigated grain boundary. In modelling, the diffusing atom continues to move over a specific period of time. The final position is recorded and assumed to be occupied and, consequently, the next atom cannot occupy this position. The movement of approximately 20 million of diffusing atoms is calculated to obtain statistically significant results. This is followed by determination of the concentration profile in the depth of penetration and the oxidation kinetics.

To describe the system completely, it is necessary to specify the appropriate parameters of the individual grain boundaries and the diffusion coefficients for them. In this case, we use the approach tested for evaluation of diffusion in the ZrO_2 (section 7.3), i.e., each boundary was allocated the diffusion coefficient in accordance with the type of boundary. The type of individual boundary was specified as random in accordance with the partial fraction of specific boundaries in the ensemble of the grain boundaries. It should be mentioned that all the boundaries were divided into three groups: $\Sigma3$ (or twins), low-angle ($\theta \leq 15°$), and high-angle arbitrary grain boundaries. These boundaries are characterised by different diffusion coefficients: the twins have the lattice diffusion coefficient; the low-angle boundaries the dislocation coefficient, and the high-angle boundaries have the grain boundary coefficient. The appropriate diffusion coefficients were taken from studies by Atkinson et al [34, 35]. For the given experimental conditions (temperature 1073 K and the external oxygen pressure

1 atm) the diffusion coefficient of nickel in the oxide film is:

$$\begin{aligned}
&D_{HAB} = 5.52 \cdot 10^{-19}\,\mathrm{m^2 \cdot s^{-1}} \\
&(D_{lattice} = 2.2 \cdot 10^{-6} \exp(-2.56\ \mathrm{keV}/kT)\ \mathrm{m^2 \cdot s^{-1}}); \\
&D_{LAB} = 3.72 \cdot 10^{-15}\ \mathrm{m^2 \cdot s^{-1}} \\
&(D_{dis} = 0{,}26 \exp(-2.00\,\mathrm{keV}/kT)\,\mathrm{m^2 \cdot s^{-1}}); \\
&D_{gb} = 7.45 \cdot 10^{-14}\,\mathrm{m^2 \cdot s^{-1}} \\
&(D_{gb} = 0.43 \exp(-1.78\ \mathrm{keV}/kT)\ \mathrm{m^2 \cdot s^{-1}}).
\end{aligned} \tag{7.23}$$

The grain boundary is assumed to be 0.7 nm wide, the size of the calculation cell is the same for all grains, and the grain size itself was determined in experiments and equals 150 nm.

Figure 7.17 shows the concentration profile for diffusion of nickel atoms through the oxide film plotted for two types of CGBS shown in Table 7.5. The graph also shows the concentration profile for lattice diffusion. It is clear that the lattice diffusion shows the smallest penetration depth. The depth of penetration of the diffusing atoms for the NiO film characterised by the maximum fraction of the twin boundaries in the microstructure is close to lattice diffusion. The oxide film with the highest fraction of the low-angle grain boundaries is most permeable for nickel. Identical results were also obtained for the oxidation kinetics. Since the fraction of the arbitrary boundaries of the grains in the NiO films, grown on both substrates, differs by approximately 7%, the main difference in the oxidation kinetics (or, in other words, the diffusion of nickel in the oxide film) is caused by

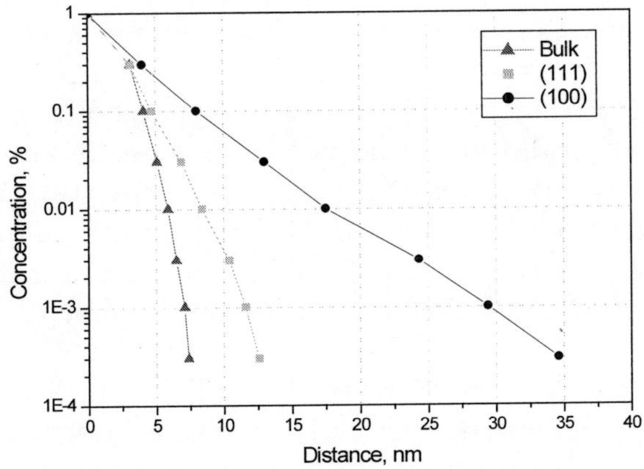

Fig. 7.17. Calculated concentration profile after 1800 s for (100)-NiO and (111)-NiO. The profile for purely bulk diffusion is also given.

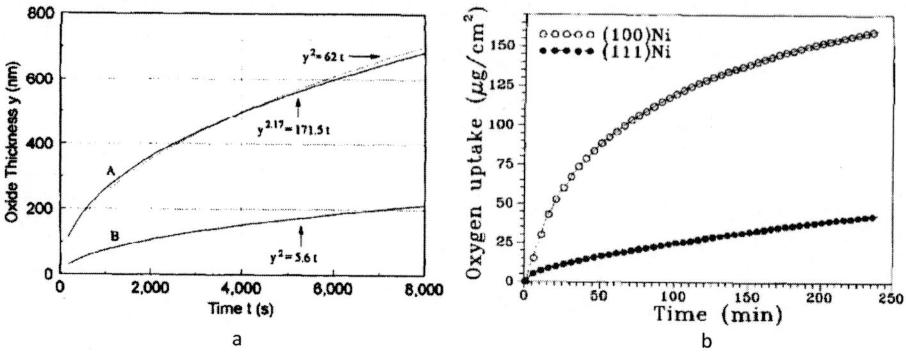

Fig. 7.18. Oxidation kinetics of the (100)-NiO (1) and (111)-NiO (2) films: a) modelling, b) experimental curves [18].

the low-angle boundaries and twins.

Figure 7.18 shows the results of modelling of the oxidation kinetics of NiO films and the appropriate experimental curves [37]. The (111)-NiO oxide film shows the parabolic law of oxidation (in the dimensionless quantities) with the equation $y^2 = 5.6t$. In the case of the (100)-NiO oxide film the coefficient of the parabolic law increases and the kinetics is described by the equation $y^2 = 171.5t$. In this case, all the modelling parameters (with the exception of the CGBS) are the same. Consequently, it can be assumed that the difference in the type of ensemble of the grain boundaries results in deviations from the oxidation kinetics. After all, complete agreement between the modelling and experimental results is not possible because in reality there are differences in the microstructure of the oxides, although the mean grain size of the oxide is approximately the same. However, modelling shows that the difference in the oxidation rate, detected in the experiments, can be explained by differences in the ensembles of the grain boundaries in the oxide films. In addition, if we calculate the ratio of the oxidation rate, using the parabolic law of oxidation for the (100)-NiO film ($y^2 = 62t$) which is close to the calculated law, we obtain the value of approximately 11. The ratio of the oxidation kinetics, determined on the basis of the experimental data, is 14. This agreement between the calculated and experimental data can be regarded as a reliable confirmation of the accuracy of the proposed model.

Another important parameter (in addition to the CGBS) which determines the oxidation kinetics is the mean grain size. In a general case, the increase of the mean grain size of the microstructure reduces the oxidation rate as a result of a reduction of the volume fraction of the grain boundaries of the function $\sim 3\delta/d$. The general equation for the oxidation kinetics (the increase of the thickness of the oxide film $y \sim \sqrt{4Dt}$) gives the dependence of the growth rate: $K \sim \frac{dy}{dt} D^{1/2}$. It is well known that in the first approximation the diffusion coefficient has the following form:

$$D = (1-x)D_V + xD_{gb}, \tag{7.24}$$

where D_V is the lattice diffusion coefficient; D_{gb} is the grain boundary diffusion coefficient; x is the volume fraction of the grain boundaries. It is clear that $x \sim 3\delta / d \ll 1$ and, consequently

$$D \sim D_V \left(1 + 3\frac{\delta}{d}\frac{D_{gb}}{D_V}\right), \quad K \sim d^{-1/2}. \tag{7.25}$$

The oxidation rate and the appropriate kinetic dependences, obtained by modelling for different values of the mean grain size, are presented in Fig. 7.19. The variation of the parameter K is governed by the law (7.25) which also confirms the accuracy of the selected model.

Cerium oxide (CeO_2) and other active additions improve the corrosion properties of many metals and alloys at elevated temperatures. In [38, 39], the Ni–NiO system was investigated in the conditions with cerium additions. The results show that the efficiency of the additions depends strongly on the type of the surface of the given substrate and treatment. The experiments were described here differ from previous ones only by the presence of active cerium additions. To introduce

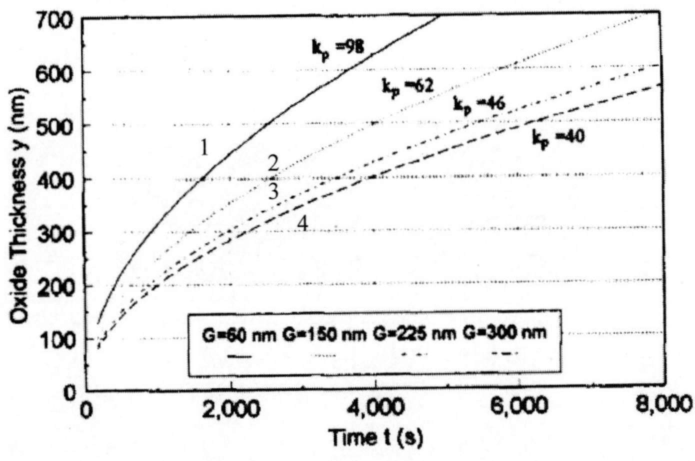

Fig. 7.19. Dependence of the oxidation rate of the NiO isotropic film on the mean grain size: G = 60 nm (1), 150 nm (2), 225 nm (3) and 300 nm (4).

additions, the surface of the specimens (a nickel single crystal with the orientations (100) and (111)) was coated with a CeO_2 solution. Subsequently, the specimens were annealed at a temperature of 1573 K for 1 h. The thickness of the coating was approximately 40 nm, and the size of the cerium oxide crystals of the order of 5 nm [40]. In addition to the x-ray investigations and recording of the texture, attention was also given to the variation of the morphology of the surface of the oxide film using a microprobe microscope (AFM – atomic force microscope) manufactured by Digital Instruments Nanoscope III. The experimental details were described in [41].

The morphology of the surface of the oxide film with both orientations, formed in the absence of cerium, has distinctive crystallographic appearance (Fig. 7.20 a–c). After coating with cerium the morphology greatly changes and becomes globular (Fig. 7.20d) and also differs in external features. The structure of the NiO oxide film on the single crystal substrate (100), coated with cerium, is of the 'cauliflower' type, and on the (111) + CeO_2 substrate it has the form of spherical grains uniformly distributed on the surface. Thus, it may be concluded that the cerium addition changes the morphology of the NiO film and reduces the rate of grain growth leading to the formation of the

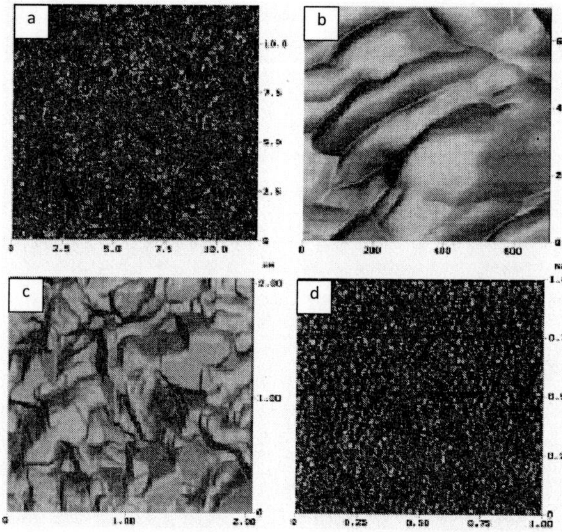

Zone	I	II	III	IV	V	VI	VII	VIII
(100)+CeO_2	1.4	1.7	9.5	8.1	4.3	0.5	1.3	72.7
(111)+CeO_2	7.6	7.8	3.8	16.1	29.6	8.8	5.9	20.4

Fig. 7.20. Morphology of NiO on the surfaces: a, b) (100) without the cerium coating; c) (111) without the cerium coating; d) (111), modified with cerium.

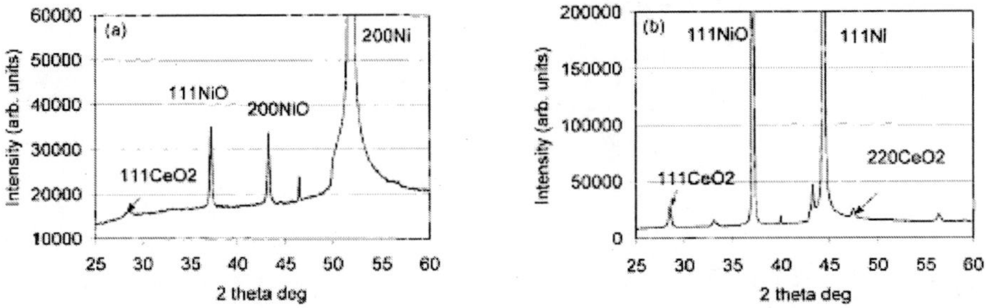

Fig. 7.21. Diffraction diagram of the nickel oxide on the cerium-modified substrates: a) (100), b) (111).

nanocrystalline structure. X-ray diffraction investigations (diffraction patterns are shown in Fig. 7.21) show that in addition to the nickel and NiO peaks, there are also CeO_2 peaks but there are no mixed phases, containing the elements Ni, O and Ce. The presence of the peaks of the nickel substrate indicates that the depth of penetration of x-ray radiation was greater than the thickness of the nickel oxide and of the cerium coating. The experimental pole figures were used for determining the orientation distribution function and consecutive modelling of the grain boundary distribution.

The modelling procedure used for the spectrum of the grain boundaries for the given case is completely identical with that described in the previous section. To shorten the explanation, we present only the distribution of the grain boundaries with respect to the inverse density of the coincident-site lattice, and also the results of investigations of the kinetics for the case of growth of a nickel oxide film on the single crystal substrates of the types (100) and (111), after preliminary modification with cerium. Table 7.6 shows the fractions of the boundaries of each type. The data described in the previous paragraph are also given for comparison. It should be mentioned that the type of spectrum given for the (100) substrate is the uncorrelated type, whereas the type *C* correlation was selected for the (111) substrates. The distribution with respect to the misorientation angles and axes for the nickel oxide film, formed on the single crystal substrate modified with Ce, is shown in Fig. 7.22a for the (100) orientation and in Fig. 7.22b for the (111) single crystal. It can be seen that the general relationships governing the formation of the spectrum of the grain boundaries depend strongly on the surface condition of the single crystal substrate, and the effect of additions greatly differs in relation to the orientation of the substrate. For example, for the (100) surface, modified with cerium, the fraction of the low-angle boundary

Table 7.6. Distribution of the grain boundaries with respect to in the (100)–NiO and (111)–NiO films

Substrate	Σ3	LAB	HAB (arbitrary)
(100)	3.1	40.4	56.6
(100)+CeO_2	0.3	69.1	30.6
(111)	27.6	6.4	66.0
(111) +CeO_2	22.2	7.6	70.2

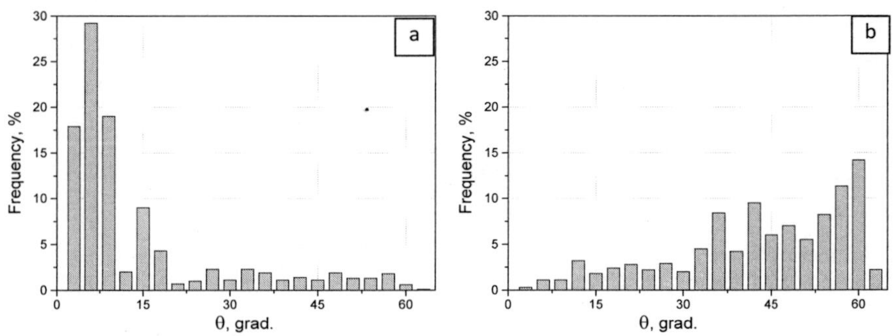

Fig. 7.22. Distribution of misorientation angles and axes in the nickel oxide films on the cerium-modified substrates: a) (100); b) (111).

is increased in comparison with the specimens with no cerium coating by almost a factor of 1.7 and equals 69.1%. The fraction of the high-angle grain boundaries decreased to 30.67. According to the model, this change should be reflected in the kinetics of growth of the oxide film. For the single crystal with the (111) orientation, coating with cerium resulted in a relatively increase of the fraction of the small angle boundaries (by approximately 20%) and a decrease of the fraction of twins. The number of the high-angle boundaries did not change. Table 7.7 shows the data on the experimental measurement of the oxidation kinetics for the investigated materials. The tendencies, detected in the variation of the misorientation distribution of the grains, reflect the tendency for the variation of the growth rate of the oxide film on the nickel single crystal substrates. Thus, the deposition of cerium on the (100) surface of nickel single crystals reduces the oxidation rate by an order of magnitude which is in agreement with the decrease of the fraction of the high-angle boundaries, acting as the so-called conductors of fast diffusion. As regards the oxide film grown on the (111) substrate modified and not modified with cerium, it may be concluded that the changes of the surface had no significant effect on the spectrum of the grain boundaries. Similarly, there were no changes in the oxidation rate of the surface (111).

Table 7.7. The oxidation rate in (100)–NiO and (111)–NiO films

Substrate	Modifying additions	Oxidation rate, $kg^2 m^{-4} s^{-2}$
(100)	–	$2.2 \cdot 10^{-10}$
(100)+CeO_2	+	$2.9 \cdot 10^{-11}$
(111)	–	$6.4 \cdot 10^{-14}$
(111) +CeO_2	+	$5.3 \cdot 10^{-14}$

To conclude this chapter, it should be mentioned that the investigations of the oxidation kinetics of zirconium and nickel single crystals confirm convincingly the conclusions on the efficiency of the proposed model, made in chapter 3. In addition, the resultant relationship between the kinetic parameters of growth of the oxide film indicates that the misorientation distribution of the grain boundaries, calculated on the basis of the texture data, describes adequately the characteristics of the grain boundary ensemble. As shown by analysis, information on some of the additional characteristics can be utilised for detailed evaluation of the grain boundary ensemble as a whole.

The experimental results also show that whole regions can form in the nanocrystalline zirconium oxide film – clusters of grains separated only by special grain boundaries. This type of organisation of the structure results in a decrease of the diffusion rate through these sections of the oxide film and improves the corrosion properties. In addition, the characteristics of these boundaries, i.e., the inverse density of the coincident-site lattice Σ, is not governed by the generally accepted law of the minimum value of Σ. It may be concluded that to determine the low-energy special boundaries, it is necessary to ensure the confidence of the sites of the lattices of the adjacent grains at the interface and not the minimum volume of the coincident-site lattice. Thus, the criterion of the 'speciality' of the grain boundaries becomes more general and the case with the cubic polycrystals is a partial case of this more general criterion.

The results of crystallographic analysis of the types of grain boundaries, carried out using the data obtained in high-resolution electron microscopy, have been used to define three regions in the ZrO_2 film in which the type of grain boundary ensemble greatly differs: they are the regions of the so-called growth and axial textures, and also an intermediate region. The grain clusters in the region of the 'growth texture' are separated only by the twin boundaries. In the regions with the 'axial texture' there is no correlation in the orientation of the adjacent grains and the texture determines the type of misorientation distribution of the grain boundaries. The resultant spectra of the grain boundaries differ for different regions and define the kinetics of oxidation and/or diffusion through the oxide film. The diffusion layers, calculated for different types of grain boundary ensemble, differ by an order of magnitude. Theoretical analysis of the diffusion problem for the zirconium oxide film subjected to the effect of stress at the oxide/metal interface (this stress may reach 1 GPa) has been used to determine the relationships governing the process, in particular, the two-stage scheme: non-stationary diffusion in the first stage and quasi-stationary diffusion after the formation of the porous layer. The results have been used to propose a possible explanation of the experimentally detected variation of the diffusion law from 'parabolic' to 'cubic'.

Identical conclusions can be made on the basis of the investigations carried out for the nickel oxide formed on the surface of the single crystal substrates with the (100) and (111) orientations. The formation of the ensemble of the grain boundaries differing in the fraction of the low-angle, special and high-angle grain boundaries, explains the difference in the oxidation kinetics of the nickel single crystal substrates which amounts to an order of magnitude. Modification of the nickel surface with the additions (Ce) resulted in the changes in the texture and, consequently, the misorientation distribution of the boundaries. In the case of the (100)+CeO_2 substrate the fraction of the low-angle boundaries increases and that of the high-angle ones decreases. The latter circumstance also explains the decrease, by an order of magnitude, of the growth rate of the oxide film on the (100) surface modified with Ce. In the case of the (111) orientation, modification did not lead to any significant changes in the spectrum of the grain boundaries nor in the growth kinetics of the oxide film.

References

1. Urbanic V. F., et al., Growth and characterization of oxide films on zirconium–niobium alloys, in: Zirconium in Nuclear Industry, 10th Int. Symp., Philadelphia, PA. 1994, 116–132.
2. Coleman C.E., et al., Journal of Nuclear Materials, 1998, vol. 257, No. 1, 35–43.
3. Gandli P., et al., Can. Metal. Q., 1995, vol. 34, 211–218.
4. Glavicic M.G., et al., Texture measurement of zirconium oxide thin films, in: Textures of Materials, Proc. ICOTOM-11, Beijing, 1996, 1137–1141.
5. Lin Y.P., Applications of analytical electron microscopy in the nuclear industry, in: Microscopy of Oxidation-3, Proc. of 3rd Int. Conf., London, The Inst. of Metals, 1996, 462–474.
6. Lin Y.P., et al., Texture and phases in oxide films on Zr–Nb alloys, in: Polycrystalline Thin Films: Structure, Properties and Applications, MRS, 1994, vol. 343, 487–493.

7. Ploc R.A., J. Nucl. Mater., 1982, vol. 110, 59–70.
8. Ploc R.A., J. Nucl. Mater., 1983, vol. 113, 75–80.
9. Ploc R.A., J. Nucl. Mater., 1983, vol. 115, 110–117.
10. Ploc R.A., J. Nucl. Mater., 1968, vol. 28, 48–60.
11. Gertsman V.Y., et al., Phil. Mag. A., 1999, vol. 79, 1567–1590.
12. Gertsman V.Y., Zhilyaev A.P., and Szpunar J., Scripta Mater., 1996, vol. 35, No.11, 1247–1251.
13. Bansal G. and Heuer A.H., Acta Met., 1972, vol., No.11, 1281–1289.
14. Bansal G. and Heuer A.H., Acta Metall., 1974, vol., No.4. 409–417.
15. Iltis X., et al., Microstructure evolution and iron redistribution in Zircaloy oxide layers: comparative effects of neutron irradiation flux and irradiation damage, in: Zirconium in the Nuclear Industry, 11th Int. Symp., 1996, ASTM STP, vol. 1295, 242–264.
16. Gertsman V.Y., et al., Modell. Simul. Mater. Sci. Eng., 1997, vol. 5, 35–52.
17. King A.H. and Singh A., J. Phys. Chem. Solids., 1994, vol. 55, No.10, 1023–1033.
18. Martinez-Fernandez J., et al., J. Am. Ceram. Soc., 1994, vol. 77(1), 57–64.
19. Van Tendello G., Acta Metall., 1983, vol. 31, No.10, 1619–1625.
20. Glavicic M.G., et al., J. Nucl. Mater., 1997, vol. 245, 147–151.
21. Nitsche R., et al., Nanostructured Mater., 1996, vol. 7, 535–546.
22. Godlewski J., How the tetragonal zirconia is stabilized in the oxide scale that is formed on a zirconium alloy corroded at 400 °C in steam, in: Zirconium in the Nuclear Industry, 10th Int. Symp., 1994, ASTM STP, vol. 1245, 663–667.
23. Beie H.J., et al., in: Zirconium in the Nuclear Industry, 10th Int. Symp., 1994, ASTM STP, vol.1245, 615–620.
24. Garcia E.A., J. Nucl. Mater., 1995, vol. 224, 299–304.
25. Garcia E.A. and Kovacs J., J. Nucl. Mater., 1994, vol. 210, 78–83.
26. Piotrowski R., et al., J. Nucl. Mater, 1993, vol. 202, 252–265.
27. Zhilyaev A.P. and Szpunar J. A., J. Nucl. Mater., 1999, vol. 264, 327–332.
28. Zirconium: Physical-Chemical Properties of its Compounds and Alloys. Atomic Energy Review, Special issue No.6, Vienna, Int. Atomic Energy Agency, 1976.
29. Moss S. J., et al., Metal. Mater. Trans. A, 1996, vol. 27, No.4, 829–837.
30. Smeltzer W.W., Mater. Sci. Forum, 1988, vol. 29, 151–172.
31. Khoi N.N., et al., J. Electrochem. Soc., 1975, vol. 122, No. 11, 1495–1503.
32. Czerwinski F. and Szpunar J.A., The microstructural characterization of nanocrystalline CeO_2 ceramics produced by the sol-gel method , in: Polycrystalline Thin Films – Structure, Texture, Properties and Applications (Eds. M. Parker, K. Barmak, R. Sinclair, D.A. Smith and J. Floro), Pittsburgh, PA, MRS, 1994, vol. 343, 131–136.
33. Zhilyaev A.P., et al., On the calculation of grain boundary character distribution in textured nickel oxide layers, in: ICOTOM-12, Proc. of 12th Int. Conf. Text. Mater., Montreal, Canada, NRC Research Press, 1999, vol. 2, 1403–1408.
34. Atkinson A. and Taylor R.I., Phil. Mag. A, 1981, vol. 43, No. 4, 979–998.
35. Atkinson A., et al., Phil. Mag. A, 1982, vol. 45, No. 5, 823–833.
36. Li H., et al., Corros. Sci, 1997, vol. 39, 1211–1219.
37. Czerwiski F. and Szpunar J.A., Acta Mater., 1998, vol. 46, 1403–1417.
38. Czerwiski F. and Smeltzer W.W., J. Electrochem. Soc., 1993, vol. 140, No.9, 2606–2615.
39. Czerwiski F., et al., Oxid. Metals, 1995, vol. 43, 25–57.
40. Czerwinski F. and Szpunar J.A., Journal of Sol-Gel Science and Technology, 1997, vol. 9, 103–114.
41. Czerwinski F., et al., Corr. Sci., 1999, vol. 41, 1703–1713.

8

MICROSTRUCTURE AND GRAIN BOUNDARY ENSEMBLES IN ULTRAFINE-GRAINED MATERIALS

Prior to discussing the main subject of this chapter, it should be mentioned that the effect of the ensemble of the grain boundaries on the mechanical and physical properties is especially strong in ultrafine-grained and nanostructured materials. The small grain size determines the greater length of the grain boundaries. In addition, the grains themselves may contain atomic defects: vacancies and complexes of vacancies, dislocations and disclinations whose density and distribution differs qualitatively from that in the coarse-grained materials. If the dimensions of the crystals are comparable with the characteristic physical parameters of length (for example, de Broglie wavelength, the free path of the electrons or the critical size of single-domain regions), then the appropriate properties are influenced by the dimensional effects. In this chapter, special attention is given to some aspects of the formation of ultrafine-grained and nanomaterials using ultrafine-grained nickel, produced by severe plastic deformation, as an example.

8.1. Methods of producing ultrafine-grained and nanostructured materials by severe plastic deformation

The large number of the methods of producing ultrafine-grained and nanostructured materials can be subdivided into two large groups: preparation of nanocrystalline powders and bulk materials with ultrafine-grained structure and nanostructure. Isolated nanoparticles can be produced by gas-phase synthesis [1], plasma chemical synthesis [2, 3], aerosol [4] and chemical [5] synthesis, thermal dissociation and reduction, mechanosynthesis [6] and electric explosion [7]. In this work, investigations were carried out on thick specimens of ultrafine-grained and nanostructured nickel and, consequently, we shall discuss in greater detail the methods of producing thick nanomaterials. The methods of producing ultrafine-grained materials, free from contamination and porosity, are based on the application of high plastic strains for the formation of a strongly fragmented and disoriented microstructure. To obtain high strength, it is necessary to use various methods: torsion under high quasi-hydrostatic pressure, equal-channel angular pressing, all-sided forging, and others. These methods are based on multiple plastic shear deformation of the processed material resulting in the true logarithmic strain ($\varepsilon_{true} \sim 4$–7). In addition to reducing the mean grain size, it is also possible to produce almost pore-free specimens. The torsional methods and equal-channel angular pressing will be investigated in greater detail.

Torsional deformation under high-pressure

Equipment in which torsional deformation is carried out under high-pressure was investigated for the first time in [8, 9]. The design of this equipment is a continuation of the well-known concept of the Bridgeman anvil [10]. In initial studies, these systems were used for investigating phase transformations under severe deformation conditions [4] and also for investigating the evolution of the structure and changes in recrystallisation temperature after high deformation [11]. The important moment was the application of this method for producing nanostructures with high-angle grain boundaries [5, 12] so that the method was regarded as a new method of producing nanostructured materials. In torsional deformation under high-pressure (Fig. 8.1), the specimens are disk-shaped. They are placed between anvils and are compressed by the applied pressure P of several giga pascals. The lower anvil rotates and the surface friction forces prevent shear deformation of the specimen. The geometrical shape of the specimens is such that the main volume of the material is deformed in the conditions of quasi-hydrostatic pressure under the effect of applied stress and the pressure from the side of the external layers of the specimen. Consequently, regardless of the high strain, the specimen does not fracture. Various methods are used to calculate the strain in torsion under high pressure. Some of them will now be discussed, together with the applicability conditions. Figure 8.1b shows the calculation of the infinitely small shear in rotation through angle $d\theta$. It is clear that $dl = rd\theta$ from which

$$d\gamma = \frac{dl}{h} = \frac{rd\theta}{h} \tag{8.1}$$

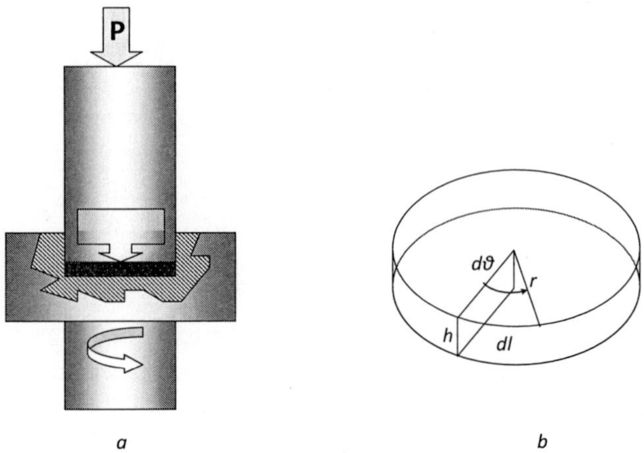

Fig. 8.1. Torsion under quasi-hydrostatic pressure: a) general scheme of equipment; b) geometrical parameters.

where h is the thickness of the disk. If it is assumed that the thickness of the disk does not depend on the angle of rotation θ, equality (8.1) can be formally integrated. However, since the tensile strain is linked with the shear strain by the relationship $\varepsilon = \gamma / \sqrt{3}$, the true logarithmic strain is expressed by the following equation:

$$\varepsilon_{true} = \ln(1+\varepsilon) = \ln\left(1+\frac{\theta \cdot r}{\sqrt{3} \cdot h}\right). \tag{8.2}$$

Since $\theta = 2\pi N$, where N is the number of revolutions, equation (8.2) can be transformed to

$$\varepsilon_{true} = \ln\left(1+\frac{2\pi N \cdot r}{\sqrt{3} \cdot h}\right). \tag{8.3}$$

In practice, this equation is further simplified [173]

$$\varepsilon_{true} = \ln\left(\frac{2\pi N \cdot r}{h}\right). \tag{8.4}$$

The equation (8.4) is identical with the relationship used for calculating the true strain of the specimens subjected to tensile loading. According to (8.3), the strain in the centre of the disk should be equal to 0. At the same time, as shown by the results of a large number of investigations, the application of this method of severe plastic deformatyion in the central part of the specimens results in the formation of the ultrafine-grained structure after several revolutions. This is also confirmed by the results of measurements of microhardness in both the centre and at the periphery of the deformed specimen. To take into account shrinkage and pressure, it was proposed to use the following equation [13]

$$\varepsilon_{true} = \ln\left[1+\left(\frac{\theta\cdot r}{h}\right)^2\right]^{1/2} + \ln\frac{h_0}{h}\ . \qquad (8.5)$$

The equations (8.2)–(8.5) will be used to estimate the strain cumulated after five complete revolutions for a disk with a radius r = 10 mm, h_0 = 0.3 mm and h = 0.1 mm, and we obtain ε_{true} ~7.5 from equation (8.3), ~8.0 from (8.4) and ~9.1 from (8.5). Thus, the difference is less than 20%, although in practice equation (8.5) is more accurate and does not give zero strain in the centre of the disk. Using this equation, we calculated the three-dimensional distribution of cumulated strain for N = 1, 5 and 7 revolutions (Fig. 8.2). Equation (8.5) is a partial case of a more general situation in which the equivalent strain at high shear strength should be calculated from the equation [14]

$$\bar{\varepsilon} = 2/\sqrt{3}\ln\left(\sqrt{1+\frac{\gamma^2}{4}}+\frac{\gamma}{2}\right)\ .$$

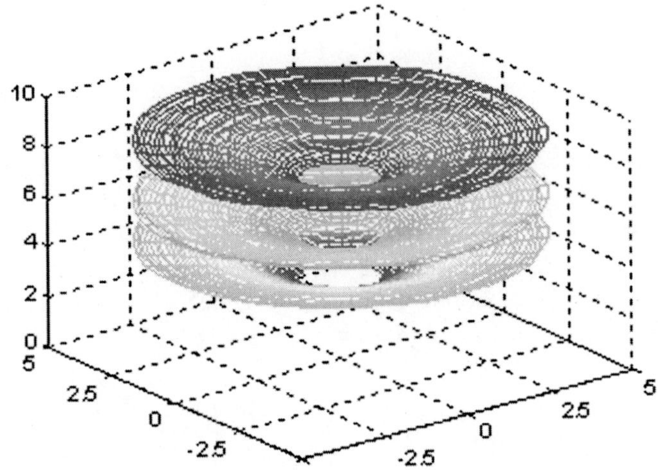

Fig. 8.2. Three-dimensional distribution of cumulated strain, calculated using equation (8.5) for N = 1, 5 and 7 (from bottom to top).

The central part of the distribution is not shown in Fig. 8.2. It can be seen that the equation (8.5) describes the gradual increase of the cumulated strain with increasing number of revolutions. In practice, the cumulated strain should be calculated using the finite element method. The viscoplastic model must take into account the variation of the friction coefficient between the anvils and the surface of the specimen during hardening of material in the course of deformation. A simplified method of calculating cumulated strain was described in [15]. However, the calculated data do not reflect the experimentally recorded facts. Two comments [16, 17] apply to the equations (8.2)–(8.5). Firstly, the calculations carried out using these equations result in the conclusions according to which the value of strain should change linearly from zero in the centre of the specimen to the maximum value at the end of the diameter of the specimen. In fact, as already mentioned previously, this is often not the case. Secondly, during deformation the initial thickness of the specimen decreases under the effect of high compressive stresses and, therefore, the application of quantity h, which is usually the case in the procedure used, reduces the calculated values of strain in comparison with true values. Both these comments indicate that the strain, calculated using these equations, is only approximately equal to the actual strain. In addition, the formation of the structure in severe plastic deformation takes place under the effect of not only external but also internal stresses. At the same time, there is no rigid link between the level of the internal stresses and the true strain. Thus, when investigating the processes of evolution of the microstructure during torsional severe plastic deformation it is often more accurate to consider the number of revolutions and not the strain calculated by analytical expressions.

Recent investigations showed that torsional severe plastic deformation can be used efficiently not only for defining the structure but also as a method of powder consolidation [18]. It was found that the high pressures, used in torsional deformation at room temperature, are capable of ensuring a sufficiently high density (close to 100%) of the produced nanostructured specimens. The specimens can be produced by consolidation by torsional severe plastic deformation using not only conventional powders but also powders processed in a ball mill. The specimens, produced by compacting in severe plastic deformation, have the form of conventional discs with a diameter of 10–20 mm and a thickness of 0.2–0.5 mm.

Deformation by equal-channel angular pressing (ECAP)

The ECAP method, based on the deformation of the specimens of thick materials by simple shear, was developed in the 70s by Segal et al. The absence of any change in the cross-section of the specimen created suitable conditions for repeated deformation [19, 20] (Fig. 8.3). At the beginning of the 90s, the method was developed further and applied for the first time to produce structures with the submicrocrystalline grain size [12]. In ECAP, the blank is pressed is several times in a special device through two channels with the same cross-section, usually intersecting under the angle of 90°. If necessary, deformation can be carried out at elevated temperatures. If the external angle $\Psi = 0^{\circ}$, and the internal angle Φ is arbitrary (Fig. 8.3b), the increase of the degree of deformation after each pass can be calculated from the following equation [20]:

$$\varepsilon_N = \frac{2N}{\sqrt{3}} \operatorname{ctg}(\Phi/2). \tag{8.6}$$

A more general relationship which can be used to calculate the degree of deformation of the specimen in ECAP in N passes, has the following form:

$$\varepsilon_N = \frac{N}{\sqrt{3}} \left[2 \operatorname{ctg}\left(\frac{\Phi}{2} + \frac{\Psi}{2} \right) + \Psi \operatorname{cosec}\left(\frac{\Phi}{2} + \frac{\Psi}{2} \right) \right]. \tag{8.7}$$

For the partial case of $\Psi = 0$ equation (8.7) is simplified to (8.6). The latter equation shows that at the angles used most frequently in

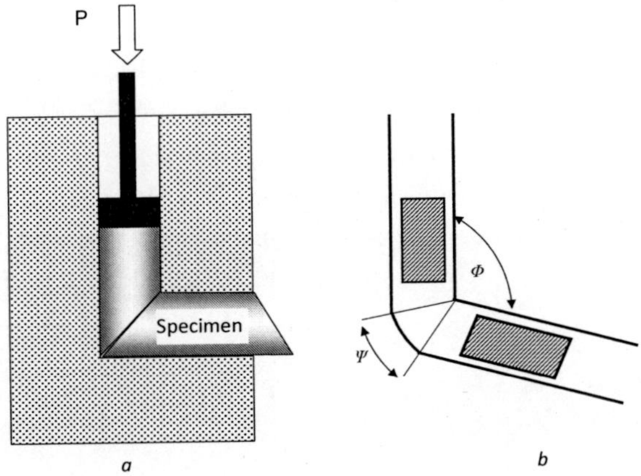

Fig. 8.3. Equal-channel angular pressing: a) schematic; b) geometrical parameters.

practice, $\Psi = 20$ and $\Phi = 90°$, each pass corresponds to the additional degree of deformation equal to approximately 1. Recently, another equation for the strain accumulated during ECAP was published in [22]

$$\varepsilon_N = \frac{N}{\sqrt{3}}\left[2 \cdot \mathrm{ctg}\left(\frac{\Phi}{2} + \frac{\Psi}{2}\right) + \Psi\right]. \tag{8.8}$$

The numerical calculations, carried out in [22], show that the equations (8.7) and (8.8) differ not more than 5% for all the values of angle Ψ.

The direction and the number of passes of the blanks in the channels are very important in ECAP. In [19, 20] and also in [23, 24] the authors studied different parts of the blanks (Fig. 8.4):

– the orientation of the blank remains unchanged in each pass (path *A*);

– after each pass the blank rotates around its longitudinal axis through the angle of 90° (path *B*);

– after each pass the blank rotates around its longitudinal axis through the angle of 180° (path *C*).

When all three paths are used, the yield and strength values of the processed material rapidly increase and reache saturation already after several passes [12]. It was shown in [25] the first three deformation cycles of Cu and Ni specimens in ECAP increase the deformation force. This is followed by the steady hardening stage and the force remains almost constant. Detailed investigations of the effect of the ECAP parameters on the microstructure and properties of aluminium alloys were described elsewhere [24, 26–30].

In addition to the discussed methods of severe plastic deformation, thick nanocrystalline materials are also produced by the electric deposition process [31, 32] which consists of the deposition of the atoms of the investigated material on a substrate from a solution. The electrochemical cell consists of a nickel anode, a cathode, a solution (Ni^{2+}, H^+, SO_4^{2-}) and a DC current source. Instead of DC, the pulsed

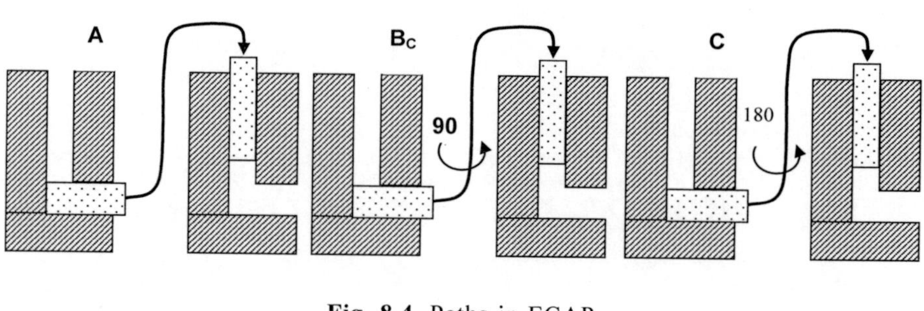

Fig. 8.4. Paths in ECAP.

regime is used to produce the nanocrystalline structure. Varying the chemical composition of the solution and the delay time of the pulses, it is possible to produce strips of nanocrystalline nickel with a thickness of up to several millimetres.

8.2. Effect of the parameters of quasi-hydrostatic pressure on the microstructure and grain boundary ensembles in nickel

A large number of investigations have been carried out to study the evolution of the structure during severe plastic deformation, in particular in torsional loading under high quasi-hydrostatic pressure. In early studies [33, 34] of loading with quasi-hydrostatic pressure in copper and nickel single crystals, special attention was given to the evolution of the microstructure from the monocrystal state to the stage of formation of the fine-grained state (with the mean grain size of approximately 100 nm). In subsequent studies of the structure of nickel and copper it was found that as the degree of deformation increases the dislocations become localised at the cell boundaries. The mean misorientation of the cells increases with increasing strain, forming ultrafine-grained fragments of the structure – a precursor of the grain structure. Deformation results in the activation of rotational modes of deformation simultaneously throughout the entire specimen, ensuring steady deformation. The decrease of the stacking fault energy results in a change of the mechanism of severe deformation in which the microstructure is refined by formation of shear bands which gradually affect the entire volume of the specimen. Electron microscope studies, carried out in [35], show that the process of formation of the microstructure in torsional severe plastic deformation in Armco iron and AISI 316L stainless steel is of a distinctive three-stage nature. According to the scheme proposed by the authors for the first stage (to complete rotation), the microstructure is characterised by a cellular structure with the mean size of fragments of approximately 400 nm and with the misorientation angle of the sub-boundaries of approximately 2–3°. With increasing strain the dislocation clusters start to occupy the entire volume of the initial gratings. The second stage (one–three revolutions) is characterised by the formation of a transitional structures with features of both the subgrain and grain structure. The mean size of the crack and the mean misorientation become greater. The third stage characterises the formation of a homogeneous ultrafine-grained structure with the mean grain size of approximately 100 nm. In this case, the grains are almost completely free from dislocations and are greatly elastically distorted. This distortion may be caused by the elastic long-range stresses generated by nonequilibrium grain boundaries.

According to the literature data, torsion under high pressure is a more efficient method for refining the structure than ECAP. For example, in an Al–3% Mg alloy quasi-hydrostatic pressure results in in the mean grain size of approximately 90 nm [36], whereas ECAP produces a microstructure with the grain size of 270 [37] on the whole, it may be assumed that the application of ECAP for many metals and alloys results in the formation of an ultrafine-grained structure with the grain size in the range 100–1000 nm, and the method of torsion under pressure – in the range 50–200 nm [38–41].

The experimental material was high purity nickel (99.99%). This material was selected on the basis of the following circumstances. In studies [42, 43] concerned with ECAP of nickel it was noted that nickel is an efficient modelling material because its stacking fault energy is lower than that of pure aluminium but higher than that of copper. Refining of the structure and size distribution of the grains are determined by the stacking fault energy. In particular, in aluminium recrystallisation starts at relatively low temperatures and it is quite difficult to produce the ultrafine-grained structure. This is possible in copper which, however, has a duplex structure: large grains are surrounded by very small grains (<100 nm). A sufficiently homogeneous fine-grained structure can be produced only in nickel. Prior to deformation the specimens were annealed in vacuum for 6 h at 700°C and the mean grain size of the specimens was greater than 100 μm and the microhardness was approximately 1.4 GPa. Subsequently, they were subjected to torsional loading under quasi-hydrostatic pressure with the variation of the torsion parameters. The data for the specimens are presented in Table 8.1. Taking into account the equations (8.2)–(8.5) it was proposed to monitor the dependence of the variation of the microstructure on the deformation parameters by measuring the errors of the entire surface of the specimen and also by electron microscopy and x-ray diffraction investigations of the centre and edges of the disk.

Table 8.1. Torsion under quasi-hydrostatic pressure

P, GPa	N, rev.				
	1/2	1	3	5	7
1					
3					
6					
9					

Microhardness measurements

After torsional deformation, the nickel specimens were disk-shaped with the diameter of approximately 10 mm, thickness approximately 0.1 mm. Assuming the axial symmetry in the microhardness distribution, we selected the measurement procedure shown in Fig. 8.5. The distances between the points were approximately 1.25 mm. Four microhardness values were measured at each point. Thus, the total number of measurements for the specimens was no less than 128. The three-dimensional pattern of the variation of microhardness variation as a function of stored energy (number of revolutions) and apply pressure for ultrafine-grained nickel is shown in Fig. 8.6 and 8.7. Figure 8.6 shows the microhardness for the applied stresses of 1, 3, 6 and 9 GPa and the total number of complete revolutions 5. Figure 8.7 shows the three-dimensional dependence of microhardness on cumulated strain, determined by the number of revolutions, at P = 6 GPa and N = 1/2, 1, 3, 7. (The specimen with N = 5 is shown in Fig. 8.6). The isolines of microhardness indicating the absolute value are shown on the XY plane.

The analysis results show that all the measured microhardness values are higher than initial microhardness. Evidently, microhardness is nonuniform along the radius of the specimens, with the smallest values recorded in the centre of the disk. As shown by Fig. 8.7, increasing load applied during torsional deformation increases microhardness (in particular in the centre). The asymmetry in the distribution of the microhardness is associated with the nonuniform development of deformation on the disk.

At a pressure of 1 GPa hardening takes place mainly at the edge of the disk and the distribution is on the whole similar to the identical strain distribution calculated from equation (4.5) and shown in Fig. 8.2. The increase of applied stress is accompanied by the formation of

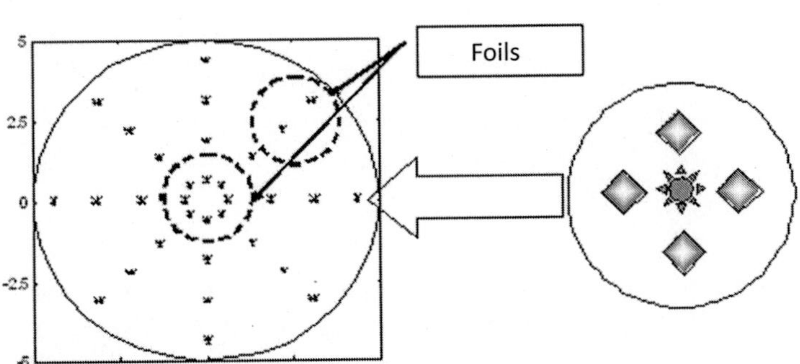

Fig. 8.5. Diagram of microhardness measurements on the disk and cutting out of blanks for transmission electron microscopy.

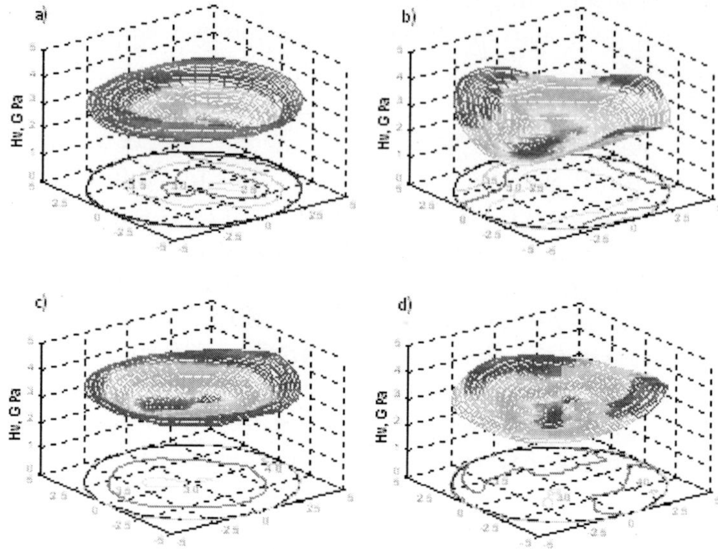

Fig. 8.6. Three-dimensional distribution of microhardness in ultrafine-grained nickel after torsional loading to $N = 5$ revolutions at applied pressure of: 1a) 1 GPa, b) 3 GPa, c) 6 GPa, d) 9 GPa.

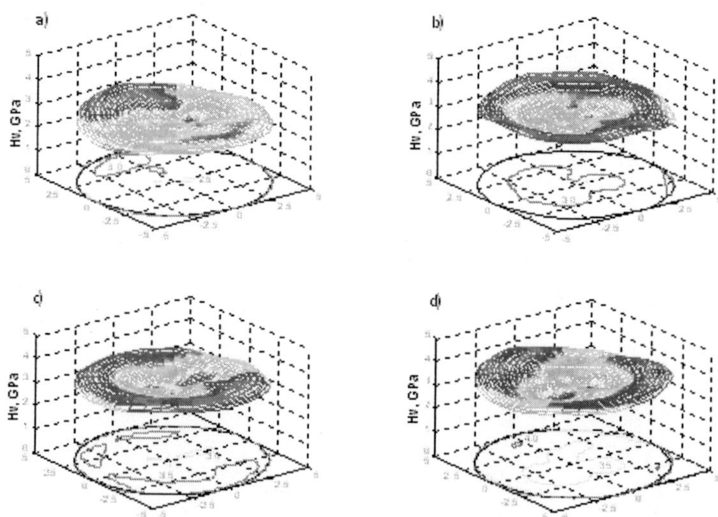

Fig. 8.7. Three-dimensional distribution of microhardness in ultrafine-grained nickel after torsional loading at applied pressure of $P = 6$ GPa and the number of revolutions: a) $N = 1/2$, b) $N = 1$, c) $N = 3$, d) $N = 7$.

the uniform distribution of microhardness for all $P > 3$ GPa. Identical conclusions can be drawn on the basis of the analysis of the dependence of microhardness distribution on cumulated strain (or the number of revolutions). At $N = 1/2$ there is a local increase of microhardness at the edge of the disk, although on the whole the level of microhardness is higher than the initial value as a result of upsetting of the specimen. With increasing number of revolutions the hardening process develops with propagation into the centre of the disk.

Discs with a diameter of 3 mm were taken from the central and peripheral parts of the disks of ultrafine-grained nickel for investigating the microstructure (Fig. 8.5). This was followed by a series of electron microscopic studies in a JEOL transmission electron microscope. Figure 8.8 shows a series of photographs of the microstructures produced in ultrafine-grained nickel after five revolutions at pressures of 1, 3 and 9 GPa. The upper row corresponds to the microstructure from the central sections, the lower – from the periphery. Analysis of the micrographs shows that the mean grain size at the periphery is smaller than in the central regions. The largest difference was recorded at a pressure of 1 GPa. The microstructures, produced at 3 GPa, have a small difference in the mean grain size but they are morphology: the grains in the central region are elongated in the tangential direction, and sub-boundaries are found inside the grains. At the highest pressure ($P = 9$ GPa) the microstructure is more homogeneous with the mean grain size of approximately 150 nm. On the whole, this is in agreement with the conclusions made in early

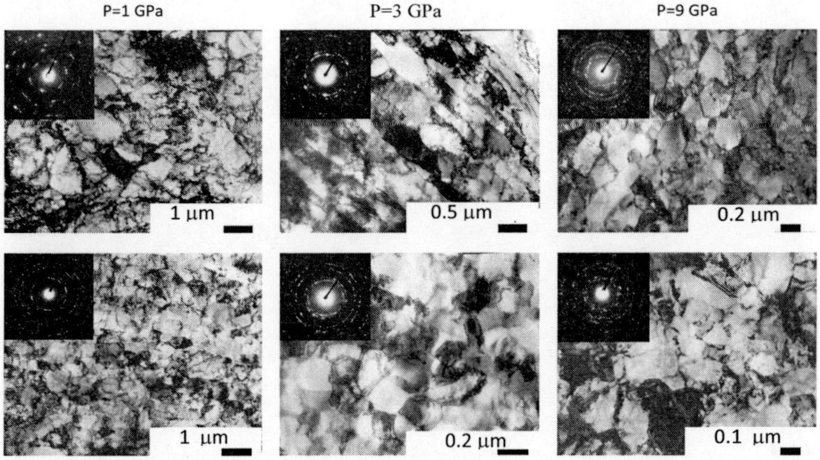

Fig. 8.8. Microstructure of ultrafine-grained nickel, produced by torsional loading at the pressure of 1 GPa (a), 3 GPa (b) and 9 GPa (c), in the centre of the disk (upper row) and at the edge of the disk (lower row) at $N = 5$.

studies [12, 34], and also in studies concerned with the investigations of ultrafine-grained iron and titanium [44]. It is important to stress two important aspects resulting from this investigation. Firstly, the method of torsional loading under quasi-hydrostatic pressure produces in fact an ultrafine-grained microstructure in metals and alloys: in pure nickel a structure with the mean grain size of approximately 170 nm is produced. Secondly, the direct measurements show that the structure is in fact uniform throughout the entire surface of the deformed discs. The process of torsional deformation takes place in a highly nonuniform fashion, and the deformed zone propagates in the form of waves from periphery to centre. This repeating process is clearly evident on the three-dimensional graphs of the dependence of microhardness on the deformation parameters (applied pressure and stored strain).

If we plot the dependence of the difference between the microhardness values averaged out at the edge and in the centre of the disk on N (Fig. 8.9), the graph shows clearly the wave-like process of deformation as a function of the number of revolutions. This parameter characterises the degree of nonuniformity. Its variations decreases with increasing number of revolutions.

8.3. Spectrum of misorientation of grain boundaries in ultrafine-grained nickel

The main problem in the production of ultrafine-grained materials is the formation of the grain structure with mostly high-angle grain boundaries. It is assumed that the set of the unusual physical mechanical properties, detected in experiments on ultrafine-grained materials [5, 6, 34, 44], is determined by the presence in the structure of mostly arbitrary high-angle grain boundaries in the metastable (or nonequilibrium) state.

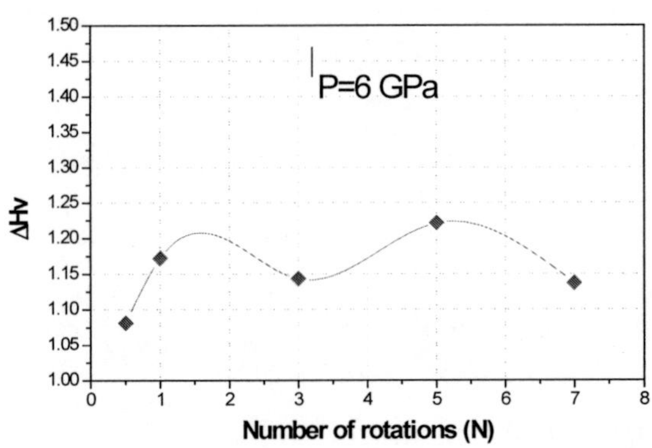

Fig. 8.9. Variation of the microhardness difference at the edge and in the centre of the disk as the function of the number of revolutions at $P = 6$ GPa.

Although the literature contains a relatively large number of data indicating that the structure of metals and alloys after severe plastic deformation is in fact characterised by the presence of an ensemble of grain boundaries with mostly high-angle boundaries; previously there were no direct experimental confirmation is of this fact. Indirect indication of the presence of high-angle grain boundaries are found in almost every study concerned with materials subjected to severe plastic deformation (see, for example, [12, 42]). The first study in which the authors mcasurcd directly the spectrum of the grain boundaries was [45]. Almost 90% of the high-angle boundaries were detected in this work. However, later the authors of [45] disawoved these results [46] and reported that the measurements were of the sampling type and, in all likelihood, the data do not correspond to the actual situation. In fact, the results obtained for aluminium alloys subjected to ECAP showed a high fraction of the low-angle grain boundaries [47] which casted doubts on the assumption on the formation of the grain (not subgrain) structure in the materials subjected to severe plastic deformation. At the same time, the authors of [48] found the tendency for the increase of the fraction of the high-angle boundaries with the increase of cumulated strain in ECAP. According to these data, the fraction of the high-angle boundaries reached 79%. The formation of the ensemble of the grain boundaries in aluminium alloys may depend strongly on the chemical composition of alloys and the deformation conditions. To explain the situation, it is necessary to carry out investigations on pure metal, for example, nickel. In addition, the problem of the type of ensemble formed in the specimens, subjected to severe plastic deformation, was unsolved until recently until investigations were carried out into the spectrum of the grain boundaries using an attachment for automatic measurement of misorientation in a Philips microscope with an auto-emission gun [17, 49].

8.4. Advanced methods of automatic measurement of the grain boundary parameters

Advanced scanning microscopes can be fitted with a special attachment for the automatic determination of the orientation of individual grains with a high degree of efficiency [50, 51]. The device of this type is referred to as orientation imaging microscopy. This method is based on indexing the channelling patterns or pseudo-Kikuchi lines formed during rocking of the electron beam falling and hitting the surface of the specimen. The geometry of these patterns of channelling and their decoding was described in detail in chapter 4. The channelling patterns are produced by focusing the electron beam followed by its precession around the axis of incidence under a small angle. As a result of beam instabilities the image is not sharp. To improve the quality of the pattern it is necessary to use another method. The specimen is tilted through approximately 70° to the horizontal plane. In this case, the electrons undergo initially inelastic scattering followed by elastic scattering, forming a diffraction pattern referred to as Kikuchi lines. The resultant diffraction pattern is transferred to a fluorescent screen and recorded using a CCD camera. The image is digitised and transferred into a computer for further processing. Figure 8.10 shows schematically the process of obtaining and indexing diffraction patterns using an OIM attachment. After recording all the essential information, the specimen is moved and the beam hits the next point. Of course, if two points belong to the same grain, the diffraction pattern does not change. The localisation of the method is thus determined by the minimum displacement of the specimen (realised by the electromechanical method) and the dimensions of the electron beam. The advanced piezoelectric motors can be used to remove the object with the accuracy of several tens of nanometres and, consequently, the localisation of the device in this stage of development is restricted only by the size of the electron beam. It is now possible to produce in microscopes diffraction patterns from regions approximately 20 nm in size so that it is possible to certify the ensemble of the grain boundaries in ultrafine-grained materials (with the grain size of 100–1000 nm). In addition to the diffraction pattern, another parameter is recorded at each point. This parameter describes the quality of the produced pattern

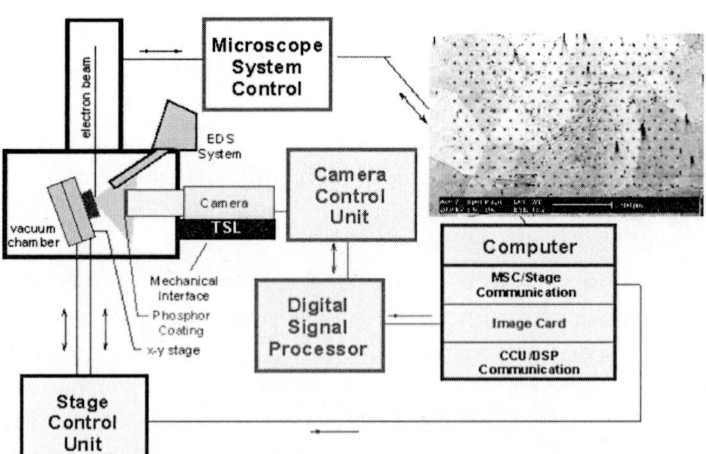

Fig. 8.10. Schematic of the method of automatic measurement of the grain boundary misorientation [50].

of electron channelling (image quality). It is determined by measuring the contrast of the diffraction pattern and reflects the imperfection of the crystal lattice. Of course, when the electron beam hits the region of the grain boundary, the IQ coefficient will be lower than in the centre of the grain. Allocating brightness values to each value of IQ on the grey scale, it is possible to form the image the microstructure as in optical microscopy. The investigations, described in this book, were carried out using an XL-30 FEG scanning electron microscope (Philips) with OIM attachment (TSL Co). This device produces both the pattern of the orientations for each grain and also the pseudo-microstructure, identical to that produced in the case of secondary electrons. The angular resolution is approximately 1°, and the average confidence index exceeds 0.4.

Careful preparation of the surface of specimens is essential to produce high-quality channelling patterns. The specimens of ultrafine-grained nickel were initially polished with diamond paste with gradually decreasing grain size and subsequently in the solution of Al_2O_3 particles with the size of 50 nm.

8.5. The misorientation distribution of the grain boundaries in ultrafine-grained nickel. Experiments and modelling

Figure 8.11 shows the microstructure of ultrafine-grained nickel, produced by torsional loading under pressure P = 6 GPa, with the total number of revolutions of N = 5. from the central section of the disk and from the peripheral region. All the grains were coloured in accordance with the scale of orientations shown in the same figure.

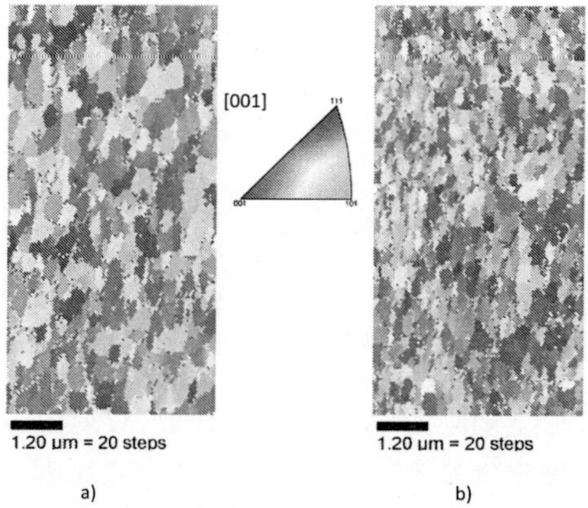

Fig. 8.11. Structure of ultrafine-grained nickel, produced by torsional loading at a pressure of P = 6 GPa (N = 5): a) central part, b) periphery.

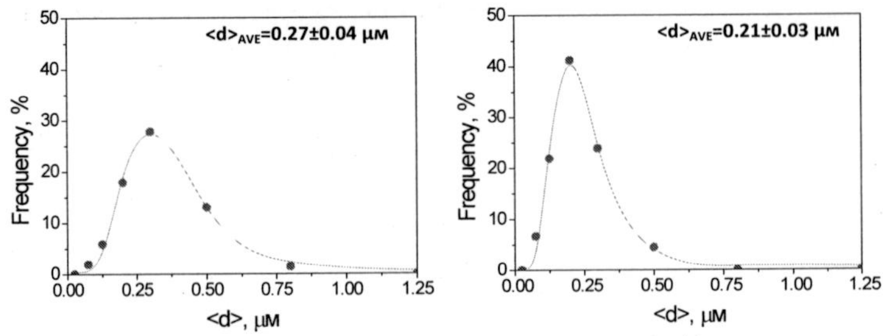

Fig. 8.12. The size distribution of the grains in the nickel specimens, subjected to quasi-hydrostatic pressure (P = 6 GPa, N = 5): a) central part ($\langle d \rangle_{mean}$ = 0.27±0.04 μm); b) periphery ($\langle d \rangle_{mean}$ = 0.21± 0.03 μm).

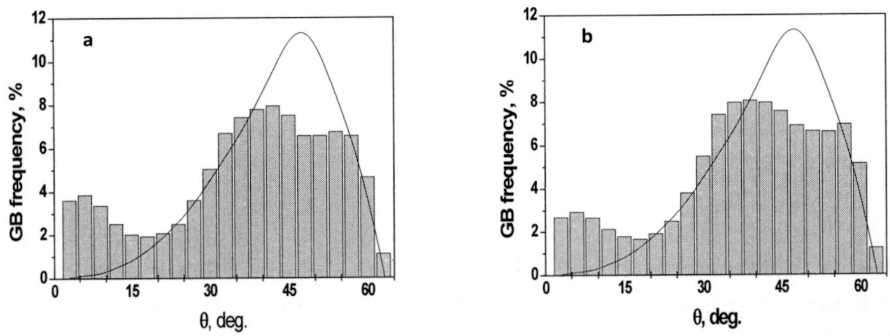

Fig. 8.13. Distribution of the misorientation angles in ultrafine-grained nickel after HPT (P = 6 GPa, N = 5), obtained from the central (a) and peripheral (b) regions of the disk.

Both photographs indicate a relatively uniform microstructure. Figure 8.12 shows the size distribution of the grains constructed for these states. The solid line indicates the log-normal distribution. The mean grain size in the central part is 0.27±0.04 μm, at the periphery 0.21±0.03 μm. The latter value is in good agreement with the results of electron microscopic studies (0.17 μm) [42], taking into account a small number of statistical data in the evaluation of the grain size for transmission electron microscopy data. The ranges of the error of the mean grain size for the centre and edge of the disk also overlap so that it may be concluded that the microstructure is relatively homogeneous and is formed by torsional loading under quasi-hydrostatic pressure at a load of $P \geq 5$ GPa, with the number of revolutions $N \geq 5$.

The total number of the grain boundary measured for producing the misorientation distribution of the grain boundaries was 1145 at the periphery and 1155 in the centre of the specimen. The appropriate distribution of the misorientation angles is shown in Fig. 8.13, and the distribution on the basis of the type of grain boundaries is in Table

Table 8.2. Distribution with respect to for ultrafine-grained nickel, produced by severe plastic deformation

Region	Σ1	Σ3	Σ5–30	HAB
Central	18.3	3.0	13.0	65.7
Peripheral	15.0	4.1	13.5	67.4
Random	2.1	1.6	7.0	89.3

8.2. The grain boundaries were divided into four categories: low angle (θ <15°, Σ1), twin (Σ3), other special (Σ5–30), and high-angle arbitrary grain boundaries. The investigated boundary related to the category then specified on the basis of the Brandon criterion [52]. The solid line in Fig. 8.13 shows the random distribution of the misorientation angles.

Comparison of the experimental data shows that the GBCD are identical from the statistical viewpoint. They are characterised by the bimodal distribution of the misorientation angles with the maximum in the low-angle range. The fraction of the low-angle grain boundaries equals 15.0% for the disk edge and 18.3% for the disk periphery of the disk. Statistically, these two values are very similar and, at the same time, they are considerably higher than the fraction of the low angle grain boundaries for the randomly disoriented polycrystal (2.1%). The fraction of the high-angle arbitrary grain boundaries (~65–67%) in both cases is lower than for the random polycrystal (almost 90%). The total number of the high-angle boundaries (including twin and other special boundaries) is more than 80%, which indicates the formation of the ultrafine-grained structure with the ensemble of the grain boundaries, consisting mainly of the high-angle boundaries. The experimental data obtained for the type of grain boundary ensemble, formed in nickel under severe plastic deformation, can be used for the verification of the proposed methods of calculating the misorientation distribution of the grain boundaries in ultrafine-grained and nanocrystalline nickel. The method of modelling GBCD was described in detail in previous chapters. Here, we present the results of investigation of the misorientation distribution of the grain boundaries in ultrafine-grained nickel, produced by ECAP, torsional loading under pressure and the combination of ECAP and loading under quasi-hydrostatic pressure. Figure 8.14 shows the brightfield images of the microstructure of all three specimens. The mean grain size, determined by transmission electron microscopy, is 0.35 μm in Ni processed by ECAP, 0.17 μm in Ni processed by quasi-hydrostatic pressure, and 0.1 μm in ECAP+HPT specimens. The most uniform structure and the smallest mean grain size were recorded in the specimens of ultrafine-grained nickel produced by

Fig. 8.14. Brightfield images of the microstructure of nickel: a) ECAP, b) HPT, c) ECAP + HPT.

the combination of the two main methods of severe plastic deformation. This is followed by HPT nickel whose microstructure is also uniform but the mean grain size is larger. The mean grain size of ECAP nickel, determined by transmission electron microscopy, is still twice as high. The azimuthal blurring of the reflection on the diffraction pattern indicates the presence of a high internal stress. Figure 8.15 shows the photographs of the microstructure of all three specimens produced using the OIM attachment to the Philips scanning electron microscope. The triangle in the figure shows the inverse pole figure as the basis for determining the crystallographic direction in the grain (crystal), parallel to the normal to the surface of the specimen. The red, green and blue colours (in the figure – various grey tones) correspond to the crystallographic directions $\langle 001 \rangle$, $\langle 110 \rangle$ and $\langle 111 \rangle$, respectively. The microstructure of ECAP nickel (Fig. 8.15a), produced using the OIM attachment, consists of elongated grains with the mean size of more than 0.5 μm. The large grains are characterised by a smooth variation of the contrast of the light reflecting the presence of sub-boundaries. The mean grain size, determined in the OIM, is 0.27 μm, which is lower than the value obtained in transmission electron microscopy (0.35 μm), and also considerably lower than the initial estimate of the size of grains with different colour. The error of the measurement of the mean grain size can be estimated as half the unit displacement during scanning (i.e. 30 nm, or 0.03 μm). Thus, taking into account the error, the mean size of the subgrain structure is 0.27 ± 0.03 μm. The difference in the estimate of the mean grain size is associated with the

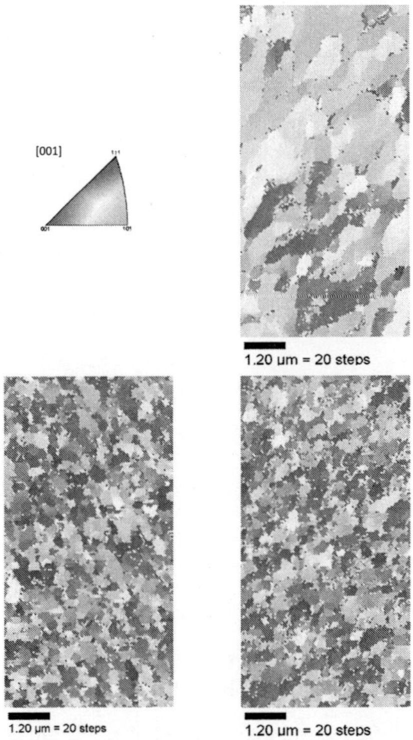

Fig. 8.15. Microstructure of nickel, produced by the OIM method: a) ECAP, b) HPT, c) ECAP + HPT.

presence of a developed structure in ECAP and with the existence of the limit of sensitivity in the determination of the misorientation in the OIM which in the given experiment was equal to 2°. The latter means that two adjacent points, having the misorientation smaller than 2° in relation to each other, are assumed to belong to the same grain. In visual examination, the microstructures of ultrafine-grained nickel, produced by torsional loading (Fig. 8.15b) and by the ECAP +HPT combination (Fig. 8.15c) are more uniform. Differences in the colour of the adjacent grains indicate the presence of the grain structure with a high fraction of the high-angle grain boundaries. The mean grain size for these two states of ultrafine-grained nickel was 0.19±0.03 μm and 0.17±0.03 μm, respectively. The estimate of the mean grain size in HPT nickel obtained in transmission electron microscopic investigations coincides with the resultant value, and for ECAP + HPT-nickel this value is slightly too high.

Figure 8.16 and Table 8.3 shows the distribution of the grains with respect to the misorientation angles and the inverse density of the coincident-site lattice Σ for the ECAP-, HPT- and ECAP + HPT

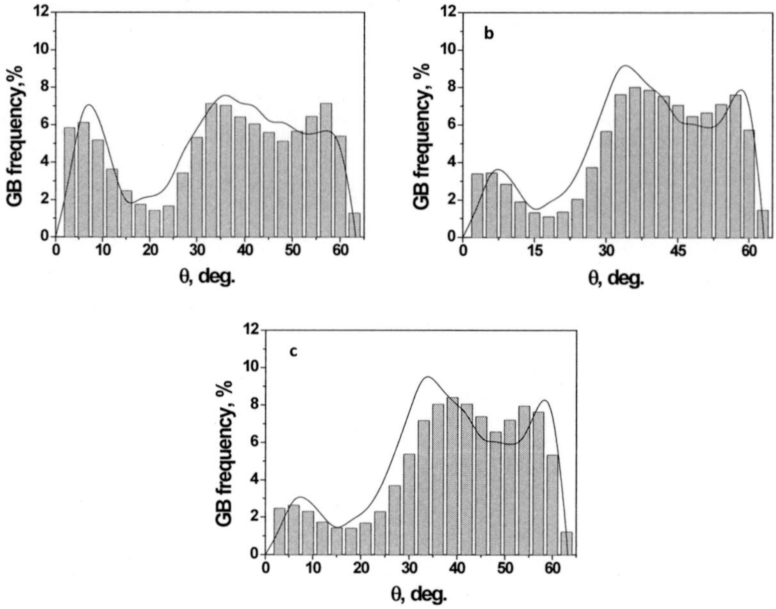

Fig. 8.16. Experimental (histogram) and modelling (solid line) distributions of the misorientation angles in ultrafine-grain nickel: a) ECAP, b) HPT, c) ECAP + HPT.

Table 8.3. Experimental and modelling distribution of the grain boundaries with respect to Σ in ultrafine-grained nickel

State	ECAP				HPT				ECAP+HPT			
	Σ1	Σ3	Σ5-30	HAB	Σ1	Σ3	Σ5-30	HAB	Σ1	Σ3	Σ5-30	HAB
Experiment	27.4	5.1	11.7	55.8	15.0	4.1	13.5	67.4	13.5	3.3	15.5	67.7
Model	25.6	2.6	13.0	58.8	14.0	3.5	10.0	72.5	12.0	3.1	10.1	74.8
Random	2.1	1.6	7.0	89.3	2.1	1.6	7.0	89.3	2.1	1.6	7.0	89.3

specimens of nickel in which the orientations of 339, 1145 and 989 grains were measured. Analysis of the data shows that the distribution with respect to the misorientation angle is bimodal for all three states of ultrafine-grained nickel. The first peak is situated in the range of low-angle misorientation, and the second one is broadened in the range 30–60°. The bimodal nature of the relationship is more evident for ECAP-nickel. This is in agreement with the results of investigation of GBCD in pure aluminium [53]. Comparison of the distribution of the misorientation angles and Σ shows that in comparison with HPT- and ECAP + HPT states the ECAP nickel is characterised by a considerably larger fraction of the low-angle boundaries. At the random distribution for the specimens of fine-grained nickel the fraction of the twins exceeds the number of the boundaries Σ3 but it is still not very high. This indicates that the processes of twining in the deformed specimens are not sufficiently extensive. Attention should be given to the unusually high fraction of other special boundaries (5–30) in all specimens of ultrafine-grained nickel. It is equal to 11.7%, 13.5% and 15.5% for ECAP-, HPT- and ECAP + HPT nickel, respectively. The question as to what is the cause of the formation of these boundaries: is it associated with special features of the deformation processes of the given material or the process of attestetion of the grain boundary ensemble?

Taking into account the results of texture analysis, described previously, we can use the modelling program for calculating the GBCD in ultrafine-grained materials. As an example, we describe the results of calculations of the GBCD for ultrafine-grained nickel produced ECAP and torsional loading under high pressure. We examine two factors which determine the formation of the misorientation distribution of the grain boundaries: texture and the type of spatial correlation. Analysing the distribution function of the grain orientation for these materials (Fig. 8.14–8.16) one can see the following special features. The texture of ECAP nickel has a distinctive maximum, whereas the texture maxima in HPT nickel were distributed more uniformly in the space of the Euler angles. Evidently, the conditions of severe plastic deformation do not lead to the formation of any specific type of correlation in the orientation of the adjacent grains. At the same time, it may be concluded that the deformation by 'pure shear' is capable of increasing the fraction of the low-angle boundaries. In fact, this was also detected in the experiments.

In general, the modelling results demonstrate the satisfactory agreement with the experiments. For example, Fig. 8.16 shows the calculated misorientation spectra of the grain boundaries in the ultrafine-grain nickel produced by ECAP pressing, torsional loading under high-pressure and the combination of these treatments. The histograms in the same figure show the experimental distribution. Table 8.3 shows the calculated data for Σ shown in the same format as previously. Comparison of the modelling data with the experiment shows that they coincide with the statistical error range. Consequently, it can be concluded that the spectra of the grain boundaries in ultrafine-grain materials, reduced by severe plastic deformation, can be modelled on the basis of the given texture.

In this chapter we examined in detail the effect of the parameters of severe plastic deformation of the formation of the microstructure and ensembles of grain boundaries in ultrafine-grained nickel, produced by ECAP pressing, torsional loading under quasi-hydrostatic pressure and the combination of these treatments. Nickel in which the stacking fault energy occupies the intermediate position between aluminium and copper is a suitable modelling material in this respect. In nickel it is possible to produce the ultrafine-grained microstructure with the grain size of approximately 0.17 μm after HPT and approximately 100 nm after torsional loading under pressure of ECAP specimens.

The experimental results were used to determine the main relationships governing the evolution of the structure with the variation of various parameters of the HPT such as the applied pressure and the stored energy (number of revolutions). It has shown that the process of torsional deformation is highly nonuniform and the deformation zone develops in the form of waves from periphery to centre. The method of measurement of microhardness of the entire surface of the deformed discs was used to investigate the evolution of microhardness (and, consequently, the evolution of the microstructure) with the increase of the load and the number of revolutions. The increase of applied pressure (to values higher than 5 GPa) leads on the whole to the formation of a homogeneous structure. The increase of the number of revolutions at P = 6 GPa results in the rapid saturation of microhardness values. The homogeneity of the microstructure from the viewpoint of formation of the ensemble of high-angle boundaries has been confirmed by direct measurements of the misorientation spectrum of the grain boundaries. For the nickel specimens, produced at P = 6 GPa and N = 5, the ensemble of the grain boundaries in the centre of the specimen and at the periphery of the specimen is statistically identical. The data obtained in the texture of nickel, formed in ECAP-, HPT- and ECAP + HPT-specimens, indicate the principal difference in the type and intensity of texture maxima. HPT nickel is characterised by the formation of the axial texture of the (100) type with the maximum three times higher than the maximum in the textureless state. The process of ECAP pressing results in the formation of a texture maximum with the coordinates (90°, 45°, 15°) on the orientation distribution function. Torsional loading of the ECAP specimens eliminates all the features of the ECAP texture. The special features of the texture formation lead to the formation of special features in the ensemble of the grain boundaries in ultrafine-grained nickel. For example, ECAP nickel is characterised by a higher fraction of the low-angle grain boundaries. This is in good agreement with the presence of a distinctive texture maximum. The highest fraction of the high-angle grain boundaries was recorded in the nickel specimens after combined deformation

(ECAP + HPT). Modelling shows the feasibility of using the method of calculating the spectrum of the grain boundaries in the ultrafine-grained materials on the basis of the texture data described in the chapters 2 and 3. Within the error range, modelling GBCD are in agreement with the experimentally measured value. Another important conclusion is the one which shows that severe plastic deformation of the materials is not accompanied by the formation of special features in the correlation in the orientation of the adjacent grains. Naturally, this conclusion can be applied to the materials subjected to severe plastic deformation at elevated temperatures when recovery and dynamic recrystallisation processes may take place.

References

1. Gleiter H., Progr. Mater. Sci., 1989, vol. 3, 223–315.
2. Morokhov I.D., et al., Physical phenomena in ultrafine media, Nauka, Moscow, 2000.
3. Gusev A.I. and Rempel' A.A., Nanocrystalline materials, Cambridge International Science Publishing, Cambridge, 2003.
4. Flagan R.C., Proc. of the NATO ASI on Nanostructured Materials: Science & Technology, Doldrech, Kluwer Acad. Publ., 1998, vol. 50, 15–30.
5. Chow G.M., II Proc. of the NATO ASI on Nanostructured Materials: Science & Technology, Doldrech, Kluwer Acad. Publ, 1998, vol. 50, 31–45.
6. Koch C.C., Materials Science and Technology, 1991, vol. 15: Processing of Metals and Alloys (R.W. Cahn, et al., eds.), New York, Wiley.
7. Burtsev V.A., al., Electrical explosion of conductors and its applications in electrophysical equipment, Energoatomizdat, Moscow, 1990.
8. Kuznetsov R.I., et al., Plastic deformation of solids under pressure, Institute of Physical Metallurgy, preprint 4/95, Sverdlovsk, 1982.
9. Zhorin V.A., et al., DAN SSSR, 1984, vol. 278, No. 1, 144–147.
10. Bridgman P.V., Investigation of high plastic strains and fracture, IL, Moscow, 1955.
11. Smirnova N.A., et al., FMM, 1986, vol. 62, No. 3, 566–570.
12. Valiev R.Z., et al., Progress Mater. Sci., 2000, vol. 45, No. 2, 103–189.
13. Efros B.M., et al., Ultrafine-grained materials (Y.T. Zhu, et al., eds.). 2002. Warrendale, PA: The Minerals, Metals and Materials Society, 193–197.
14. Polakowski N.H. and Ripling E.J. Strength and structure of engineering materials, Prentice-Hall, Englewood Cliffs, N.J.
15. Kim H.S., J. Mater. Process. Techn., 2001, vol. 113, 617–621.
16. Valiev R.Z., et al., Acta Mater, 1996, vol. 44, 4705–4712.
17. Zhilyaev A.P., et al., Acta Materialia, 2003, vol. 51, No. 3, 753–765.
18. Mishra R.S., et al., Materials Sci. Forum, 1996, vol. 225–227, 605–610.
19. Segal V.M., et al., Izv. AN SSSR, Metally, 1981, No. 1, 115–120.
20. Segal V.M., et al., Processes of plastic structure formation of metals, Navuka i Tekhnika, Minsk, 1994.
21. Iwahashi Y., et al., Acta Mater., 1997, vol. 45, 4733–4741.
22. Aida T., et al., Scripta Mater, 2001, vol. 44, 575–579.
23. Nakashima K., et al., Mat. Sci. & Eng., A, 2000, vol. 281, 82–87.
24. Langdon T.G., et al., JOM, 2000, vol. 52(4), 30–33.
25. Akhmadeev N.A., et al., Izv. VUZ, Metally, 1992, No. 5, 96–102.

26. Nakashima K., Horita Z., Nemoto M., Langdon T.G., Mat. Sci. & Eng. A. 2000. vol. 281, 82–87.
27. Horita Z., et al., Acta Mater., 2000, vol. 48, 3633–3640.
28. Akamutsu H., et al., Scripta Mater., 2001, vol. 44, 759–764.
29. Komura Sh., et al., Mat. Sci. & Eng. A., 2001, vol. 297, 111–118.
30. Horita Z., et al., Metal. Mater. Trans. A, 2000, vol. 31, 691–701.
31. Erb U., et al., Processing and properties of nanocrystalline materials (C. Suryanarayana, J. Singh, F. H. Froes, eds.) Warrendale, PA, TMS, 1996, 93–110.
32. Natter H., et al., J. Mater. Res., 1998, vol. 13, vol. 5, 1186–1197.
33. Horita Z., et al., J. Mater. Res., 1996, vol. 13, 446–450.
34. Valiev R.Z., et al., FMM, 1992, No. 4, 70–86.
35. Dobatkin S., Ultrafine Grained Materials II (Y. T. Zhu, et al., eds.), 2002, Warrendale, PA, The Minerals, Metals and Materials Society, 183–192.
36. Iwahashi Y., et al., Metall. Mater. Trans., A, 1998, vol. 29, 2245–2252.
37. Iwahashi Y., et al., Metall. Mater. Trans., A, 1998, vol. 29, 2503–2510.
38. Korznikov A.V., et al., Acta Mater., 1999, vol. 47 (11), 3301–3311.
39. Islamgaliev R.K., et al., Mater. Sci. Eng., A, 2001, vol. 319–321, 872–876.
40. Oh-ishi K., et al., J. Mater. Res., 1999, vol. 14, 4200–4207.
41. Dudova N., et al., Ultrafine Grained Materials II (Y.T. Zhu, et al., eds.) 2002. Warrendale, PA, The Minerals, Metals and Materials Society, 75–80.
42. Zhilyaev A.P., et al., Scripta Mater., 2001, vol. 44, 2753–2758.
43. Neishi K., et al., Mater. Sci. Eng. A, 2002, vol. 325, 54–58.
44. Valiev R.Z., et al., J. Mater. Res., 2002, vol. 17, 5–8.
45. Mishin O.V., et al., Scripta Mater., 1996, vol. 35(7), 873–878.
46. Mishin O.V., et al., Science of metastable and nanocrystalline alloys: structure, properties, modelling, Proc. of the 22nd Risø Intern. Symp. (A.R. Dinesen et al., eds.), Roskilde, Denmark, Risø National Laboratory, 2001, 335–340.
47. Gholina A., et al., Acta Mater., 2000, vol. 48, 1115–1130.
48. Huang J.C., et al., Scripta Mater., 2000, vol. 43, 213×220.
49. Kirn B.-K., et al., Materials Science Forum, 408–412, 943–948.
50. Dingley D.J. and Field D.P., Mat. Sci. and Technology, 1996, vol. 12, 1–9.
51. Field D.P., Ultramicroscopy, 1997, vol. 67, 1–9.
52. Brandon D.G., Acta Metall., 1966, vol. 14, 1479.
53. Terhune S.D., et al., Metall. Mater. Trans., A, 2002, vol. 33, 2173–2184.

9

GRAIN BOUNDARY PROCESSES IN ULTRAFINE-GRAINED NICKEL AND NANONICKEL

The problem of thermal stability of ultrafine-grained and nanostructured materials is very important because of the high degree of development of the ensemble of the grain boundaries. Grain growth in such materials starts already at temperatures $0.4T_m$ or even lower [1]. It is assumed that this is associated primarily with the reduced activation energy of the grain boundary processes controlled by grain boundary diffusion [2, 3]. The estimates of the activation energy of grain growth in ultrafine-grained copper, published in [4], give values almost 50 lower than the activation energy of grain boundary dislocations. This should accelerate the kinetic processes by several orders of magnitude. In [5] it is reported that the rate of grain boundary diffusion of copper in ultrafine-grained nickel increases by five orders of magnitude. However, the results of these experiments, carried out in the same group, have not as yet been confirmed in other investigations. In addition, in the studies [6, 7] of diffusion in nanocrystalline materials produced by gas phase synthesis there was no acceleration of the diffusion processes. It is therefore interesting to study the evolution of the microstructure and the ensemble of the grain boundaries during annealing. This chapter describes the results of investigations of the kinetics of grain growth in isothermal annealing of ultrafine-grained nickel [8] produced by equal-channel angular pressing (ECAP) and also the results of differential scanning calorimetry of ultrafine-grained nickel after ECAP, HPT and combination of these treatments [9]. The evolution of the microstructure, texture and the ensemble of the grain boundaries were investigated in HPT nickel [10, 11]. In addition, the results are presented of the investigation of the low-temperature superplasticity in nanocrystalline nickel produced by electrodeposition [12–14].

9.1. Grain growth kinetics in ECAP specimens

To investigate the relaxation processes leading to the gradual transition of the ultrafine-grained structure to the conventional coarse-grained condition, it is logical to investigate the simple single-phase materials. Firstly, after explaining the types of relaxation processes and investigating the kinetics of these processes in pure deformed metals, we develop a structural model to predict and explain the unusual physical–mechanical properties of more complicated alloys and composites. Secondly, the investigations of the evolution of the structures in heating helps to understand their complicated nature and examine the physical principle of the thermal instability of the ultrafine-grained materials. This is important for developing scientific fundamentals of reducing this instability. For example, the thermal instability of ultrafine-grained pure metals, produced by severe torsional plastic deformation under high pressure, was investigated in copper [15, 16] and nickel [15, 17]. In addition, experiments were carried out with nanocrystalline copper produced by electrodeposition [18]. These studies were concerned with the investigation of the recovery stage and determination of the temperature ranges corresponding to the recrystallisation processes. At the same time, the evolution of the microstructure and changes in the microhardness under the thermal effect on copper, deformed by ECAP, have not been explained. In particular, in recently published dissertations [4, 19, 20] it is claimed that the relaxation processes in ultrafine-grained materials have three stages and grain growth is characterised by a reduced activation energy. In this case, the activation energy is almost halved in comparison with the activation energy of grain boundary self-diffusion. This reduction should accelerate the processes associated with grain boundary dislocations (for example, at room temperature (300 K) by approximately ten orders of magnitude) which undoubtably can be recorded in the experiments. Although the recent studies concerned with the investigation of grain boundary dislocations in copper and ultrafine-grained nickel contain data on the acceleration of the diffusion processes by 5–6 orders of magnitude [5], no additional confirmation was presented in studies by other investigator groups. Thus, the relaxation processes taking place in ultrafine-grained materials during thermal annealing have their special features which have been studied insufficiently and require additional investigations. This is associated with the high level of the strain stored in the material. Heating of the deformed material is accompanied by complicated structural changes associated with recovery and recrystallisation processes. The nature of recovery is determined primarily by the redistribution and annihilation of the dislocations at the grain boundaries and in the body of the grains leading to a decrease of internal elastic stresses. Recrystallisation consists of the nucleation and growth of new, perfect grains as a result of the deformed matrix in which recovery takes place. The investigations of the evolution of the structure during annealing helps to understand better the nature of these phenomena, their relationship with the lattice defects and the nonequilibrium state of the grain boundaries, the relationship of the formation of the crystallographic texture and other special features of the behaviour of ultrafine-grained structures during annealing. The effect of temperature and annealing time on the microstructure of ECAP nickel will be investigated.

The investigated material was high-purity nickel (99.99%) in which the ultrafine-grained state was produced by ECAP at room temperature. The specimens were taken from the central part of the blanks and annealed in air in a muffle furnace. Variations of annealing temperature did not exceed ±5°C. The nickel produced by ECAP pressing (at pressures of 700–800 MPa), with the mean grain size of 0.4–0.5 μm and a microhardness of 2.6 GPa was annealed for 1 h at temperatures of 100, 200, 300, 400 and 500°C. In addition, the annealing time at 250°C was varied from 1 to 12 hours.

Specimens for optical metallography were etched by the standard procedure. The investigations were carried out in a Aksiolab polarisation microscope with a Vx–44 camera (PSO Inc., Germany), with the resolution of 756 × 591 pixels and in a JEOL transmission electron microscope. TEM foils were prepared by the standard method. Microhardness was measured by the conventional procedure.

Figure 9.1 shows the data for the variation of the microhardness and mean grain size as a function of annealing temperature. The results indicate grain growth at 300–500°C and, consequently, the microhardness decreases. The behaviour of microhardness and the evolution of the microstructure during annealing of ultrafine-grained nickel produced by ECAP is governed on the whole by the already mentioned relationships described in [17]. There are three stages of relaxation of the structural state with increasing temperature. The following temperature ranges were defined: 20–175°C, 175–250°C and 250–400°C (annealing time 30 min). In the temperature range 175–250°C the microhardness rapidly decreases from 2.5 to 1.4 GPa. It should be mentioned that in [17] the investigations were carried out on ultrafine-grained nickel produced by torsion under high pressure (7 GPa). According to the data obtained from dark field images, the mean grain size in the initial condition was approximately 100 nm.

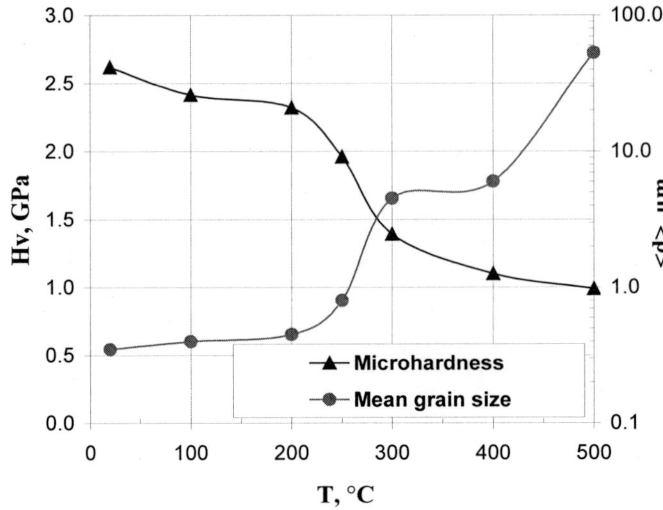

Fig. 9.1. Dependence of the mean grain size and microhardness on annealing temperature.

In this case, the mean grain size in the nickel specimens after ECAP was 0.4–0.5 µm. It may be assumed that at the same cumulated strain in severe plastic deformation the applied pressure influences the grain size: with increasing pressure the grain size decreases (from 0.5 µm up to 0.7–0.8 GPa to 0.1 µm at 7 GPa).

Figure 9.2 shows the histograms of the grain size distribution after annealing at 300°C, 400°C and 500°C for 1 h, obtained on the basis of the results of optical metallographic examination. The mean grain size is 4.7 µm after annealing at 300°C and increases to 6 µm after annealing at 400°C. In both cases, the form of the distribution is close to log-normal, although the maximum is displaced in the direction of larger dimensions. This indicates that the grain growth is normal. Analysis of the distribution for annealing at 500°C shows that this temperature is in fact the start of anomalous grain growth. This is confirmed by the bimodal size distribution of the grains: the first maximum is found in the range 20–40 µm and, in addition to this, there is a second small maximum in the range 120–140 µm.

The results of electron microscopic studies can be used to investigate the process of variation of the microstructure of the material (Fig. 9.3). The microdiffraction pattern shows a large number of point reflections uniformly distributed around the circumference which evidently indicates the presence of high-angle grain boundaries. The considerable azimuthal diffusion of the reflections indicates the high level of elastic internal stresses. Transmission electron microscopy results show that after annealing at 100 and 200°C there is no significant grain

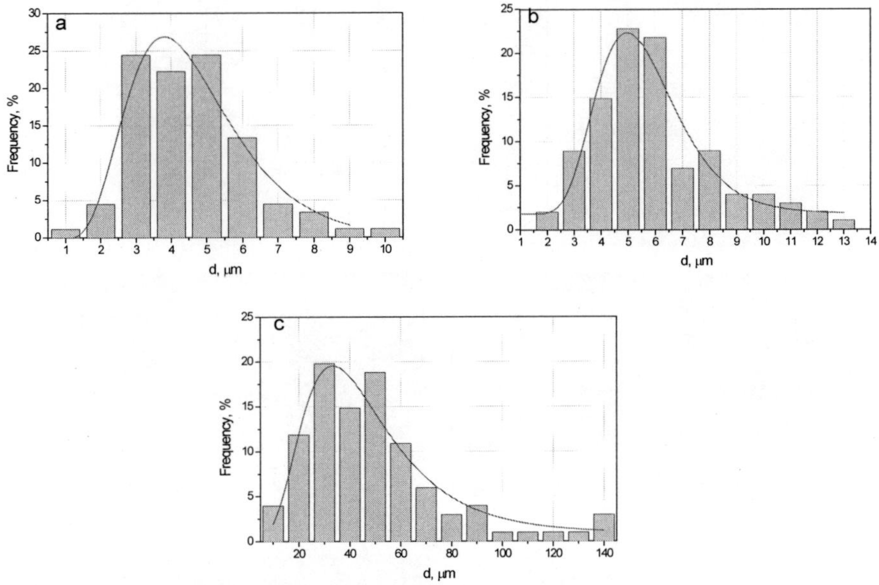

Fig. 9.2. Size distribution of grains in ECAP nickel, annealed for 1 h at temperatures of: a) 300°C ($\langle d\rangle_{\text{dist}}$ = 3.8±0.1 μm); b) 400°C ($\langle d\rangle_{\text{dist}}$ = 5.0±0.2 μ); c) 500°C ($\langle d\rangle_{\text{dist}}$ = 33±0.2 μm).

growth in comparison with the initial condition but the relaxation of microdistortions (sink of vacancies and dislocations) starts to take place and slightly reduces the microhardness values. Thus, in the initial stage in heating the dislocations found inside the grains disappear, i.e., recovery takes place. Annealing at 200–250°C (1 h) is accompanied by a rapid decrease of microhardness (from 2.3 to 1.4 GPa) caused by grain growth. This is indicated by the electron microscopy results: on the background of the majority of the grains with the size of 0.6–0.7 μm there are large grains with the size of 1–4 μm. Dislocation pile-ups are found at the grain boundaries and in the body of small grains. For more detailed examination of the process taking place in this temperature range the nickel specimens were annealed for different periods of time (1, 6 and 12 h) at a temperature of 250°C. An increase of the annealing time from 1 h to 6 h increases the number of large grains and the mean size of the grains increases to 3–4 μm. In addition, the body of the large grains shows the formation of annealing twins and the boundaries of the grains contain a large number of dislocations. Annealing for 12 h at a temperature of 250°C is characterised by further grain growth, with the mean size of these grains being 4–5 μm, whereas the fraction of the small grains rapidly decreases (to 30%). Straight grain boundaries appear. At the same time the degree of azimuthal

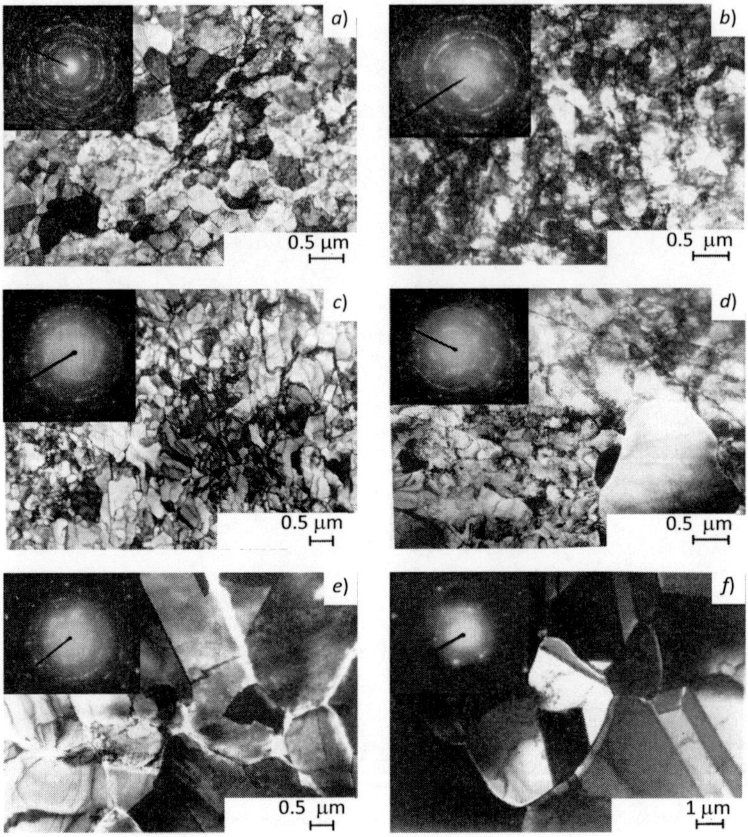

Fig. 9.3. Microstructure of nickel produced by ECAP pressing and annealed for 1 h at different temperatures: a) initial condition, b) 100°C, c) 200°C, d) 250°C, e) 300°C, f) 400°C.

diffusion of the reflections on the microdiffraction patterns decreases.

The specimens annealed at 300°C showed large grains with the size of 5 µm, with small grains are distributed along the boundaries of the large grains. The number of twins with straight equilibrium boundaries increases. After annealing at 400°C the fraction of the small grains (the size of approximately 2.5 µm) rapidly decreases (to 20%) and the size of the large grains reaches 8 µm. The mean grain size is 5–7 µm. The number of the straight grain boundaries increases and many of these boundaries are characterised by a typical banded contrast (Fig. 9.3) indicating the recovery of the structure of the grain boundaries. The specimens, annealed at 500°C, showed large changes in the microstructure, with the mean size of these grains being 40–50 µm; the majority of the grain boundaries and the body of the

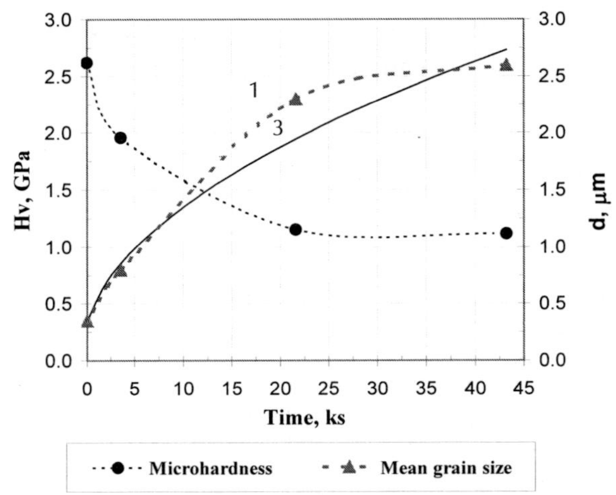

Fig. 9.4. Dependence of the mean grain size on annealing time at T = 250°C: 1) microhardness, 2) mean grain size, 3) parabola from equation (9.1).

grains are free from dislocation networks, and the structure is fully recrystallised, and the grain size further increases. The dependence of the mean grain size on annealing time at T = 250°C is shown in Fig. 9.4. The kinetic parameters of the grain growth process will be evaluated taking into account the results of isothermal annealing investigations. The dependence of the mean grain size on the annealing time at the given temperature is described by the relationship

$$d^2 - d_0^2 = k \cdot t^n \tag{9.1}$$

where d_0 is the initial grain size; k is a constant which depends on the driving force of grain growth and mobility of the grain boundary; n is the exponent (usually equal to 0.5–1). The kinetics at n = 1 corresponds to the parabolic law and normal grain growth. Interpolating the experimental points (Fig. 9.4) by equation (9.1) and using the method of least squares, it can be determined that the coefficient k = 0.6165 μm^2 · h^{-1} or 170 nm^2 · s^{-1}, where k is described by the equation

$$k = k_0 \exp\left(-\frac{Q}{RT}\right).$$

Assuming that the grain growth processes in annealing at temperatures higher than 250°C also satisfy equation (9.1), the data obtained for isochronic annealing at temperatures in the range 300–350°C can be used to estimate the activation energy of grain growth in ultrafine-grained nickel produced by ECAP. The results show that k_0 = 3.9·10^{-6} m^2 · s^{-1}

and Q = 103 kJ · mole^{-1}. This value is very close to the value of grain boundary self-diffusion of nickel: 115 kJ · mole^{-1} [21], 131 kJ · mole^{-1} [22], and 108.8 kJ · mole^{-1} [23]. Thus, it may be assumed that grain growth is controlled by grain boundary self-diffusion and, consequently, the kinetic parameter k_0 can be determined from the equation:

$$k_0 = \frac{4 \cdot \gamma \cdot \Omega}{\delta^2} \cdot \frac{(\delta \cdot D_{0b})}{kT} , \tag{9.2}$$

where γ is the surface tension of the grain boundaries in nickel (~0.7 J·m^{-2}); Ω is the atomic volume ($1.1 \cdot 10^{-29}$ m^3); δ is the width of the grain boundary (~$0.7 \cdot 10^{-9}$ m); $\delta \times D_{0b}$ is the pre-exponential multiplier of grain boundary diffusion multiplied by the width of the grain boundary (~$3.5 \cdot 10^{-15}$ m^3 s^{-1}); k is the Boltzmann constant; T is absolute temperature. Equation (9.2) for T = 500 K gives $(k_0)_{calc} \approx 4.6 \cdot 10^{-5}$ m^2 · s^{-1}. The order of magnitude of this value is identical with that obtained from equation (9.1) which indicates that the estimate is accurate. It should be stressed that to analyse the kinetic parameters it is necessary to carry out more detailed investigations. However, here it is attempted to evaluate only the order of these parameters to verify the facts of the larger decrease of the activation energy in the ultrafine-grained materials reported in the literature. As shown by the experimental results, the normal grain growth is governed by the conventional law and is controlled by grain boundary processes.

The following conclusions can be drawn here. Firstly, the nickel structure, produced by ECAP, is characterised by a high level of stored strain, as confirmed by high microhardness values (~2.6 GPa), a high dislocation density and extensive azimuthal diffusion of the reflections on the microdiffraction pattern. Secondly, the evolution of the microstructure and the variation of the microhardness of nickel under the thermal effect take place in three stages: 100–200°C, 200–300°C and 300–500°C. The first stage is characterised by the constant grain size, the almost complete relief of internal stresses and a monotonic decrease of microhardness. The second stage is characterised by the rapid grain growth and a large decrease of microhardness in the absence of internal stresses. In the third stage, the grains continue to grow and the microhardness monotonically decreases. Thirdly, the second stage for ultrafine-grained nickel, produced by torsional loading at a specific pressure of 7 GPa, starts at temperatures approximately 40–50°C lower than for ultrafine-grained nickel produced by ECAP at a pressure of 700–800 MPa. Finally, the ultrafine-grained nickel with the mean grain size of 0.4–0.5 μm, produced by ECAP pressing, can be regarded as thermally stable at temperatures up to approximately 200°C.

9.2. Activation energy and stored enthalpy in ultrafine-grained nickel

One of the efficient methods of investigating thermal stability (and generally relaxation processes) is differential scanning calorimetry (DSC). In this section, we present the results of DSC analysis of the specimens of ultrafine-grained nickel and nanonickel produced by the methods of severe plastic deformation. The investigated conditions are shown in Table 9.1. Two main states of ultrafine-grained nickel were selected for investigations: ECAP nickel (eight passes; mean grain size 0.37 μm), HPT nickel (6 GPa, five revolutions), and also a nickel specimen produced by cold rolling (room temperature) with a reduction of approximately 85% (mean grain size approximately 0.3 μm). Another pair of specimens was produced by torsional loading under high-pressure of blanks after ECAP and blanks produced by ECAP + rolling. To check the experimental data, experiments were also carried out with electrodeposited nickel with the mean grain size of 35 nm. The results of electron microscopic investigations show that the mean grain size in the ECAP + rolling nickel is approximately 0.3 μm,

Table 9.1. Investigated specimens of nickel prepared by severe plastic deformation and DSC data

Heating rate, K·min^{-1}	ECAP	ECAP +rolling	HPT	ECAP+ HPT	ECAP +rolling +HPT	eNi	eNi [22]
	Peak temperature T_p, K						
20	652.0	630.4	525.9	554.9	552.0	595.4	587
40	672.9	646.5	542.2	570.3	563.2	609.5	606
60	681.2	659.2	549.0	578.6	575.7	620.7	–
d_0, nm	350 [3]	300	170 [2]	140	100	35 [20]	20 [6]
Δf_{IC}, 10^3	8.6	9.4	17.6	21.4	30.0	85.7	150.0
E_{gr}, kJ·mole^{-1}	107.9± ±5.4	105.8±5.3	89.2± ±4.5	98.7± ±4.9	97.9±4.9	107.8± ±5.4	131.5
ΔH, J·mole^{-1}	59.3	75.1	102.7	187.9	229.0	610.6	415.7

in ECAP + HPT 0.14 μm, and in ECAP + rolling + HPT approximately 100 nm.

Investigations of ultrafine-grained nickel by differential scanning calorimetry were carried out in equipment Perkin-Elmer DSC7 with the accuracy not lower than 2%. Figure 9.5 shows the typical DSC curves, produced for ECAP nickel at a heating rate of 20, 40 and 60 K/min. The curves were plotted by deducting DSC signals, measured in the first and second pass.

The theoretical principles used in analysis of the DSC curves will be briefly analysed. Two main parameters are measured in the experiments of this type: activation energy and released (or absorbed) enthalpy (heat release). Activation energy can be determined by the standard Kissinger [24] or Ozawa [25] methods or from the Kolmogorov–Johnson–Mehl–Avrami equations [26]. Regardless of the fact that these methods are based on different physical assumptions [9], they lead to identical results within the range of the measurement error. The Kissinger and Ozawa methods were developed for non-isothermal heating conditions. The method is based on the equation of the type of chemical reaction:

$$\frac{dx}{dt} = A\left((1-x)^n\right)\exp\left(-\frac{E}{RT}\right) \tag{9.3}$$

where dx/dt is the reaction rate; x is the amount of the reacted substance; n is the reaction order; E is activation energy; T is absolute temperature. The reaction rate is maximum when $d(dx/dt)/dt = 0$. If the time dependence of temperature in heating is linear (i.e. T =

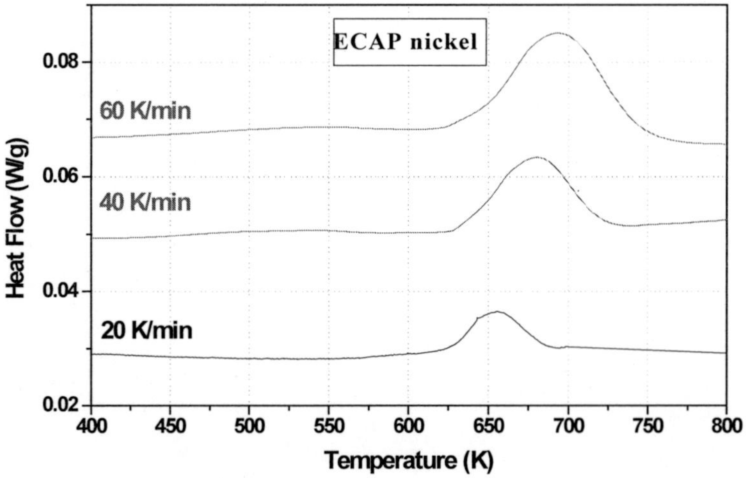

Fig. 9.5. Characteristic DSC-curves for ECAP nickel for three heating rates.

$T_0 + \beta t$), then, differentiating (9.3), we obtain

$$\frac{d}{dt}\left(\frac{dx}{dt}\right) = \frac{dx}{dt}\left[\frac{E\beta}{RT^2} - An(1-x)^{n-1}\exp\left(-\frac{E}{RT}\right)\right]. \tag{9.4}$$

Equating (9.4) to 0, we obtain

$$\frac{E\beta}{RT_m^2} = An(1-x)_m^{n-1}\exp\left(-\frac{E}{RT_m}\right), \tag{9.5}$$

where T_m is the temperature of the peak on the DSC curve; $(1-x)_m^{n-1}$ is the amount of the untransformed material. It may be shown that the product $(1-x)_m^{n-1}$ does not depend on heating rate β and is close to 1. Consequently, activation energy is determined by differentiating (9.5):

$$\frac{d\left[\ln\frac{\beta}{T_m^2}\right]}{d\left(\frac{1}{T}\right)} = -\frac{E}{R} \quad . \tag{9.6}$$

In the Ozawa method, equation (9.6) has the following form

$$\frac{d(\ln\beta)}{d\left(\frac{1}{T}\right)} = -\frac{E}{R} \quad . \tag{9.7}$$

Both equations, (9.6) and (9.7), are independent of *n*.

Recently, a large number of attempts have been made to increase the accuracy or modify the expression for the determination of activation energy with special reference to the recrystallisation of ultrafine-grained and nanocrystalline materials [22]. However, as shown by the experimental results, the experimental errors are higher than the resultant improvement of accuracy. The suitability of using Kissinger analysis for determining the kinetic parameters of recrystallisation of nanocrystalline and ultrafine-grained materials was investigated in detail in [27]. The effective energy of activation, measured in DSC experiments, consists of two parts and can be written in the following form:

$$E \approx \frac{E_N + mE_{gr}}{m+1} \tag{9.8}$$

where E_N and E_{gr} is the activation energy of nucleation and growth of grains, respectively; *m* is the integer or odd half-integer which depends on the mechanisms of nucleation and growth of the grains. For the three-dimensional case of grain growth controlled by grain boundary processes, $m = 1$. The activation energy of new nuclei in the investigated temperature range is negligible in comparison with the activation energy of grain growth (i.e., $E_N \ll E_{gr}$). In this case

$$E_{gr} = 2 \cdot E. \tag{9.9}$$

Heating at different rates can be used to determine the displacement of the maximum peaks of the grain growth and find the effective activation energy and activation energy of grain growth from the equations (9.5) and (9.9). Figure 9.6 shows the Kissinger graph for analysis of all the investigated states of ultrafine-grained and nanostructured nickel. The appropriate values of the parameters are presented in Table 9.2. Analysis shows that in the nickel specimens subjected to torsional loading at a high quasi-hydrostatic pressure, grain growth starts at lower temperatures. In this set, the ECAP nickel has the lowest temperature of the start of grain growth processes. The electrodeposited nickel, characterised by the smallest mean grain size in the group of the investigated specimens, occupies an intermediate position. For comparison, Fig. 9.6 shows the data taken from [22]. It can be seen that the results of these investigations coincide with the already

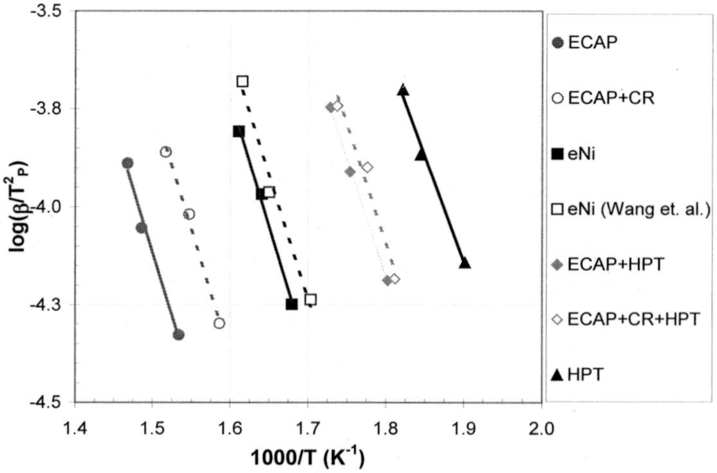

Fig. 9.6. Kissinger graph for ultrafine-grained and nanostructured nickel: 1) ECAP, 2) ECAP + rolling, 6) the electrodeposited nickel (eNi), 4) the same eNi by the according to the data in [22]; 5) ECAP + HPT; 6) ECAP + rolling + HPT; 7) PHP.

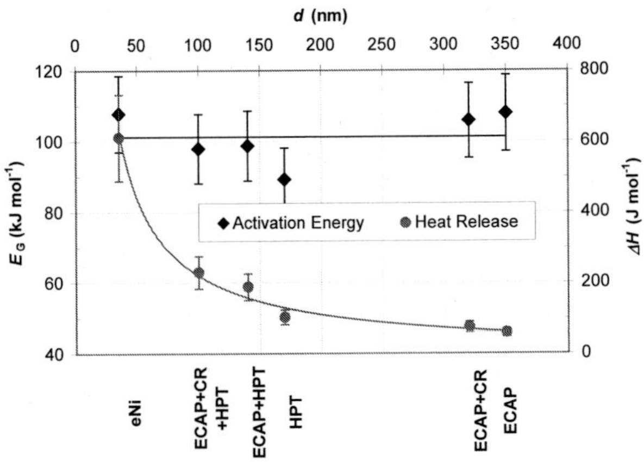

Fig. 9.7. Heat release and activation energy (2) as a function of the grain size.

published experimental data. Calculations of the activation energy give the values for ultrafine-grained nickel, produced by ECAP pressing, and nanostructured nickel, produced by electrodeposition, close to the activation energy of grain boundary diffusion. For the ultrafine-grained nickel produced by HPT or by ECAP + rolling + HPT combination, the activation energy is on average 10% lower (~89.2±0.5 kJ/mole). The latter may be used to explain the thermal stability of HPT nickel. Figure 9.7 shows the activation energy and heat release values as a function of the state or volume fraction of the grain boundaries. In the model of the 'spherical' grains, the variation of the volume, occupied by the grain boundaries, has the following form

$$\Delta f_{IC} = \left(1 - \frac{\delta}{d}\right)^3 - \left(1 - \frac{\delta}{d_0}\right)^3 \tag{9.10}$$

where $\delta \approx (0.8\text{–}1)$ nm is the thickness of the grain boundaries, d_0 and d is the initial and final grain size. Since the inequality $\delta/d \ll \delta/d_0 \ll 1$ is fulfilled with higher accuracy, equation (9.11), can be presented in the following form

$$\Delta f_{IC} = \frac{3\,\delta}{d_0} \quad . \tag{9.11}$$

The energy generated during a decrease of the volume fraction of the grain boundaries during grain growth, will be estimated. The surface density of energy in nickel γ~0.7 J m^{-2} which corresponds to the volume density $\alpha\gamma/d_0$, where α is the shape coefficient. For the model of spheres $\alpha = 3$. The excess energy of the grain boundaries (in J $mole^{-1}$) is expressed by the relationship

$$\Delta H_{gb} = 3\frac{\gamma}{d_0}\Omega \cdot N_A \qquad (9.12)$$

where $\Omega = 1.09 \cdot 10^{-29}$ m^3, N_A is the Avogadro number. The enthalpy, measured in the DSC experiments, and the excess energy calculated from equation (9.12) are presented in Table 9.2. The measured and calculated values differ by a factor of 2. It may be assumed that the source of a similar difference is the elastic energy of the dislocations, introduced into the material during severe plastic deformation or electrodeposition. Using the simplifying assumption according to which the dislocations are uniformly distributed in the volume of the specimen, it will be attempted to estimate the dislocation density in ultrafine-grained and nanostructured nickel. The elastic energy of a dislocation ensemble is expressed as [28]:

$$E_{EL} = A^* G b^2 \rho \ln \frac{R_C}{r_0} \qquad (9.13)$$

where coefficient A^* is equal to $1/(4\pi)$ for screw dislocations and $1/(4\pi\,(1-\nu))$ for edge dislocations; G is the shear modulus for nickel ($7.89 \cdot 10^{10}$ Pa); ν is the Poisson coefficient; b is the modulus of the Burgers vector ($2.49 \cdot 10^{-10}$ m); R_C is the external radius of truncation; $r_0 \approx b$ is the radius of the dislocation core. In many theoretical studies the external radius of truncation R_C is equated to the grain size for nanocrystalline materials. The physical meaning of this approach is reduced to the fact that the elastic fields of the dislocations cannot spread outside the limits of the nanograin. Thus, the grain boundaries are regarded as some sort of screening for the elastic fields of the

Table 9.2. Heat release and density of 'effective' dislocations in ultrafine-grained nickel

Parameter	Grain size, nm					
	35	100	140	170	300	350
ΔH, J/mol	610.6	229.0	187.9	102.7	75.1	59.3
ΔH_{gb}, J/mol	265.0	92.7	66.2	54.5	30.9	26.5
E_{EL}, J/mol	345.6	136.3	121.7	48.1	44.2	32.8
$x = 4b^2\rho$, 10^{-4}	6.613	2.607	2.328	0.921	0.845	0.628
ρ, m^{-2}	$6.0 \cdot 10^{14}$	$2.1 \cdot 10^{14}$	$1.9 \cdot 10^{14}$	$6.8 \cdot 10^{13}$	$6.1 \cdot 10^{13}$	$4.0 \cdot 10^{13}$

dislocations. However, since the logarithmic function smoothly increases with increasing R_C, this approach is also extended to the ultrafine-grained materials. In this case we shall try to estimate the external radius of truncation as the function of the dislocation density ρ. If the ensemble of the dislocations is treated as a square network (Fig. 9.8), then R_C is described by the equation

$$R_C = \frac{1}{2\sqrt{\rho}} \tag{9.14}$$

and the energy associated with the ensemble of the dislocations is:

$$E^* = \frac{4E_{EL}}{A^*G} = x\ln\left(1/\sqrt{x}\right) \tag{9.15}$$

where $x = 4b^2\rho$ is a dimensionless parameter. The type of function $x\ln(1/\sqrt{x})$ in the interval [0, 1] is shown in Fig. 9.9. The function has a maximum at $x^* = 1/e = 0.3679$, and decreases with the increase of x at $x > x^*$. Since the decrease of the elastic energy with increasing x (and, consequently, with the increase of dislocation density) has no physical meaning, it is assumed that the value $x^* = 1/e$ corresponds to the maximum possible dislocation density for a given material. For nickel $\rho_{max} = 1/(4b^2e) = 1.5\cdot10^{18}$ m^{-2}, which is a reasonable value. Now, the dislocation density in ultrafine-grained nickel can be estimated from the equation

$$\frac{4\left(\Delta H - 3\gamma\Omega N_A / d_0\right)}{A^*G} = x\ln\left(1/\sqrt{x}\right). \tag{9.16}$$

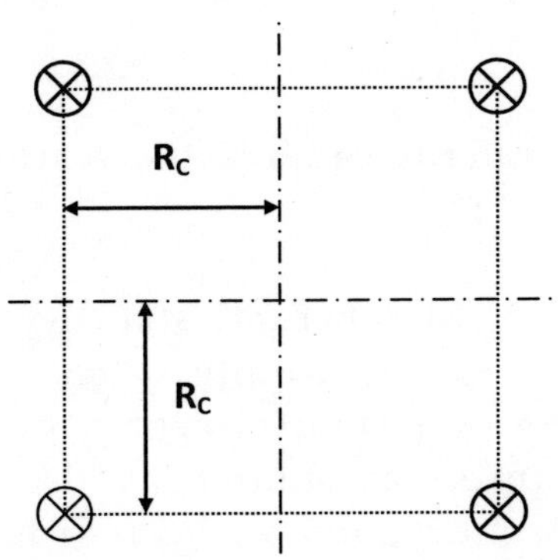

Fig. 9.8. Network of 'effective' dislocations.

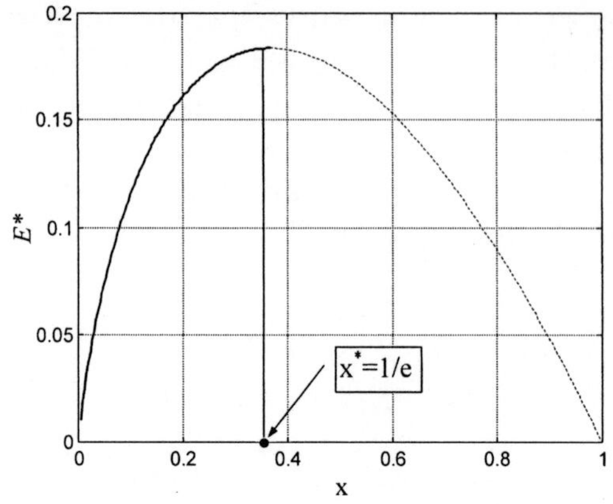

Fig. 9.9. Function $x \ln (1/\sqrt{x})$.

All the calculations carried out using this approximation are presented in Table 9.2. Dislocation density varies in the range from $6{\cdot}10^{14}$ m^{-2} for electrodeposited nickel to $4{\cdot}10^{13}$ m^{-2} for ECAP nickel. These values are in good agreement with the values obtained from x-ray measurements carried out for nanocrystalline and ultrafine-grained materials [29]. It should be mentioned that these dislocation densities lead to mean distances between them in the range from 20 nm for electrodeposited nickel to ~60 and 80 nm for HPT and ECAP nickel, respectively. The distance of 20 nm is comparable with the mean grain size in electrodeposited nickel (35 nm) and, consequently, no more than one dislocation can be present inside the individual grains. In the case of HPT and ECAP nickel the body of the grain may contain on average 3–5 dislocations. These estimates are capable of explaining why it is not possible to observe individual dislocations in nanocrystalline materials (with the grain size smaller than 100 nm) when the grains either does not contain dislocations or it is not in the reflecting position.

9.3. Evolution of the microstructure and texture in HPT nickel in annealing

As already mentioned, the thermal stability of ultrafine-grained materials has been investigated quite extensively mainly owing to the fact that it has an important effect in the application of these materials. The structure was studied by the methods of electron microscopy and x-ray diffraction analysis to investigate the thermal stability in pure ultrafine-grained metals: copper [16, 30], cobalt [31], nickel [15], and also in some industrial alloys [32], etc. In these studies, special attention was given to changes in the structure on the microscopic level (i.e., annihilation of the vacancies and dislocations) on the basis of changes of the electrical resistance and by analysis of diffraction contrast. However, the evolution of the ensemble of the grain boundaries and texture in annealing of ultrafine-grained nickel, produced by torsional loading under quasi-hydrostatic pressure, has not been investigated. Recently, we also carried out investigations of this type [11]. These results are presented in this chapter. The investigated materials were pure nickel in which the microstructure formed by means of quasi-hydrostatic pressure with the applied pressure of 6 GPa and the total number of revolutions of $N = 6$. The specimens were deformed at room temperature. Annealing was carried out at a temperature of 300°C which results, according to the data in [9, 17], in a rapid decrease of microhardness and increase of the mean grain size. The annealing time was 1, 10 and 30 min, and also 5 and 20 h. Texture measurements were taken in a Siemens diffractometer. The evolution of the microstructure and of the ensemble of the grain boundaries was investigated in a Philips scanning electron microscope. The pole figures and ODF of the initial specimens were described in chapter 4. It is well known that pure shear in FCC metals results in the formation of the texture component $\{001\}\langle 110\rangle$ which was found in the initial condition [33]. Annealing resulted in changes in the texture. The evolution of the texture can be investigated by monitoring the texture components (001), (111), (112) and (103) whose volume fraction was calculated on the basis of experimental ODFs. Figure 9.10 shows the variation of each components with increasing annealing temperature in comparison with the textureless state. It can be seen the fraction of the

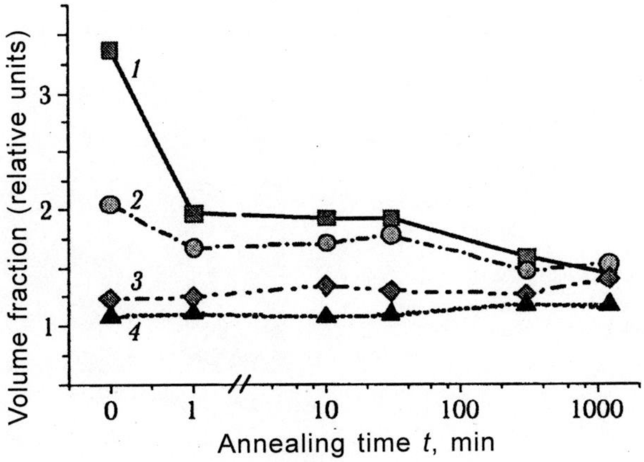

Fig. 9.10. Variation of the intensity of texture components (001) (1), (111) (2), (112) (3) and (103) (4) in annealing of HPT nickel.

components (100) and (111) greatly decreases already after annealing for 1 min. This is followed by a small change of all components with increasing annealing time and the intensity of the components (100) and (111) continues to decrease, whereas the sharpness of other texture components smoothly increases. After long-term annealing the intensity of all the investigated texture components is approximately the same. The most important characteristics of the microstructure are the size distribution of the grains (and also the mean grain size) and the misorientation distribution of the grain boundaries. The variation of the mean grain size during annealing of ultrafine-grained materials has been studied in considerable detail. At the same time, there is only a very small number of experimental data on the size distribution of the grains and evolution of the ensemble of the grain boundaries during annealing. There are no data for the evolution of the GBCD in annealing of ultrafine-grained materials. Figure 9.11 shows the dependence of the mean grain size on annealing time. The same figure shows the microstructure produced by the OIM method. It can be seen that in the first minutes the grain growth rate is high and the mean size of the grains increases by an order of magnitude: from 0.17 to 1.5 μm. It may be concluded that at a temperature of 300°C and the annealing time longer than 30 min the microstructure becomes almost completely stable. The mean grain size in annealing from 30 min to 20 h increases from 1.5 to 2.3 μm.

It is very important to investigate the evolution of the ensemble of the grain boundaries in annealing by determining the variation of the fraction of low-angle, special and high-angle arbitrary grain boundaries. The variation of the fractions of the various types of the

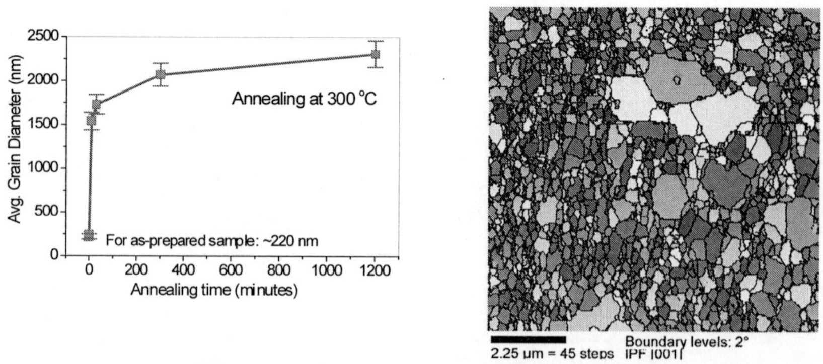

a b

Fig. 9.11. Dependence of the mean grain size on annealing time at T = 300°C for ultrafine-grained nickel, produced by torsional loading under pressure (a), and the microstructure after annealing at 300°C for 1 min (b).

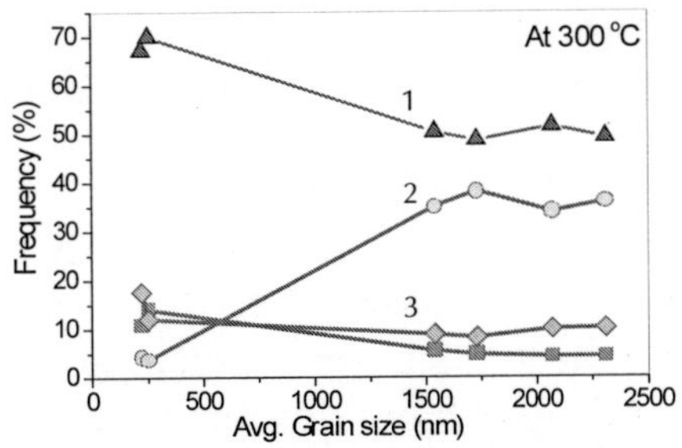

Fig. 9.12. Variation of the fraction of the high-angle grain boundaries (1), the boundaries $\Sigma3$ (2), $\Sigma5$–29 (3) and low-angle boundaries (4) in the general spectrum of the boundaries as a function of the grain size (or annealing time at 300°C) for ultrafine-grained nickel produced by torsional loading under pressure.

grain boundaries for the HPT nickel, annealed at 300°C, annealing time 1, 10 and 30 min and also 5 and 20 h, is shown in Fig. 9.12. It can be seen that with increasing annealing time only two components change greatly: the fraction of the twin boundaries $\Sigma3$ increases as a result of a decrease of the fraction of the high angle arbitrary boundaries. It should be mentioned that the fraction of the low- angle grain boundaries and the $\Sigma5$–30 boundaries also decreases, although not so greatly as the fraction of the high-angle grain boundaries. The identical trend was found in investigating the evoluation of the microstructure of the nanocrystalline nickel produced by electrodeposition, which is characterised by the presence of a sharp (001) texture. Annealing at 300°C was accompanied by the nucleation and growth of the (111) texture component. This is a clear indication of the nucleation and growth of the twin boundaries $\Sigma3$. Thus, the evolution of the ensemble of the grain boundaries in the ultrafine-grained and nanocrystalline materials takes place in the manner similar to the process of multiple twinning detected in FCC materials with low and mean stacking fault energies.

9.4. Superplasticity of nanocrystalline nickel

Superplasticity can be described by the generalised equation linking the flow stress σ and strain rate $\dot{\varepsilon}$ [34–36]:

$$\dot{\varepsilon} = A\frac{DGb}{kT}\left(\frac{b}{d}\right)^p\left(\frac{\sigma}{G}\right)^n, \tag{9.17}$$

where D is the appropriate diffusion coefficient (volume or grain boundary); G is the shear modulus; b is the Burgers vector; T is deformation temperature; d is the mean grain size; σ is the flow stress; p is the exponent of the dependence on the grain size; $n = 1/m$ is a quantity inversely proportional to the rate sensitivity coefficient [35, 36]. The results of a large number of experiments show that in the metals, intermetallics and ceramics with a nanostructure the value of the exponent p is close to 2, if the bulk diffusion is the main accommodation mechanism, and equal to 3 for grain boundary diffusion [37]. As indicated by equation (9.18), the decrease of the grain size should result in a decrease of the temperature of superplastic deformation and/or increase of strain rate. Theoretical predictions of this type were made by Gleiter [38] and confirmed by experiments published in [6, 13, 14]. However, the nanostructured materials may show grain growth even at reduced temperatures and this growth leads to structural instability. The experiments show that in many cases the superplasticity of the nanostructured materials is in fact accompanied by grain growth. In intermetallics, grain growth has a destructive role, and in pure nickel it may lead to accommodation of grain boundary sliding by 'sweeping up' triple junctions [13, 14]. In our view, grain growth has a positive role only when the mean grain size is lower than some critical size.

In early studies of superplasticity of nickel in experiments with uniaxial tensile loading special attention was given to the effect of small grains (~4 μm) on the superplastic behaviour of the material [39]. For comparison, experiments were carried out with coarse-crystalline nickel produced by annealing at 1173 K for 1 h. The mean grain size after this treatment was approximately 100 μm. The strongest effect of the initial microstructure was found at elevated temperatures. The results show that at temperatures of 723 K the flow stress of fine-grained nickel is higher than that of coarse-crystalline nickel; however, with a further increase of temperature in deformation the situation was reversed: coarse-grained nickel was stronger. In fine-grained nickel, plasticity increased with increasing temperature, reaching a maximum of approximately 180% at 1073 K. The typical diagram of superplastic flow was obtained in this case. For nickel with larger grains the behaviour of the specimens during deformation was different: hardening takes place up to fracture. The initial microstructure has a strong effect on both the rate dependence of the flow stress and relative elongation. For low strain rates the flow stress for fine-grained nickel was lower than for coarse-grained nickel. The situation changed at rates higher than $5\cdot10^{-2}$ s^{-1}. For the strain rate of $5\cdot10^{-3}$ s^{-1} the coefficient of rate sensitivity was equal to 0.38 for fine-grained nickel and 0.22 for coarse-grained. With increasing strain rate the plasticity of coarse-

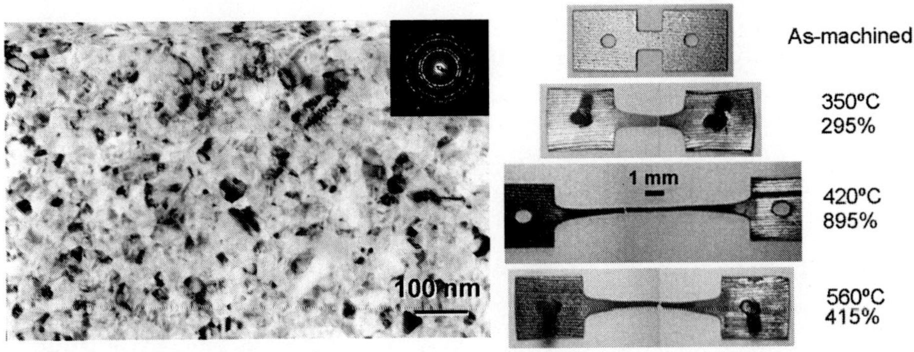

a b

Fig. 9.13. Brightfield images and diffraction pattern of electrodeposited nickel (a). The external appearance of the specimen in the tensile test (b).

grained nickel increased monotonically and that of fine-grained nickel showed a maximum at $5 \cdot 10^{-2}$ s^{-1} and then rapidly decreased.

Of special interest is the investigation of the superplastic properties of nickel with the ultrafine-grained structure. These investigations were carried out on nickel specimens produced by electrodeposition [13, 14]. The brightfield image of the microstructure and the microdiffraction pattern of nanocrystalline nickel are shown in Fig. 9.13a. The mean grain size of the initial structure was approximately 35 nm. The mean size of the grains, produced on the dark field images, is in good agreement with the results obtained in x-ray diffraction measurements. Experiments with uniaxial tensile loading were carried out in a mini-machine. The external appearance of the minispecimens prior to and after tensile loading is shown in Fig. 9.13b. Elongation was measured directly in the gauge length of the specimens (not on the basis of the displacement of thc clamps of the machine). The shape of the gauge part indicates the uniform elongation without formation of a localised neck. Maximum elongation was recorded at 693 K and a strain rate of $1 \cdot 10^{-3}$ s^{-1}. The stress–strain curves, measured at a rate of $1 \cdot 10^{-3}$ s^{-1}, for three different temperatures (453, 623 and 693 K) are shown in Fig. 9.14. At temperatures higher than 453 K the flow stress rapidly decreases and already at 623 K the specimens of nanocrystalline nickel show plasticity higher than 200%. A further increase of temperature increases the plasticity of the specimens, with the maximum plasticity recorded at 693 K equalling 895%, which is the highest recorded elongation for pure nickel. The transition of the nickel specimens from the state with low plasticity to the state without plasticity coincides with the isothermal peak on the curve plotted in the differential scanning

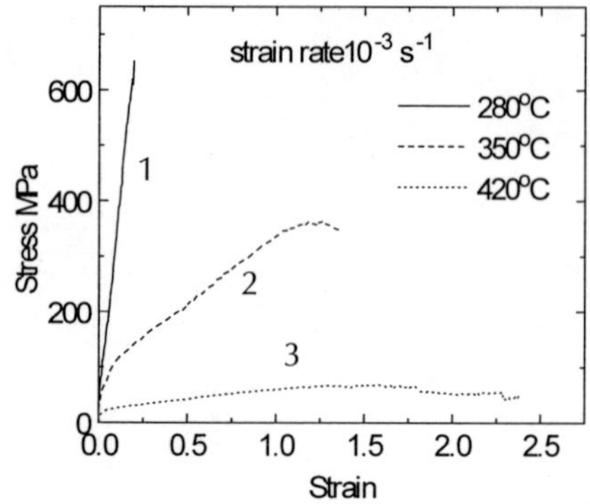

Fig. 9.14. True dependence of the flow stress on strain for the specimens taken from electrodeposited is equal, at a speed of movement of the clamps of the machine of 10^{-3} s^{-1} and temperature T = 453 K (1), 623 K (2) and 693 K (3).

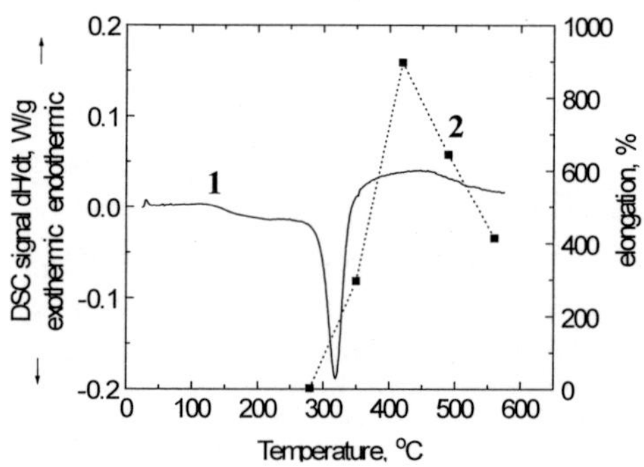

Fig. 9.15. The curve, recorded in DS C (1; heating rate 40 K/min), combined with the dependence of maximum elongation on temperature (2)

calorimeter. Figure 9.15 shows the signal from DSC at a heating rate of 40 K/min and the elongation values recorded at the given deformation temperature. The nanocrystalline nickel specimens were annealed to determine the microstructure and the start of tensile loading on the basis of the temperature profile of the mechanical tests. Figure 9.16a shows the microstructure of the specimen after heating from room temperature to 623 K at the same rate as in the mechanical tests (approximately 40–50 K/min). The microstructure remained ultrafine-grained, although the size of the individual grains reached 0.3 μm. In heating to 693 K (experiments in which the highest elongation values were recorded) the microstructure had a more uniform grain structure with the mean grain size of approximately 0.5 μm (Fig. 9.16b). Annealing twins were found in the structure. The results of transmission electron microscopy of thin foils taken from the gauge part of the specimens deformed at 623 K, show that the mean grain size in the gauge part was 1.3 μm

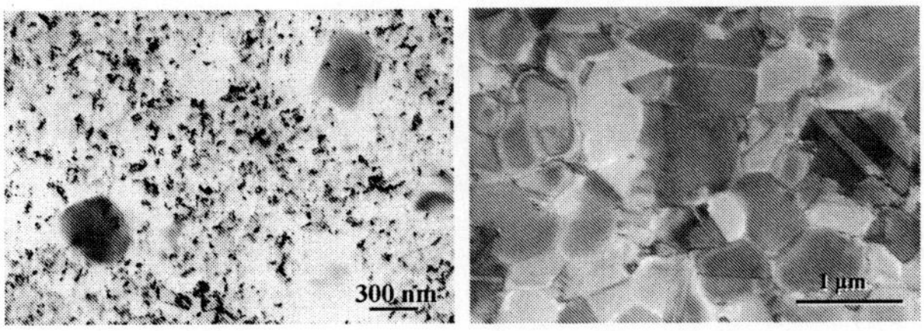

a b

Fig. 9.16. Microstructure of nickel specimens after heating to 623 K (a) and 693 K (b) at a rate of 40–50 K/min.

along the tensile loading axis, and 0.6–0.7 μm in the cross-section. For the specimens, deformed at 693 K, the mean grain size was approximately 2 μm in the direction of the tensile loading axis and 1 μm across this axis. Thus, the rate of grain growth was high in heating to the temperature of the start of deformation and the mean size of the grains increased by three orders of magnitude or more (from 35 nm to 0.3–0.5 μm). Nevertheless, the structure of nickel just prior to the start of testing remained ultrafine-grained with the mean grain size considerably smaller than that recorded in the experiments in earlier studies [39].

The most important characteristic of the superplastic deformation process is the rate sensitivity exponent n which can be measured in experiments by varying the rate of superplastic deformation. The value n can be calculated from the equation

$$n = \frac{\ln\left(\dot{\varepsilon}_2 / \dot{\varepsilon}_1\right)}{\ln\left(\sigma_2 / \sigma_1\right)}. \tag{9.18}$$

The rate sensitivity parameter was measured in [13]. Tests were carried out at two temperatures (623 and 693 K). Strain rate was changed in 'jumps' from $1 \cdot 10^{-3}$ s^{-1} to $2 \cdot 10^{-3}$ s^{-1} for T = 623 K, and from $1 \cdot 10^{-3}$ s^{-1} to $5 \cdot 10^{-2}$ s^{-1} for 693 K. In both cases the rate sensitivity decreases with increasing strain. These experimental results are very unusual because up to now it has not been possible to reach such high elongation values into pure metals because of grain growth at test temperatures. An exception is only pure aluminium – the highly plastic material on its own.

The transition of nanocrystalline nickel from the state with low plasticity to the one with high plasticity, detected at a test temperature higher than 453 K, corresponds to the transition from the purely dislocation mechanism of deformation of the thermally activated mechanism (grain boundary sliding). A similar behaviour is characteristic of nickel [39, 40] but in this case the transition temperature is considerably lower and the 'jump' in the flow stress is considerably greater. This is a consequence of the small grain size: follow temperatures (and, consequently, the dislocation mechanism of deformation) the high flow stress is caused by the Hall–Petch effect, and is the crease at elevated temperatures is caused by the small mean grain size (see equation (9.16)). In the materials with a small grain size in transition to the thermally activated deformation process (grain boundary sliding, diffusion and dislocation accommodation) the thermal energy is distributed between the processes of superplastic flow and grain growth. Both these processes have approximately the same activation energy which is equal to the activation energy of grain boundary diffusion. The phenomenon of grain growth in superplastic deformation was observed a long time ago but it has not been investigated for the materials with the ultrafine-grained structure. It would appear that only the driving force of grain growth greatly decreases with the increase of the mean grain size to approximately 1 μm, including the superplasticity mechanism, and grain growth is accompanied by grain boundary sliding. The rate of grain growth decreases in this case. The rate sensitivity parameter is an indicator of the type of controlling mechanism. For both deformation temperatures the rate sensitivity coefficient tends to decrease from 4 to 2 with increasing strain. This is in agreement with the results published in [39] in which the value of $n \approx 2.5$ was obtained for fine-grained nickel deformed at 1073 K and a strain rate of $5 \cdot 10^{-3}$ s. The variation of the rate sensitivity parameter with increasing strain indicates the evolution of the microstructure which takes place parallel with the superplastic flow.

In conclusion, it should be mentioned that the problems of the type of ensemble of the grain boundaries have not been examined in the phenomenological theory of superplasticity. We believe that this is not completely correct. According to the current views, superplasticity is a cooperative process in which grain boundary sliding affects not individual grains but groups of grains (pile-ups). It may be assumed that the boundaries of these pile-ups are represented by high-angle arbitrary boundaries and the grains inside the pile-ups are separated by low-angle and/or special grain boundaries, with sliding along these grain boundaries being difficult or completely impossible [40]. This is discussed in chapter 10.

In this chapter, attention was given to the phenomena directly associated with the ensemble of the grain boundaries in the ultrafine-grained and nanocrystalline nickel specimens. In cases in which the evolution of GBCD was not discussed, it was assumed that the investigated phenomena are caused by high-angle grain boundaries. Analysis of the results of investigation of the grain growth in ultrafine-grained nickel, produced by ECAP, confirmed the assumption that this process is controlled by grain boundary self-diffusion. The recently published data on the large decrease of the activation energy for ECAP nickel have not been confirmed. To evaluate the activation energy, experiments were carried out using differential scanning calorimetry for a set of ultrafine-grained nickel specimens produced by different methods of severe plastic deformation with the mean grain size in the range 0.35 μm to 35 nm. In all cases DSC yielded the activation energy values close to the values for grain boundary diffusion. This conclusion is also supported by the experiments with annealing of ultrafine-grained nickel produced by torsional loading under quasi-hydrostatic pressure. The evolution of the ensemble of the grain boundaries in ultrafine-grained nickel was investigated in the given experimental range. In particular, it was found that only two components of GBCD change greatly with increasing annealing time. The fraction of the twin boundaries $\Sigma 3$ increases as a result of a decrease of the fraction of high-angle arbitrary boundaries. It should be mentioned that the fraction of the low-angle grain boundaries and of the $\Sigma 5$–30 boundaries decreases, although not so greatly as the fraction of the high-angle boundaries. The identical situation was observed when investigating the evolution of the microstructure of nanocrystalline nickel produced by electrodeposition and characterised by the presence of the sharp texture (001). Annealing at 300°C was accompanied by the nucleation and growth of the (111) texture which is a clear indication of the nucleation and growth of the twin boundaries $\Sigma 3$. Thus, the evolution of the ensemble of the grain boundaries in the ultrafine and nanocrystalline materials takes place by the same mechanism as the process of multiple twinning detected in the FCC materials with low and mean stacking fault energies.

In the final part of the chapter we present the results of investigations of the superplastic behaviour of nanocrystalline nickel. Nickel is nanocrystalline only in the initial condition and at the start of tensile loading its mean grain size was in the range 0.3–0.5 μm and in fact, nickel was ultrafine-grained. The highest elongation value, recorded in these investigations, was approximately 900%. At the moment this is the highest elongation known in the literature. There are two hypotheses regarding the mechanism of superplastic deformation of nanonickel. The first of them is associated with the presence of the residual impurity of sulphur, found in nickel produced by electrodeposition [41]. The sulphur atoms, dissolved in the matrix, segregate in the vicinity of the migrating grain boundaries, reduce their mobility and, consequently, delay grain growth. This hypothesis is supported by the investigations of the effect of small sulphur additions on the high-angle grain boundaries in nickel. However, in our studies and also in some other investigations annealing at temperatures corresponding to the superplastic deformation temperature did not result in the precipitation of nickel sulphides at the grain boundaries. Similarly, no significant increase of the sulphur concentration along the grain boundaries in annealing of the electrodeposited metal was recorded in [42]. At the same time, in [43] the authors reported a large increase of the fraction of the grain boundaries in the spectrum of the grain boundaries in nanonickel produced by electrodeposition. It is well known that the twin boundaries have low mobility and may act as obstacles in the path of migrating grain boundaries.

References

1. Ultrafine-grained materials prepared by severe plastic deformation (ed. by R.Z. Valiev, Ann. De Chimie, 1996, vol. 21, No. 6–7.
2. Kolobov Yu.R., et al., Grain boundary diffusion and properties of nanocrystalline materials, Cambridge International Science Publishing, Cambridge, 2006.
3. Valiev R.Z. and Aleksandrov N.V., Nanostructured materials produced by severe plastic deformation, Logos, Moscow, 2000.
4. Amirkhanov N., Dissertation, IFMK, Ufa, 2002.
5. Kolobov Y.R., et al., Scripta Mater., 2001, vol. 44, No. 6, 873–878.
6. Würschum R., et al., Phil. Mag. B, 1997, vol. 76, 407–417.
7. Würschum R., et al., Ann. Chim. Fr., 1996, vol. 21, 471–482.
8. Zhilyaev A.P., et al., Metall. Mater. Trans., A, 2002, vol. 33, 1865–1868.
9. Zhilyaev A.P., et al., Mater. Phys. Mech., 2002, vol. 5, No. 1, 1–8.
10. Zhilyaev A.P., et al., Acta Materialia, 2003, vol. 51, No. 3, 753–765.
11. Kim B.-K., et al., Materials Science Forum, vol. 408–412, 943–948.
12. McFadden S.X., et al., Nature, 1999, vol. 398, No. 6729, 684–686.
13. McFadden S.X.,et al., Materials Letters, 2000, vol. 45, No. 6, 345–349.
14. Zhilyaev A.P., Mater. Phys. and Mech., 2000, vol. 1, 98–102.
15. Islamgaliev R.K., et al., Mat. Sci. Eng., A, 1997, vol. 237, 43–51.
16. Amirkhanov N.M., et al., Structure and properties of nanocrystalline materials, Proceedings, ed. by N.I., Noskova and G.G. Talutsa, Ekaterinburg, 1999, 333–343.
17. Korznikov A.V., et al., FMM, 1997, vol. 84, No. 4, 413–417.
18. Lu L., et al., Mat. Sci. Eng., A, 2000, vol. 286, 125–129.
19. Islamgaleev R.K., Dissertation, IFMK, Ufa, 2000.
20. Korznikov A.V., Dissertation, IMSP, Ufa, 2001.
21. Kaur I., et al., Handbook of grain and interphase boundary diffusion, Stuttgart, Ziegler, 1989.
22. Wang N., et al., Acta Mater., 1997, vol. 45, 1655–1669.
23. Iordache M.C., et al., Nanostr. Mater., 1999, vol. 11, 1343–1349.

24. Kissinger H.E., Ann. Chem, 1957, vol. 29, 1702–1706.
25. Ozawa T.J., J. Thermal Anal., 1970, vol. 2(2), 201–205.
26. Belen'kii G.Z., Geometrico-probability models of crystallisation. Phenomenological aproach, Nauka, Moscow, 1980.
27. Chen L.C. and Spaepen F., J. Appl. Phys., 1991, vol. 69 (2), 679–688.
28. Krivoglaz M.A., X-ray and neutron diffraction in non-ideal crystals, Berlin, Springer Verlag, 357–420.
29. Zhilyaev A.P., et al., Phys. Stat. Sol., (a), 2003, vol. 198, No. 2, 263–271.
30. Gertsman V.Yu., et al., Scripta Metal. Mater., 1994, vol. 28, 229–235.
31. Mulyukov Kh.Ya., et al., Phys. Stat. Sol. A, 1991, vol. 125, No. 2, 609–614.
32. Korznikov A.V., et al., Acta Met. Mater., 1991, vol. 39, No. 12, 3193 3197.
33. Holscher M., et al., Acta Metall. Mater., 1994, vol. 42, 879–886.
34. Kaibyshev O.A., Superplasticity of industrial alloys, Metallurgiya, Moscow, 1984.
35. Nieh T.G., et al., Superplasticity in Metals and Ceramics, Cambridge University Press, 1997.
36. Mukherjee A.K., Materials Science and Technology, vol, 6, ed. R. Arsenault, New York, Academic Press, 1994, 164.
37. Mukherjee A.K., Materials Science and Technology, ed. R.W. Cahn, et.al., New York, VCH Publishers, 1993, 407.
38. Gleiter H., J. Appl. Cryst., A, 1991, vol. 24, 79–90.
39. Kaibyshev O.A. and Markelov A.A., Izv. VUZ, Chern. Met., 1973, vol 7, 119–124.
40. Zhilyaev A.P. Vestnik UGATU, 2000, No. 2, 1–9.
41. Hibbard G., Mater. Sci. Forum, 2002, vol. 386–388, 387–396.
42. Abraham M., et al., Mater. Sci. Forum, 2002, vol. 386–388, 397–402.
43. Kim B.-K. and Szpunar J.A., Proceedings of the First Joint International Conference, eds. G. Gottstein and D.A. Molodov, Springer Verlag, 2001, 525–531.

10

DURATION OF THE STABLE FLOW STAGE IN SUPERPLASTIC DEFORMATION

10.1. Superplastic capacity and the rate sensitivity parameter

High deformation to failure is the distinctive feature referring to plasticity as a physical phenomenon. Starting with a study by Hart [1], the magnitude of deformation to failure directly depends on the value of the strain rate sensitivity coefficient $m = d\ln\sigma/d\ln\dot{\varepsilon}$. In [1], deformation to failure is regarded as the formation of a microscopic neck from the initial geometrical imperfection of the specimen – a prototype of the neck. At high values of m the initial nonuniformity does not develop into the neck for a long period of time. This results in high elongation of specimens. In [2] using a large number of experimental results it was shown that there is a correlation between the superplastic capacity and the value of m but the functional relationship between them could not be determined. The dependence cannot be described on the basis of the single parameter m. Formally, Hart's analysis permits the introduction of further two parameters: strain hardening exponent n and the parameter of the geometrical defect

$$f = \frac{\delta A_0}{A_0},$$

where A_0 is the cross-sectional area of the specimen; δA_0 is the deviation of the cross-sectional area from its value A_0 in the vicinity of the defect. It should be mentioned that the localised geometrical imperfection typical of the specimen in the initial stage is difficult to reconcile in the experiments with tensile loading of the specimens with the phenomenon of the running neck. This explains the attempts for formal introduction of the deformation defect of the unknown nature instead of the geometrical one [3]. In order to estimate the role of the parameters f and n in the formation of the limiting elogations, we shall discuss several equations which improve the accuracy and develop further the Hart approach but do not extend outside the limits of general ideology. The following equations were obtained in [4]

$$e_f = \left[1-(1-f)^{1/m}\right]^{-m} - 1 \tag{10.1}$$

$$e_f = (1+e_0)\eta^{m/(1-\gamma)} - 1, \quad \eta = 67, \tag{10.2}$$

where γ is the strain hardening parameter linked with n by the relationship $\gamma = n/\varepsilon_f$. Here and later e_f and ε_f are two known deformation measures: $\varepsilon_f = \ln(1+e_f)$. Using the parameters n, m and f, the following relationship was obtained in [6]:

$$e_f = \frac{e^n}{f^m} - 1. \tag{10.3}$$

The generalising equation (10.1) for limiting deformation which is very similar to the equation derived in an early study by Ghosh, was published by Baudelet [3]:

$$e_f = e^n\left[1-(1-f)^{1/m}\right]^{-m} - 1 \tag{10.4}$$

$$e_f = e^n\left(\frac{m}{f}\right)^m - 1\ . \tag{10.5}$$

This expression also resembles the result of an earlier study by Nicholson [7]:

$$e_f = \left(\frac{1}{f}\right)^{m/(1-\gamma)} - \gamma. \tag{10.6}$$

Finally, there are two known results obtained in describing the dependence of ε_f on all these parameters in the implicit form. These equations were published by Hutchinson and Neale [8]

$$\exp(-s\varepsilon_f)\sum_{k=0}^{p}\frac{(s\varepsilon_f)^k}{k!} = (1-(1-f)^s), \quad s = 1/m,\ p = n/m \tag{10.7}$$

and also Baudelet [3]:

$$\varepsilon_f = n - m \ln\left[1 - \exp\left(-\varepsilon_0 \frac{\varepsilon_f - n}{m\varepsilon_f}\right)\right]. \qquad (10.8)$$

The dependences of the limiting strain on the rate sensitivity parameter, determined using one of these relationships, are presented in Fig. 10.1. Regardless of the large external difference between the equations (10.4) and (10.5), the values determined using these equations do not differ by more than 1% and the curves on the graphs coincide almost completely. It should be mentioned that all these relationships lead to unlikely high strain values at high values of m. The value corresponding to ε_f ~5 is f ~15 000%. Of course, such elongation have never been experienced.

The presence of the inflection point on the $\sigma - \dot{\varepsilon}$ sigmoidal dependence determines the non-monotonic relationship between the rate sensitivity parameter and the strain rate. If we select two identical values of m on both sides of the maximum, these values will correspond to different limiting strain values. In addition, the maximum elongation is not obtained at the rate which corresponds to the maximum value of m [9].

The role of m in the formation of limiting strains changed unexpectedly after the discovery of high-strain rate superplasticity. It was found that high elongation values can also be reached at low values of m (less than 0.3). Thus, the high (at least in the initial meaning) value of m is not essential for obtaining high strains. The results obtained in recent experiments on ceramics were even more

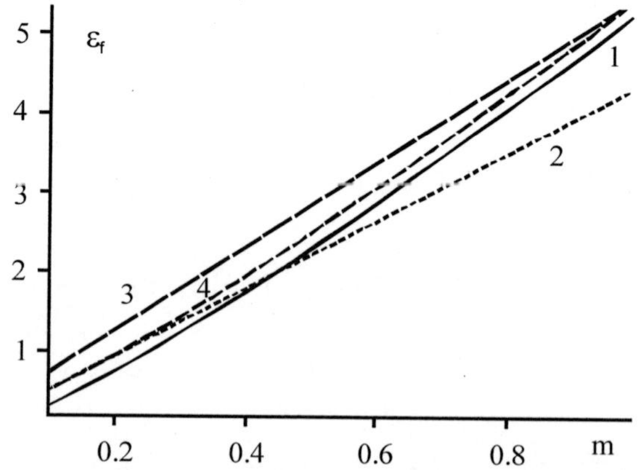

Fig. 10.1. Dependence of strength and fracture on the strain rate sensitivity parameter, calculated from the equations: 1) (10.1); 2) (10.2); 3) (10.3); 4) (10.4). The following values of the parameters were used in the calculations: $f = 0.005$, $e_0 = 0.1$, $\gamma = 0.01$, $n = 0.2$.

surprising [10]. It was established that even higher values (~1) of m do not guarantee high strains, i.e., the condition $m > m_c$ appears to have lost its meaning even as a sufficient condition of superplasticity. Although the limiting strains are determined by the formation of the neck, the phenomenological description of the development of the neck becomes unconvincing: high values of m do not guarantee high values of elongation. The essential and sufficient condition for superplasticity in the proposed model is the formation of bands of cooperative grain boundary sliding (CGBS). Taking these considerations into account leads to the task of calculating limiting strains [11].

10.2. Description of thickness differences of a flat specimen in tensile deformation

A flat specimen is regarded as a two-dimensional region defined at the initial moment of time in the coordinates x and z by the inequalities $|x| \le r_0$, $|z| \le l_0$ (where r_0 is the half width of the specimen, l_0 is the half length of the specimen). The tensile deformation along the axis z takes place by shearing along two systems of the CGBS bands, oriented under the angle of 45° to the tensile loading axis. If the total number of the bands in the specimen at the given moment of time t is equal to $M(t)$, and V is the shear rate in an individual band, then the instantaneous speed with which the length of the specimen increases is determined by the obvious relationship:

$$\frac{dl(t)}{dt} = \frac{V}{\sqrt{2}} M(t). \tag{10.9}$$

The specimen fails in localised deformation in some cross-section and irreversible growth of the neck. Thus, the duration of the process is determined by the degree of uniformity of the profile of the cross-section of the specimen – thickness difference. The rate of variation of any cross-section in complete analogy with the equation (10.9) is determined by the number of CGBS bands intersecting the given cross-section. The spatial distribution of the bands along the z axis is defined by the function $n(z,t)$, normalised with respect to the total number of the bands $M(t)$

$$M(t) = \int_{-l(t)}^{l(t)} n(z,t)dz.$$

Let $r(z,t)$ be the transverse dimension of the specimen in the cross-section with the coordinate z at the moment of time t. Consequently, by analogy with (10.9) we have

$$\frac{dr(z,t)}{dt} = -\frac{V_B}{\sqrt{2}} \int_{z-r(z,t)}^{z+r(z,t)} n(z',t)\,dz', \qquad r(z,0) = r_0 \cdot \tag{10.10}$$

Here the integral in the right-hand part also determines the number of bands intersecting the cross-section with the coordinate z. Equation (10.10) describes the correlated behaviour of closely spaced sections: a number of bands, intersecting such cross sections, is common in this case. In the uniform distribution of the bands along the length of the specimen the rate of variation of the transverse dimensions does not depend on the selected cross-section and the neck does not form. In this case, $n\ (z,t) = M\ (t)/(2l(t))$, and the equation (10.10) is reduced to the incompressibility condition, $d[l(t)r(z,t)]/dt = 0$, and describes the uniform variation of the cross-section. If the scale of the nonuniformity of the distribution of the bands is considerably smaller than the actual width then, since the length of the integration interval in (10.10) is equal to $2r\ (z,t)$, such nonuniformity will be removed by integration and has only a slight effect on the value of $r\ (z,t)$. Significant localisation of the flow occurs only in the case in which the characteristic scale of the nonuniformity with respect to the order of magnitude is comparable with the actual width. Thus, at high strains (or the initial narrow specimen), fine-scale nonuniformities in the band distribution may lead to large thickness differences. In addition, with increasing length of the specimen the probability of the nonuniform distribution of the bands becomes higher (both as a result of the increase of the base in which the bands are distributed and as a result of the possible decrease of the absolute number of the bands due to the buildup of damage in the material) and the probability of necking increases.

10.3. Formation of thickness difference as a random process

We consider equation (10.10) as a stochastic equation determined by the random process $n(z,t)$. From the incompressibility condition

$$\int_{-l(t)}^{l(t)} r(z,t)dz = 2r_0 l_0 \,.$$

Consequently, the spatial mean of the random quantity $r\ (z,t)$ is:

$$\overline{r}(t)=\frac{1}{2l(t)}\int_{-l(t)}^{l(t)} r(z,t)dz=r_0\frac{l_0}{l(t)}$$

and is unambiguously determined through the tensile loading kinetics by the experiment conditions. At a constant speed of the grips of the machine $\dot{l}(t) = V_0$ ($l(t) = l_0 + V_0 t$); at a constant strain rate $\dot{l}(t)/l(t) = \dot{\varepsilon} =$ const ($l(t) = l_0 \exp(\dot{\varepsilon} t) = l_0 \exp \varepsilon(t)$). On the other hand, the dependence $l(t)$ is determined by the total number of acting bands and by the shear rate in these bands (see (10.9)). We introduce the probability $p(t)$ of the formation of the band at time t. The total number of the bands $M(t)$ is assumed to be a determined value which does not differ from the mean value. If $d(t)$ is the grain size at time t, the number of potentially possible sites for the bands is $2l(t)/d(t)$ and the mean number of the bands on the entire length is

$$M(t)=\frac{2l(t)}{d(t)}p(t)\,.$$

We transfer from the independent variable t to strain ε. Their relationship is determined by kinematics. In a general case

$$\frac{\dot{l}(t)}{l(t)}=\dot{\varepsilon}(t),\quad l(t)=l_0\exp[\int_0^t\dot{\varepsilon}(t')dt']=l_0\exp(\varepsilon(t)),\quad l(\varepsilon)=l_0\exp(\varepsilon)\,.$$

After transition to variable ε, the equations (10.9) and (10.8) have the following form

$$\dot{\varepsilon}(\varepsilon)=\sqrt{2}V_B\frac{p(\varepsilon)}{d(\varepsilon)} \tag{10.11}$$

$$\frac{dr(z,\varepsilon)}{d\varepsilon}=-\frac{d(\varepsilon)}{2p(\varepsilon)}\int_{z-r(z,\varepsilon)}^{z+r(z,\varepsilon)} n(z',\varepsilon)\,dz'. \tag{10.12}$$

Let $\int_{z-r(z,\varepsilon)}^{z+r(z,\varepsilon)} n(z',\varepsilon)\,dz'=\xi\left[z\middle|2r(z,\varepsilon)\right]$ be the number of bands in the section z at the given with $r(z,\varepsilon)$, i.e., the random quantity with the average value

$$\langle\xi\rangle = \frac{2r(z,\varepsilon)}{d(\varepsilon)} p(\varepsilon) .$$

Consequently, because of (10.12), the ensemble average is defined by

$$\frac{d\langle r(z,\varepsilon)\rangle}{d\varepsilon} = -\langle r(z,\varepsilon)\rangle \qquad \text{or} \qquad \langle r(z,\varepsilon)\rangle = r_0 \exp(-\varepsilon) .$$

Thus, the mean ensembe average coincides with the spatial average. Generally speaking, it is necessary to determine the conditions (or the moment of arrest in the variable ε) at which the random realisation of the transverse size of the specimen in some cross-section differs so much from the spatial average that the shear stress in this section is outside the superplasticity range and the material is deformed in the conditions of classic diffusion (wide cross-section) or dislocation (narrow cross-section) creep. If the ensemble average and the spatial average coincide, it is much easier from the technical viewpoint to monitor the scatter of realisations in respect of the ensemble average. Here z is an inessential parameter and the equation with respect to the independent variable ε which, to shorten considerations, will be referred to as time, has the final form:

$$\frac{dr(\varepsilon)}{d\varepsilon} = -\frac{d(\varepsilon)}{2p(\varepsilon)} \xi\left[r(\varepsilon)\right], \tag{10.13}$$

where function $d(\varepsilon)$ is determined by the grain growth mechanism in deformation; $p(\varepsilon)$ is the probability of formation of a CGBS band which depends on the buildup of damage in deformation. Random quantity ξ is determined as follows. At time $\varepsilon = 0$ there is some random number of bands ξ_0 determined by the quantities $r(0)$, $d(0)$ and $p(0)$; during time Δ_0 the cross-section decreases at a constant rate. Correlated switching (taking into account the changing base $r(\Delta_0)$) takes place at time $\varepsilon = \Delta_0$, another system of bands forms, etc. After integration of (10.13) in the range $\varepsilon \in \left[\varepsilon_k, \varepsilon_k + \Delta_k\right]$, where $\varepsilon_k = \Delta_0 + \ldots + \Delta_{k-1}$, and after introducing the notation $r(\varepsilon_k) = r_k$ we obtain a random process

$$r_{k+1} = r_k - \frac{d_k \Delta_k}{2p_k} \xi_k \left[r_k\right]; \quad \varepsilon_{k+1} = \varepsilon_k + \Delta_k$$

$$P_0(\xi_k \,|\, r_k) = C_{[2r_k/d_k]}^{\xi_k} p_k^{\xi_k} q_k^{[2r_k/d_k]-\xi_k}$$

with the binomial distribution ξ_k and the evident transition probability

$$P(r_{k+1}\Big|r_k)=\sum_{\xi_k=0}^{[2r_k/d_k]}P_0(\xi_k\Big|r_k)\delta\left[r_{k+1}-r_k+\frac{d_k\Delta_k}{2p_k}\xi_k\right].$$

We construct the characteristic function of the random quantity r_{k+1} [12]

$$\langle\exp(iu_{k+1}r_{k+1})\rangle=\left\langle\exp(iu_{k+1}(r_k-\frac{d_k\Delta_k}{2p_k}\xi_k))\right\rangle=\left\langle\exp(iu_{k+1}r_k)\left\langle\exp(-iu_{k+1}\frac{d_k\Delta_k}{2p_k}\xi_k)\right\rangle\right\rangle$$

where the internal angular bracket denotes averaging with respect to the distribution of random quantity ξ_k. Consequently

$$\left\langle\exp(-iu_{k+1}\frac{d_k\Delta_k}{2p_k}\xi_k)\right\rangle=\sum_{l=0}^{[2r_k/d_k]}\exp(-iu_{k+1}\frac{d_k\Delta_k}{2p_k}l)\,C^l_{[2r_k/d_k]}p^l_kq_k^{[2r_k/d_k]-l}=$$

$$=\left[p_k\exp(-iu_{k+1}\frac{d_k\Delta_k}{2p_k})+q_k\right]^{2r_k/d_k}.$$

The characteristic function of random quantity r_{k+1} is linked with the characteristic function of quantity r_k by the relationship

$$\langle\exp(iu_{k+1}r_{k+1})\rangle=\langle\exp(iu_kr_k)\rangle$$

where

$$u_k=u_{k+1}-\frac{2i}{d_k}\ln(p_k\exp(-iu_{k+1}\frac{d_k\Delta_k}{2p_k})+q_k);$$

$$\left\langle r^n_{k+1}\right\rangle=\left(\frac{1}{i}\frac{d}{du_{k+1}}\right)^n\langle\exp(iu_{k+1}r_{k+1}\rangle;$$

$$\langle r_{k+1}\rangle=(1-\Delta_k)(1-\Delta_{k-1})...(1-\Delta_0)r_0=r_0\prod_{i=0}^{k}(1-\Delta_i);$$

$$\sigma^2_{k+1}=\left\langle r^2_{k+1}\right\rangle-\langle r_{k+1}\rangle^2=\sum_{i=0}^{k}\frac{q_i}{2p_i}\Delta_i^2d_i\langle r_i\rangle;$$

$$\langle r_kr_n\rangle=\frac{\langle r_k\rangle}{\langle r_n\rangle}\left\langle r_n^2\right\rangle,\quad k\geq n.$$

Since $r_0\gg d_0$, the distribution of r_i at small values of i is described with sufficient accuracy by the Gauss distribution (the well-known conditions of approximation of the binomial distribution are fulfilled).

It is assumed that the combined density of probability for the set $\{r_1,r_2,..., r_n\}$ is defined by the n-dimensional Gauss distribution

$$P(r_1,r_2,...,r_n)=\exp(-\frac{1}{2}\sum_{i=1}^{n}\sum_{k=1}^{n}\Lambda_{ik}(r_i-\langle r_i\rangle)(r_k-\langle r_k\rangle))/\sqrt{(2\pi)^n\det[\lambda_{ik}]};$$

$$\Lambda_{ik}=\Lambda_{kk}\delta_{i,k}+\Lambda_{k,k+1}\delta_{i,k+1}+\Lambda_{k-1,k}\delta_{i,k-1};$$

$$\Lambda_{kk}=\frac{\alpha_{k+1}-\alpha_{k-1}}{(\alpha_{k+1}-\alpha_k)(\alpha_k-\alpha_{k-1})\langle r_k\rangle^2},\qquad k=1,2,..,n-1,\quad(\alpha_0\equiv 0);$$

$$\Lambda_{k,k+1}=\frac{1}{(\alpha_k-\alpha_{k+1})\langle r_k\rangle\langle r_{k+1}\rangle},\qquad k=1,2,...,n-1;$$

$$\Lambda_{n,n}=\frac{1}{(\alpha_n-\alpha_{n-1})\langle r_n\rangle^2},\qquad \Lambda_{ik}=\Lambda_{ki},\quad \alpha_k=\frac{\sigma_k^2}{\langle r_k\rangle^2}.$$

The determinant of the covariance matrix is equal to

$$\det\hat{\lambda}=[\langle r_1\rangle\langle r_2\rangle...\langle r_n\rangle]^2(\alpha_n-\alpha_{n-1})(\alpha_{n-1}-\alpha_{n-2})...(\alpha_2-\alpha_1)\alpha_1$$

and the sum

$$\sum_{i,k=1}^{n}\Lambda_{ik}(r_k-\langle r_k\rangle)(r_i-\langle r_i\rangle)=\sum_{i=1}^{n}\frac{(\nu_i-\nu_{i-1})^2}{\alpha_i-\alpha_{i-1}},$$

where to shorten the equations, it is assumed that

$$\nu_k=r_k/\langle r_k\rangle\quad(\nu_0\equiv 1).$$

10.4. Absorption condition and the equation for limiting strain

Plotting the combined distribution of realisations r_i for the arbitrary number of steps n, we can determine the conditions of absorption of the process as the conditions of formation of a thickness difference at which the extreme cross-sections (thinnest or widest) are no longer in the range of the superplasticity conditions. Consequently, the local regions are transferred to the classic plasticity regime with rapid failure. Let the range of superplasticity in the stress variable be given by the boundaries $\tau_1<\tau<\tau_2$, i.e., τ_1 is the threshold stress, τ_2 is the transition stress to the dislocation creep regime. The stress $\tilde{\tau}$ in the section r_k

is linked with the macroscopic stresses τ (generally speaking, with $\tau(\varepsilon)$) by the relationship $\tau\langle r_k\rangle = \tilde{\tau}\, r_k$. It is assumed that if $\tilde{\tau}$ does not fit in the range $[\tau_1, \tau_2]$, absorption has taken place. Thus, the conditions of continuation of the process in the k-th step are reduced to fulfilling the inequalities restricting the difference between the random realisation r_k and its mean value:

$$a_k \equiv \frac{\tau}{\tau_2}\langle r_k\rangle < r_k < \frac{\tau}{\tau_1}\langle r_k\rangle \equiv b_k \ .$$

It is now quite easy to write the equation for the probability of absorption of the random process in the k-th step:

$$Q_k = \int_{-\infty}^{a_k}\int_{a_{k-1}}^{b_{k-1}}\cdots\int_{a_1}^{b_1} P(r_k, r_{k-1},\ldots,r_1)dr_1\ldots dr_k + \int_{b_k}^{\infty}\int_{a_{k-1}}^{b_{k-1}}\cdots\int_{a_1}^{b_1} P(r_k, r_{k-1},\ldots,r_1)dr_1\ldots dr_k = P_{k-1} - P_k$$

where

$$P_k = \int_{a_k}^{b_k}\int_{a_{k-1}}^{b_{k-1}}\cdots\int_{a_1}^{b_1} P(r_k, r_{k-1},\ldots,r_1)dr_1\ldots dr_k \ .$$

Since Q_k is the normalised (taking into account the equality $P_0 = 1$) density of probability of the discrete random quantity with the values 1, 2,..., the mean number of steps to absorption is:

$$\langle k\rangle = \sum_{k=1}^{\infty} kQ_k = \sum_{k=1}^{\infty} k(P_{k-1} - P_k) = \sum_{k=0}^{\infty} P_k = 1 + \sum_{k=1}^{\infty} P_k \, ,$$

where the first equality is definition and the remaining ones are identical transformations. The calculation of the probabilities P_k is reduced to calculating the integral with respect to the k-dimensional cube:

$$P_k = \int_{-1}^{1}\ldots\int_{-1}^{1}\exp\Big(-\sum_{i=1}^{k}\frac{(x_i - x_{i-1})^2}{2\gamma_i}\Big)\frac{dx_1\ldots dx_k}{\sqrt{(2\pi)^k\gamma_1\ldots\gamma_k}}$$

$$\gamma_i = \left(\frac{2}{t_1 + t_2}\right)^2(\alpha_i - \alpha_{i-1}) \, , \ x_0 = \frac{t_2 - t_1}{t_2 + t_1} \, , t_1 = \frac{\tau}{\tau_1} - 1, \quad t_2 = 1 - \frac{\tau}{\tau_2} \ .$$

As a result of the evident substitution of the variables, P_k can be written in the form of the integral of the isotropic function:

$$P_k = \int\limits_{\Omega_k} \exp(-y_1^2 - ... - y_k^2) \frac{dy_1 ... dy_k}{\pi^{k/2}}$$

where the integration range is given by the inequalities

$$-(1+x_0) \le \sum_{l=1}^{i} \sqrt{2\gamma_l} y_l \le 1 - x_0, \quad i = 1,2,...,k\ .$$

Rotating the coordinate system with $\{y_1\}$, the oblique prism Ω_k can be transformed to the form

$$-\frac{1+x_0}{\sqrt{2(\gamma_1 + ... + \gamma_i)}} \le \sum_{j=i}^{k-1} \sqrt{\frac{\gamma_1 + ... + \gamma_i}{\gamma_1 + ... + \gamma_j}} \sqrt{\frac{\gamma_{j+1}}{\gamma_1 + ... + \gamma_{j+1}}} z_j + \sqrt{\frac{\gamma_1 + ... + \gamma_i}{\gamma_1 + ... + \gamma_k}} z_k \le$$

$$\le \frac{1-x_0}{\sqrt{2(\gamma_1 + ... + \gamma_i)}}$$

$$-\frac{1+x_0}{\sqrt{2(\gamma_1 + ... + \gamma_k)}} \le z_k \le \frac{1-x_0}{\sqrt{2(\gamma_1 + ... + \gamma_k)}}, \quad i = 1,2,...,k-1.$$

It is now quite easy to find the maximum parallelepiped inscribed in Ω_k (considering successively the inequalities, starting with older numbers):

$$-\frac{1+x_0}{2} \frac{t_1 + t_2}{\sqrt{2\alpha_i}} \sqrt{1 - \frac{\alpha_i}{\alpha_{i+1}}} \le z_i \le \frac{1-x_0}{2} \frac{t_1 + t_2}{\sqrt{2\alpha_k}} \sqrt{1 - \frac{\alpha_i}{\alpha_{i+1}}}$$

$$-\frac{1+x_0}{2} \frac{t_1 + t_2}{\sqrt{2\alpha_k}} \le z_k \le \frac{1-x_0}{2} \frac{t_1 + t_2}{\sqrt{2\alpha_k}}, \quad i = 1,2,...,k-1.$$

Thus, cutting off from the oblique prism the external regions in which the subintegrand function is small, we split the k-fold integral into k independent one-dimensional integrals and the required probability can be easily calculated:

$$P_k = 2^{-k} \left(\mathrm{erf}\left[\left(1 - \frac{\tau}{\tau_2}\right) \frac{\langle r_k \rangle}{\sqrt{2\sigma_k^2}} \right] + \mathrm{erf}\left[\left(\frac{\tau}{\tau_1} - 1\right) \frac{\langle r_k \rangle}{\sqrt{2\sigma_k^2}} \right] \right) \times$$

$$\times \prod_{i=1}^{k-1} \left(\mathrm{erf}\left[\left(1 - \frac{\tau}{\tau_2}\right) \sqrt{\frac{\langle r_i \rangle^2}{2\sigma_i^2} - \frac{\langle r_{i+1} \rangle^2}{2\sigma_{i+1}^2}} \right] + \mathrm{erf}\left[\left(\frac{\tau}{\tau_1} - 1\right) \sqrt{\frac{\langle r_i \rangle^2}{2\sigma_i^2} - \frac{\langle r_{i+1} \rangle^2}{2\sigma_{i+1}^2}} \right] \right)$$

The accurate upper estimate of $\langle k\rangle$ for the stresses coinciding with the boundaries of the superplasticity range can be obtained directly from the last equation. Thus, at $\tau = \tau_1$

$$P_k = 2^{-k}\left(erf\left[\left(1-\frac{\tau_1}{\tau_2}\right)\frac{\langle r_k\rangle}{\sqrt{2\sigma_k^2}}\right]\right)\prod_{i=1}^{k-1}\left(erf\left[\left(1-\frac{\tau_1}{\tau_2}\right)\sqrt{\frac{\langle r_i\rangle^2}{2\sigma_i^2}-\frac{\langle r_{i+1}\rangle^2}{2\sigma_{i+1}^2}}\right]\right)<2^{-k}$$

which shows that $\langle k\rangle \le 2$. This also applies to the case in which $\tau = \tau_2$. Thus, only two switches of the system of the CGBS bands take place on average at the boundaries of the superplasticity range.

To calculate $\langle k\rangle$ at arbitrary τ, the equation for P_k will be simplified. If all the quantities which determine the random process are independent of the number of the step (there is no grain growth, no buildup of damage), it is quite easy to obtain the equality

$$\frac{\langle r_i\rangle^2}{2\sigma_i^2}=\frac{pr_0}{qd\Delta}\frac{(1-\Delta)^{2i}}{1-(1-\Delta)^i}$$

At low k because of the evident inequality $r_0/d >>1$, all the arguments of the erf function are high and the values P_k are almost equal to unity. With increasing k the values of $\langle r_k\rangle$ decrease and the dispersion increases resulting in a decrease of P_k to 0 at k tending to infinity. The inflection point on the dependence of probability of the number can be determined as the solution of the equation

$$P_{n+1}-2P_n+P_{n-1}=0$$

If n is the required solution, then in calculating $\langle k\rangle$ it may be concluded with good approximation that all the values of P_k with the number $k < n$ are equal to unity and all 'older' probabilities are equal to 0.

Initially, we obtain the estimate of n for a simpler case: $\tau = 2\tau_1\tau_2/(\tau_1+\tau_2)$. In this case

$$P_k = \left(\text{erf}\left[t\frac{\langle r_k\rangle}{\sqrt{2\sigma_k^2}}\right]\right)\prod_{i=1}^{k-1}\left(\text{erf}\left[t\sqrt{\frac{\langle r_i\rangle^2}{2\sigma_i^2}-\frac{\langle r_{i+1}\rangle^2}{2\sigma_{i+1}^2}}\right]\right)$$

where $t = (\tau_2-\tau_1)/(\tau_2+\tau_1)$. Since the argument of the multiplier standing inside the general product is far greater than the argument of the multiplier with the number $k-1$, it is different from unity and can be ignored. Consequently, the equation for n has the form

$$\operatorname{erf}\left[t\sqrt{\alpha_n^{-2}-\alpha_{n+1}^{-2}}\right]+\left\{\operatorname{erf}\left[t\sqrt{\alpha_{n-1}^{-2}-\alpha_n^{-2}}\right]\right\}^{-1}=2.$$

In the notations $S=(\alpha_{n-1}^{-2}-\alpha_{n+1}^{-2})/2$, $T=(\alpha_{n-1}^{-2}-2\alpha_n^{-2}+\alpha_{n+1}^{-2})/2$ we have

$$\operatorname{erf}\left(t\sqrt{S-T}\right)+\left[\operatorname{erf}\left(t\sqrt{S+T}\right)\right]^{-1}=2$$

For constant *p, d*, and Δ, the quantities S and T are linked by an approximate (its accuracy improves with increasing n) relationship: $T = \Delta S$ (i.e., $T \ll S$). Using the approximation $\operatorname{erf}(x)\approx\sqrt{1-\exp(-4x^2/\pi)}$, we obtain the approximate solution:

$$S=\frac{\pi}{4t^2}L\left(\frac{1}{4\Delta}\right)$$

where $L(x)$ is the Lambert function determined as the solution of the equation $L\exp(L) = x$. At high values of n (dispersion at saturation), quantity S is described by the equation

$$S=(\Delta_{n-1}+\Delta_n)\alpha_{n+1}^{-2}$$

and at constant Δ in the form

$$S=2(1-\Delta)^{2n}\left\{\sum_{i=0}^{n}\frac{q_i}{p_i}\frac{d_i}{r_0}(1-\Delta)^i\Delta\right\}^{-1}.$$

Thus, the equation for n has the form

$$2(1-\Delta)^{2n}\left\{\sum_{i=0}^{n}\frac{q_i}{p_i}\frac{d_i}{r_0}(1-\Delta)^i\Delta\right\}^{-1}=\frac{\pi}{4t^2}L\left(\frac{1}{4\Delta}\right)\qquad(10.14)$$

and can be solved by defining the dependence p_i and the law of grain growth d_i. Consequently, because of the accepted dependence of probability of index, $\langle k\rangle = n$ and the problem is solved. It should be mentioned that at the given number $\langle k\rangle$ of the steps to failure, the appropriate strain is determined as $\Delta\langle k\rangle$. In this case, the equation

(10.14) can be transformed immediately to strain to fracture, using the equality

$$(1-\Delta)^n = \left((1-\Delta)^{1/\Delta}\right)^{n\Delta} \approx \exp(-n\Delta) = \exp(-\varepsilon)$$

with the sum in the denominator of the left-hand part of the equation (10.6) treated as the integral sum corresponding to the integral

$$\int_0^{\varepsilon} \frac{q(\varepsilon')}{p(\varepsilon')} \frac{d(\varepsilon')}{r_0} \exp(-\varepsilon')\, d\varepsilon'$$

We examine the quantity S introduced into the variable ε and defined by the equation

$$S = 2\exp(-2\varepsilon) \Big/ \int_0^{\varepsilon} \frac{q(\varepsilon')}{p(\varepsilon')} \frac{d(\varepsilon')}{r_0} \exp(-\varepsilon')\, d\varepsilon'.$$

To derive the equation for determining the strain to fracture for the arbitrary value of stress τ, it is necessary to find the inflection point on the dependence P_k, taking into account the more complicated equation:

$$\operatorname{erf}\left(t_1\sqrt{\alpha_n^{-2} - \alpha_{n+1}^{-2}}\right)/2 + \operatorname{erf}\left(t_2\sqrt{\alpha_n^{-2} - \alpha_{n+1}^{-2}}\right)/2 +$$

$$+2\left[\operatorname{erf}\left(t_1\sqrt{\alpha_{n-1}^{-2} - \alpha_n^{-2}}\right) + \operatorname{erf}\left(t_2\sqrt{\alpha_{n-1}^{-2} - \alpha_n^{-2}}\right)\right]^{-1} = 2$$

where the notations $t_1 = \tau/\tau_1 - 1$, $t_2 = 1 - \tau/\tau_2$ are used to shorten the equations. The solution will be determined in the form

$$S = \frac{c_1}{t_1^2} + \frac{c_2}{t_2^2},$$

where the constants c_1 and c_2 are determined on the basis of the already known solution corresponding to the condition $t_1 = t_2$, and the estimated solution for the edges of the superplasticity range (t_1 tends to 0 – $S \sim c_1/t_1^2$, t_2 to 0 – $S \sim c_2/t_2^2$). Consequently, we obtain

$$c_1 = \frac{\pi}{4}\left[\left(1-\sqrt{2\Delta}\right)^2 + 2\Delta\right];\quad c_1 + c_2 = \frac{\pi}{4} L\left(\frac{1}{4\Delta}\right).$$

The equation for limiting elongation ε_f has the following form

$$\exp(2\varepsilon_f)\int_0^{\varepsilon_f}\frac{q(\varepsilon')}{p(\varepsilon')}\frac{d(\varepsilon')}{r_0}\exp(-\varepsilon')\,d\varepsilon' = \frac{2}{c_1(\Delta)\left[\frac{\tau}{\tau_1}-1\right]^{-2}+c_2(\Delta)\left[1-\frac{\tau}{\tau_2}\right]^{-2}},$$

where τ is the stress at strain equal to ε_f. If the strain rate in the experiments is maintained constant $\dot{\varepsilon}$, and the hardening in the stable flow stage is negligible, then τ can be regarded as a quantity independent of ε and determined by the value of $\dot{\varepsilon}$ (the appropriate point on the sigmoidal curve). However, if the experiments are carried out with the constant speed of movement of the grips V_0, ε_f is determined by the equation

$$\exp(2\varepsilon_f)\int_0^{\varepsilon_f}\frac{q(\varepsilon')}{p(\varepsilon')}\frac{d(\varepsilon')}{r_0}\exp(-\varepsilon')\,d\varepsilon' =$$

$$= \frac{2}{c_1(\Delta)\left[\frac{\tau(\dot{\varepsilon}_0 e^{-\varepsilon_f})}{\tau_1}-1\right]^{-2}+c_2(\Delta)\left[1-\frac{\tau(\dot{\varepsilon}_0 e^{-\varepsilon_f})}{\tau_2}\right]^{-2}},$$

where the initial strain rate $\dot{\varepsilon} = V_0/L_0$, L_0 is the initial length of the specimen. In this case, as a result of the variation of the strain rate during loading the transition from one loading curve to another one (movement along the sigmoidal curve in the direction of lower stresses) takes place even in the absence of hardening in the stable flow stage. This greatly complicates the solution.

10.5. Some properties of limiting strain

Ignoring the grain growth at deformation ($d(\varepsilon) = d_0$) and the buildup of damage ($p(\varepsilon) = p_0$), and studying the deformation variant with a constant rate, it is quite easy to derive an explicit equation for the dependence of limiting strain on applied stress:

$$\varepsilon_f = \ln\left(\sqrt{R(\tau)+1/4}+1/2\right) \qquad (10.15)$$

where

$$R(\tau) = 2\frac{r_0}{d_0}\frac{p_0}{q_0}\left\{c_1(\Delta)\left[\frac{\tau}{\tau_1}-1\right]^{-2}+c_2(\Delta)\left[1-\frac{\tau}{\tau_2}\right]^{-2}\right\}^{-1}.$$

The dependence (10.15) inside the superplasticity range, $[\tau_1, \tau_2]$ is

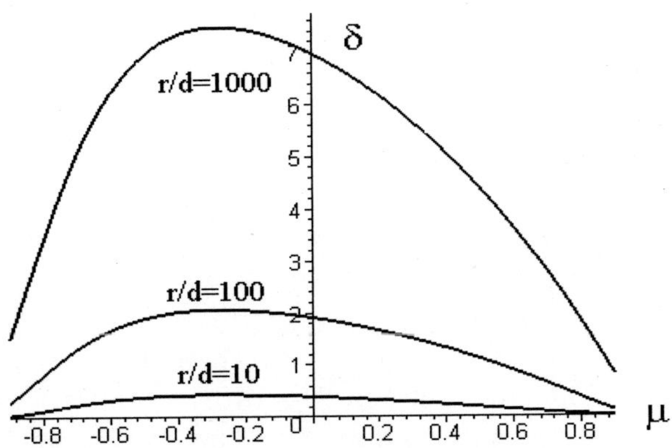

Fig. 10.2. Dependence of strain δ on the reduced logarithm of stresses μ for three values of the width of the specimen: 1) r/d = 10; 2) r/d = 100; 3) r/d = 1000.

described by a curve with a maximum at the optimum stress

$$\tau_0 = \tau_1 \frac{1+\left(\dfrac{c_1\tau_2}{c_2\tau_1}\right)^{1/3}}{1+\left(\dfrac{c_1\tau_1^2}{c_2\tau_2^2}\right)^{1/3}}$$

determined by the boundaries of the superplasticity range and the rate of switching the CGBS bands (through the quantities c_1, c_2). The dependence on the grain size in the vicinity of the optimum deformation conditions and at the boundary of the superplasticity range is described by different relationships. For example, in the vicinity of the optimum value the limiting strain increases logarithmically with decreasing grain size and at the boundaries of the range $\varepsilon_f \sim d_0^{-1}$. This is in agreement with the estimate obtained in [13] in which the limiting plasticity is regarded as the strain restricted by the buildup of some critical damage – micropores.

An example of the dependence of limiting strain $\delta = \exp \varepsilon_f - 1$ (where ε_f is determined by the expression (10.15)) and the acting stress for three values of the width of the specimen differing by an order of magnitude is shown in Fig. 10.2. The graph shows the role of the geometrical dimensions of the specimen (scale factor in the terminology used in [14]). Calculations were carried out for the values p_0 = 0.1 and Δ = 0.01. The range of the stresses $\tau \in [\tau_1, \tau_2]$ is converted into the range of the reduced logarithmic stresses $\mu \in [-1, 1]$ in accordance with the relationship

$$\frac{1+\mu}{2} = \frac{\ln \frac{\tau}{\tau_1}}{\ln \frac{\tau_2}{\tau_1}}.$$

Thus, the superplastic capacity can be investigated on the basis of considerations regarding the deformation mechanism on the mesolevel. The nonuniformity of the spatial distribution of the CGBS band formed as a fluctuation manner during the formation pistol localisation of the flow on the macrolevel. No initial macroheterogeneity needs to be defined for the specimens. The model reproduces the non-monotonic dependence of the limiting strain on the strain rate inside the superplastic range. It is possible to determine the optimum deformation conditions for the arbitrary loading law on the basis of the input parameters of the model: the probability of formation of the band p_0, the rate of switching bands Δ, the boundaries τ_1 and τ_2 of the superplasticity range.

References

1. Hart E.W., Acta Metall., 1967, vol. 15, 351-355.
2. Langdon T.G., Scripta Metall., 1977, vol. 11, 997-1000.
3. Lian J. and Baudelet B., Mater. Sci. Eng., 1986, vol. 84, 157-162.
4. Ghosh A.K. and Ayres R.A., Met. Trans., A, 1976, vol. 7, No. 10, 1589-1591.
5. Lin I.H., et al., Acta Metall., 1981, vol. 29, No. 5, 819-827.
6. Semiatin S.L., et al., Acta Metall. Mater., 1984, vol. 32, No. 9, 1347-1354.
7. Nichols F.A., Acta Metall., 1980, vol. 28, No. 6, 663-673.
8. Hutchinson J.W. and Neale K.W., Acta Metall. Mater., 1977, vol. 25, No. 8, 839-846.
9. Mohamed F.A. and Langdon T.G., Acta Metall. Mater., 1981, vol. 29, No. 5, 911-920.
10. Zaripov N.G. Kaibyshev, in: Proc. Int. scientific conference: Current state of the theory and practice of superplasticity of materials, Ufa, 2000, 174-179.
11. Pshenichnyuk A.I., Electronic journal Issledovano v Rossii, 2001, No. 25, 264-272, http://zhurnal.ape.relarn.ru/articles/2001/025.pdf.
12. Klyatskin V.I., Stochastic equations and waves in random-nonuniform media, Nauka, Moscow, 1980.
13. Khannanov Sh.Kh., in: Proc. Int. scientific conference: Current state of the theory and practice of superplasticity of materials, Ufa, 2000, 157-161.
14. Astanin V.V., FMM, 1995, vol. 79, No. 3, 166-173.

11

DERIVATION OF CONSTITUTIVE EQUATIONS IN MULTICOMPONENT LOADING CONDITIONS

11.1. From the deformation mechanism to constitutive equations

The constitutive equations (CE) are the expressions describing the relationship between the measures of the stress state and the measures of deformation [1–3]. These relationships are required for solving the problem of predicting the behaviour of material in the given loading conditions. This may include the task of calculating structures, calculations and optimisation of the technological processes of treatment of materials, etc. The constitutive equations allow to close the dynamic equations and the boundary-value conditions to an unambiguously defined boundary-value problem. Specific methods of constructing the CEs have been developed in the mechanics of solids. Some of them use, as a prototype, the Hooke law which defines the linear relationship between the components of the strain tensor and the components of the stress tensor in the elastic deformation range. When working outside this range it is immediately necessary to face a large number of problems associated with the need to investigate the strains which cannot any longer be regarded as low. It is essential to develop adequate deformation measures and the appropriate non-linear generalisation of the relationship of these measures with the measures of the stress state. The clue which produces results in this approach is the requirement for maintaining the sign of the corresponding thermodynamic functionals which greatly reduces the number of potentially possible variants of the CE. These results smoothly (or not very smoothly) continue into the plastic deformation range. In another approach to the construction of the CE of the plastic deformation processes the problem is approached from the opposite side – using the physical models describing the flow of viscous fluids. The result is the construction of a non-linear-viscous medium with rheological characteristics restored usually from some set of experimental relationships. Calculation of the rheological characteristics from the first principles can be carried out if we have some microscopic model of the flow and are capable of transferring its properties to the macrolevel. The model of the first type describes the relationship between the stress and strain tensors, the models of the second type links the stress tensor and the strain rate tensor. The connecting link of these limiting variants in the linear case was proposed by Maxwell in the form that is not resolved with respect to the stress tensor components:

$$\frac{d\sigma_{ik}}{dt}+\frac{1}{\tau}\sigma_{ik}=2\mu\frac{du_{ik}}{dt},$$

where σ_{ik} is the stress tensor, u_{ik} is the strain tensor, μ is the shear modulus, τ is the Maxwell relaxation time. If may easily be seen that for the case of slow loading this relationship is equivalent to the expression for the linear fluid with the viscosity $\mu\tau$ and reproduces the Hooke law for high loading rates. It should be mentioned that if the last relationship is regarded as a differential equation with respect to σ_{ik}, and in solving this equation with some initial conditions we obtain the CE in the form of the functional of the loading process. In addition to these relationships, the structural–mechanical models of plasticity, which are combinations of elastic, viscous (including non-linearly viscous) and plastic (dry friction) elements, are used. The number of variants formed in this case is described in detail in [1], and the CE variants, formed within the framework of creep theory, and also the so-called CE with internal variables, are also discussed ther. Studies have been published in which the construction of CE for high elastic strains in polycrystalline materials is based on assumptions regarding the colliding planes specified in each grain independently of others, i.e., it would appear they are based on the assumptions matched with the microscopic pattern of deformation [4]. However, the program cannot be realised gradually; at some stage it is necessary to introduce phenomenological structures describing the hardening and determine their characteristics using experimental results. In addition, the CE derived in this case cannot be used to describe superplastic deformation. Their main fault is that the deformation of the specimen is determined by the formation of the grains in the specimen and this is in direct contradiction with the experimental results obtained in investigations of superplastic deformation. These approaches were developed mainly in studies by main users, i.e., mechanical engineers, given the tasks to solve the appropriate boundary-value problems.

The structural ideology of construction of the CE can be developed only under the condition of understanding the processes accompanying deformation rearrangement, on a relatively low structural level – the level of special features of the crystalline structure. However, these formulations are the prerogative of another region of investigations – physics.

The physical models of superplastic deformation could not even be used for formulating the problem of construction of the CE because the models are restricted only to the structural level whose scale is given by the mean grain size whereas the CEs are used for macroscopic description. If in constructing the CE for the regime of classic plasticity of coarse-grained materials the transfer of the relationships from the level of the grain to the level of specimen is ensured using the assumptions forming the principle of the Taylor model, in investigating superplastic deformation this possibility is not at our disposal. The previously described model of superplastic deformation, based on the considerations on the cooperative mechanism of deformation uses the completely macroscopic formation, i.e. CGBS band. Knowing the special features of the large-scale organisation of the flow we can construct a scheme of the structural approach to the derivation of the CE.

The formalism, used in chapter 10, is based on the two-dimensional pattern of the flow characteristics of flat specimens. We determine the kinematics of the flow of polycrystalline continuum, restricted by the condition of integrity of the material, in the conditions of a general multicomponent stress state.

It is not justify to assume that the mechanism of superplastic deformation in the conditions of uniform multiaxial loading differs from the mechanism described for the conditions of uniaxial loading. The same processes lead to the formation of CGBS bands and the formation of a similar large-scale flow pattern. Since the orientation of the bands coincides with the orientation of the planes of the maximum tangential stresses, the problem is reduced to calculating these orientations for an arbitrary given stress state and to constructing a geometrical pattern of the appropriate flow. Consequently, we obtain a relationship between the components of the stress tensor and the components of the strain rate tensor. *A priori* it is not possible to say whether the assumption on the coaxiality of the tensors, which is always used in applied calculations, is fulfilled. This circumstance is the main intrigue of the proposed formulation [5–7].

11.2. Kinematics of polycrystalline continuum

We start with the description of the kinematics of a polycrystalline continuum restricted only by the conditions of integrity of the material. It should be noted that under the effect of only diffusion deformation mechanisms this condition was used as the basis of the formal scheme in the well-known study by Lifshits [8]. The investigated volume is regarded as the region $\Omega(t)$ in R^3, being the union of the regions Ω^α:

$$\Omega = \bigcup_\alpha \Omega^\alpha,$$

where $\Omega^\alpha(t)$ is the region occupied by the grain with the number $\alpha \subset (1, ..., N)$ at time t. It is assumed that the union occupies the region without 'holes' i.e., if some point x_i from R^3 belongs to Ω, then this point also belongs to some grain Ω^α or the general boundary of the adjacent grains. The total number of the grains N in the representative volume Ω should be sufficiently large to enable us to consider the co-operative mechanism of formation of the CGBS bands. On the other hand, the region Ω should be sufficiently small to fulfil the condition of uniformity of the stress state. Of course, in the superplastic deformation conditions (small grain size) it is quite easy to satisfy these conditions. The kinematics in Ω is given by the time dependence of carriers, i.e., the function is determined by the condition

$$\chi^\alpha(\mathbf{x},t) = \begin{cases} 1, & \text{at } \mathbf{x} \subset \Omega^\alpha(t) \\ 0, & \text{at } \mathbf{x} \not\subset \Omega^\alpha(t) \end{cases} .$$

The conditions of integrity of the material in deformation show that the functions χ^α cannot be defined arbitrarily and independently of each other. The instantaneous fields of speeds in the α-th grain is denoted by $v_i^\alpha(\mathbf{x},t)$. Therefore, the field of speeds in the region $\Omega(t)$ can be described by the equation:

$$v_i(\mathbf{x},t) = \sum_\alpha v_i^\alpha(\mathbf{x},t)\chi^\alpha(\mathbf{x},t) \quad . \tag{11.1}$$

To calculate the field of the strain rate tensor, we determine the derivative of the components of the field of speed v_i with respect to the spatial coordinates x_j

$$\frac{\partial v_i}{\partial x_j} = \sum_\alpha \left[\frac{\partial v_i^\alpha}{\partial x_j} \chi^\alpha + v_i^\alpha \frac{\partial \chi^\alpha}{\partial x_j} \right]. \tag{11.2}$$

Taking into account the equality

$$\frac{\partial \chi^\alpha}{\partial x_j} = -\sum_{\nu_\alpha} n_j^{\alpha_\nu} \delta(\mathbf{x} | \nu_\alpha),$$

where $\mathbf{n}^{\alpha_\nu}$ is the external normal to the ν-th boundary of the α-th grain; $\delta(\mathbf{x}|\nu_\alpha)$ is the delta function with the carrier on the ν-th face of the α-th grain, we obtain

$$\frac{\partial v_i}{\partial x_j} = \sum_\alpha \frac{\partial v_i^\alpha}{\partial x_j} \chi^\alpha - \sum_\alpha v_i^\alpha \sum_{\nu_\alpha} n_j^{\nu_\alpha} \delta(\mathbf{x} | \nu^\alpha).$$

The same face is included twice in the second term of the above equation in summation with respect to all grains: as the face of some grain and as the face of the grain adjacent to the former. Since the external normals at the general point of two adjacent grains have opposite directions (the line of triple junctions is ignored), we have

$$\frac{\partial v_i}{\partial x_j} = \sum_\alpha \frac{\partial v_i^\alpha}{\partial x_j} \chi^\alpha + \sum_{(\alpha\beta)} (v_i^\alpha - v_i^\beta) n_j^\beta \delta(\mathbf{x} | \Gamma^{(\alpha\beta)}) \tag{11.3}$$

where $\Gamma^{(\alpha\beta)}$ is the common face of two grains with the numbers α and β, and the second summation is carried out over all grain boundaries. The speed at some point of the arbitrary boundary is expanded with respect to the orthonormalised base represented by the following vectors: **n** – the vector of the normal to the boundary, $\boldsymbol{\tau}^{(1)}$ and $\boldsymbol{\tau}^{(2)}$ – the vectors tangent to the boundary forming a right-handed vector with **n**

$$\mathbf{v} = (\mathbf{v}\cdot\mathbf{n})\mathbf{n} + (\mathbf{v}\cdot\boldsymbol{\tau}^{(1)})\boldsymbol{\tau}^{(1)} + (\mathbf{v}\cdot\boldsymbol{\tau}^{(2)})\boldsymbol{\tau}^{(2)}.$$

According to the integrity condition, the normal components of the speed is continuous in transition through the boundary:

$$(\mathbf{v}^\alpha - \mathbf{v}^\beta)\cdot\mathbf{n} \equiv 0$$

$$V_{ij}(\mathbf{x},t) = \sum_\alpha V_{ij}^\alpha(\mathbf{x},t)\chi^\alpha(\mathbf{x},t) + \sum_{(\alpha\beta)} (\mathbf{v}^\alpha - \mathbf{v}^\beta)\cdot\tau \frac{\tau_i n_j^\beta + \tau_j n_i^\beta}{2} \delta\left(\mathbf{x} \middle| \Gamma^{(\alpha\beta)}\right) \tag{11.4}$$

where $\boldsymbol{\tau}$ is the unit vector normal to the vector of the normal at the investigated boundary and is such that the maximum shear stress acts along this vector. Since the shear stress for the given deviator of stress tensor σ_{lm}, normal $\mathbf{n}$ which determines the area, and the direction $\boldsymbol{\tau}$ has the form

$$T = \sum_{l=1}^{3} \sum_{m=1}^{3} \tau_l \sigma_{lm} n_m \,, \tag{11.5}$$

the value $\boldsymbol{\tau}$ should be determined from the condition of the maximum of this linear form. The projection of the speed difference $\mathbf{v}^{\alpha} - \mathbf{v}^{\beta}$ to the direction of the unit vector $\boldsymbol{\tau}$ is in fact the speed of grain boundary sliding at the boundary $(\alpha\beta)$. It is denoted by $\Delta V^{(\alpha\beta)}$.

Since we investigated the uniform stress state, the latter should be characterised by the appropriate uniform macroscopic tensor of the strain rate and not by the field of the tensor of the strain rate reflecting the complicated pattern of the layers and of the grains on the microlevel. We determine the means as conventional spatial means:

$$\left\langle V_{ij}(t) \right\rangle = \frac{1}{|\Omega|} \int\limits_{\Omega(t)} V_{ij}(\mathbf{x}, t) \, d\mathbf{x} \,, \tag{11.6}$$

where $|\Omega|$ is the volume of the averaging region which can be regarded as independent of time in the incompressibility conditions. Strictly speaking, the volume of the α-th grain cannot be regarded as independent of time if deformation is accompanied by the increase of the mean grain size. However, this factor will be ignored. Consequently

$$\left\langle V_{ij} \right\rangle = \sum_{\alpha} \frac{|\Omega^{\alpha}|}{|\Omega|} \left\langle V_{ij}^{\alpha} \right\rangle + \sum_{(\alpha\beta)} \frac{S_{\alpha\beta}}{|\Omega|} \Delta V^{(\alpha\beta)} \frac{\tau_i n_j^{\beta} + \tau_j n_i^{\beta}}{2} \,, \tag{11.7}$$

where $S_{\alpha\beta}$ is the area of the face $(\alpha\beta)$. Thus, we have removed part of the degrees of freedom of kinematics of the general type, but neither the set $\left\langle V_{ij}^{\alpha} \right\rangle$ nor even the first and second terms of the above equation can be defined independently of each other. At the same time, although formally, we have already separated the contribution from intragrain deformation (the first term) and from grain boundary sliding (the second term). In the approximation which forms the basis of the Taylor model (deformation of all the grains is the same and coincides with the deformation of the representative volume), the second term is also equal to 0. Thus, the Taylor model completely ignores the contribution of grain boundary sliding to the deformation processes and in this sense is the antipode of the models which claim to describe superplastic deformation. In analysis of the uniaxial experiments we formulated the result in the bicrystal approximation, i.e., in fact, we completely ignored the contribution of the first term. This is justified for describing the stationary flow stage. In a similar case, the CGBS bands are already formed, their contribution to the total deformation is large, and the deformation of the grains is only added to this shear. Here, we have the time dependence $\left\langle V_{ij}^{\alpha} \right\rangle \approx \exp\,(i\nu t)$ with the characteristic frequency $\nu \approx V/d$ (where V is the shear rate along the CGBS band, d is the mean grain size). However, the stage of exit to the stationary flow regime (when the bands have not as yet been formed and the contribution of the second term is very small) requires an accurate examination of intragrain deformation. This is very important in the analysis of the multiaxial loading.

11.3. Strain rate tensor determined by shear along the CGBS bands

We used the standard notations for the macroscopic tensor is $\left\langle V_{ij}^{\alpha} \right\rangle = \dot{\varepsilon}_{ij}^{\alpha}$ denoting the second term in (11.7) by $\dot{\varepsilon}_{ij}^{GBS}$ and introducing the tensor

$$m_{ij} = (\tau_i n_j + \tau_j n_i)/2\,. \tag{11.8}$$

We examined the second term in (11.7). Summation is transformed with respect to the arbitrarily renumerated faces $(\alpha\beta)$ in summation with respect to faces $(\alpha\beta)_k$, belonging to the k-th CGBS band, and with respect to all bands. Consequently

$$\dot{\varepsilon}_{ij}^{GBS} = \sum_{k=0}^{M} \sum_{(\alpha\beta)_k} \frac{S_{(\alpha\beta)_k}}{|\Omega|} \Delta V_{(\alpha\beta)_k} m_{ij}^{(\alpha\beta)_k}\,, \tag{11.9}$$

where M is the number of the CGBS bands form under the given loading conditions. We shall not go outside the limits of the bicrystal approximation which is justified when describing the experiments with uniaxial tensile loading, and it will be assumed that the shear rate is constant $\Delta V_{(\alpha\beta)_k}$ along the band and the same for all bands. In addition, the difference in the orientation of the faces belonging to the band is ignored, i.e., the deviation of some number of the faces from the most favourable faces oriented in the plane of the effect of the maximum shear stresses is ignored. The latter causes that tensor m_{ij} no longer depends on the index of the face and is a global characteristic of the geometrical pattern of the flow. This is natural for the uniform stress state. To calculate the remaining sum it should be mentioned that

$$\sum_{(\alpha\beta)_k} S_{(\alpha\beta)_k} = S_k .$$

where S_k is the surface area of the k-th band. If H is the mean distance between the bands, then

$$H\sum_{k=0}^{M} S_k \cong \Omega .$$

The last equality is approximate because the surface of the CGBS is not a plane. Finally, since

$$H \cdot M_0 = d \cdot M$$

where M_0 is the maximum possible number of bands in the specimen, we obtain

$$\dot{\varepsilon}_{ij}^{GBS} = \frac{M}{M_0}\frac{\Delta V}{d} m_{ij} . \tag{11.10}$$

Thus, the difference from the case of uniaxial tensile loading is reduced to substitution of the constant of the geometrical origin by the tensor m_{ij} which also characterises the flow geometry. To estimate this tensor, it is necessary to determine the area with the maximum shear stress and calculate the appropriate direction of this area. The scalar multiplier $(M\Delta V)/(M_0\, d)$ in equation (11.10) can be restored from the experiments with uniaxial tensile loading. The multiplier is determined unambiguously by the same maximum shear stress.

The geometrical characteristics of the flow pattern and the orientation of the sliding systems will be investigated. On the area given by the vector of the normal **n**, the shear stress in direction $\boldsymbol{\tau}$ is described by equation (11.5). The maximum is realised for vector $\boldsymbol{\tau}$ determined by the condition

$$\overline{\tau} = \frac{\boldsymbol{\omega} \times \mathbf{n}}{|\boldsymbol{\omega}|}, \tag{11.11}$$

where the numerator contains cross products, and

$$\varpi = \mathbf{n} \times \hat{\sigma}\mathbf{n} \tag{11.12}$$

and $|\boldsymbol{\omega}|$ defines the shear stress. The term $\hat{\sigma}\,\mathbf{n}$ denotes the result of multiplying the matrix $\hat{\sigma}$ by the vector $\mathbf{n}$ (also a vector). It may easily be seen that the shear stress in the direction normal to the vectors $\boldsymbol{\tau}$ and $\mathbf{n}$ is equal to 0. In the group of the arbitrarily oriented areas we select the facet with the maximum shear stress, i.e., construct the vector $\mathbf{n}$ which ensures the maximum value $|\boldsymbol{\omega}|$. $\mathbf{n}$ will be determined in the form of expansion with respect to the eigenvectors of the matrix $\hat{\sigma}$:

$$(\hat{\sigma} - \lambda \hat{E}) \cdot \mathbf{n}^* = 0 \,. \tag{11.13}$$

The equation for the eigenvalues (in the present case these are the main stresses) has the form

$$\lambda^3 - (Sp\hat{\sigma}) \cdot \lambda^2 - \frac{1}{2} I_0^2 \cdot \lambda - \det(\hat{\sigma}) = 0 \tag{11.14}$$

where according to the definition of the deviator, the trace is equal to 0:

$$Sp\hat{\sigma} \equiv \sigma_{11} + \sigma_{22} + \sigma_{33} = 0$$

and the intensity of the deviator of the stress tensor is given by the expression

$$I_0^2 = \sigma_{11}^2 + \sigma_{22}^2 + \sigma_{33}^2 + 2 \cdot (\sigma_{12}^2 + \sigma_{13}^2 + \sigma_{23}^2) \,. \tag{11.15}$$

Taking into account the properties of the roots of the cubic equation, the following relationships can be written:

$$\lambda_1 + \lambda_2 + \lambda_3 = Sp\hat{\sigma}$$

$$\lambda_1^2 + \lambda_2^2 + \lambda_3^2 = I_0^2$$

$$\lambda_1 \lambda_2 \lambda_3 = \det \hat{\sigma} \equiv D \,.$$

We consider a non-degenerate case and arrange the solution of the cubic equation in the following manner: $\lambda_1 > \lambda_2 > \lambda_3$. We assume that the appropriate eigenvectors, $\mathbf{n}^{(1)}$, $\mathbf{n}^{(2)}$ and $\mathbf{n}^{(3)}$, form a right-handed vector. Then

$$\mathbf{n} = \alpha_1 \mathbf{n}^{(1)} + \alpha_2 \mathbf{n}^{(2)} + \alpha_3 \mathbf{n}^{(3)}$$

where according to the condition of unit normalisation $\alpha_1^2+\alpha_2^2+\alpha_3^2=1$. Equation (11.12) is transformed to the form

$$\boldsymbol{\varpi}=\alpha_2\alpha_3(\lambda_3-\lambda_2)\mathbf{n}^{(1)}+\alpha_1\alpha_3(\lambda_1-\lambda_3)\mathbf{n}^{(2)}+\alpha_1\alpha_2(\lambda_2-\lambda_1)\mathbf{n}^{(3)}$$

from which

$$|\boldsymbol{\varpi}|^2=\alpha_2^2\alpha_3^2(\lambda_2-\lambda_3)^2+\alpha_1^2\alpha_3^2(\lambda_1-\lambda_3)^2+\alpha_1^2\alpha_2^2(\lambda_1-\lambda_2)^2. \tag{11.16}$$

The maximum value $|\boldsymbol{\varpi}|^2$ is determined by the term with the maximum value $(\lambda_k-\lambda_m)^2$, and since $\lambda_1>\lambda_2>\lambda_3$, we have $\alpha_1^2=\alpha_3^2=1/2,\ \ \alpha_2=0$. Thus, for the given stress state σ_{ij}, the maximum shear stress is

$$T=(\lambda_1-\lambda_3)/2\,. \tag{11.17}$$

This stress acts only on the facet given by the vector
$\mathbf{n}=\alpha_1\mathbf{n}^{(1)}+\alpha_3\mathbf{n}^{(3)}$
in the direction
$\overline{\tau}=\alpha_1\mathbf{n}^{(1)}-\alpha_3\mathbf{n}^{(3)}$.
Substituting **n** and $\boldsymbol{\tau}$ into equation (11.8), we obtain

$$m_{ij}=(n_i^{(1)}n_j^{(1)}-n_i^{(3)}n_j^{(3)})/2\,. \tag{11.18}$$

It should be mentioned that for **n** (and $\boldsymbol{\tau}$) we obtain two mutually perpendicular vectors differing in the selection of the signs α_1 and α_3. This defines two sliding systems. However, tensor m_{ij} for both systems is the same because it is determined by the squares of the values α_1 and α_3. The eigenvectors are represented by the normalised columns of the adjoint matrix of the deviator [9] which is determined unambiguously by the coefficients of the characteristic equation (11.14)

$$t_{ij}=\lambda^2\delta_{ij}+\lambda\sigma_{ij}+\sigma_{il}\sigma_{lj}-\delta_{ij}I_0^2/2\,.$$

Here usual summation is carried out with respect to the repeating index. Because of the properties of the deviator it is quite simple to confirm the validity of the identity

$$\sigma_{il}\sigma_{lj}-\delta_{ij}I_0^2/2=S_{ij} \tag{11.19}$$

where S_{ij} is the algebraic cofactor of the element σ_{ij} in the deviator matrix. Consequently, the adjoint matrix has the form

$$t_{ij} = \lambda^2 \delta_{ij} + \lambda \sigma_{ij} + S_{ij}$$

and the j-th eigenvector is in fact the j-th column (with the substitution $\lambda = \lambda_j$). This is verified by the following chain of equalities:

$$\sigma_{li} t_{ij} = \lambda^2 \sigma_{li} \delta_{ij} + \lambda \sigma_{li} \sigma_{ij} + \sigma_{li} S_{ij} = \left| \hat{S} = D\hat{\sigma}^{-1} \right| =$$
$$= \lambda^2 \sigma_{lj} + \lambda \sigma_{li} \sigma_{ij} + D\delta_{lj} = \lambda^2 \sigma_{lj} + \lambda \sigma_{li} \sigma_{ij} + \lambda_1 \lambda_2 \lambda_3 \delta_{lj} = \left| \lambda = \lambda_1 \right| =$$
$$= \lambda_1 \left[\lambda_2 \lambda_3 \delta_{lj} + \lambda_1 \sigma_{lj} + \sigma_{li} \sigma_{ij} \right] = \lambda_1 \left[\lambda_2 \lambda_3 \delta_{lj} + \lambda_1 \sigma_{lj} + \delta_{lj} I_0^2 / 2 + S_{lj} \right].$$

When deriving the last equality we used the identity (11.19). Finally, because of the properties of the roots

$$\lambda_2 \lambda_3 + I_0^2 / 2 = \lambda_1^2$$

we obtain

$$\sigma_{li} t_{ij} (\lambda = \lambda_1) = \lambda_1 t_{ij} (\lambda = \lambda_1)$$

i.e., the first column of the adjoint matrix is the eigenvector corresponding to the first eigenvalue. The two remaining vectors are processed in the same manner. The squares of the norms of the eigenvectors are factorised as follows:

$$\begin{aligned} &(\lambda_1 - \lambda_2)(\lambda_1 - \lambda_3)(\lambda_1^2 + \sigma_{11}\lambda_1 + S_{11}) \\ &-(\lambda_1 - \lambda_2)(\lambda_2 - \lambda_3)(\lambda_2^2 + \sigma_{22}\lambda_2 + S_{22}) \\ &(\lambda_1 - \lambda_3)(\lambda_2 - \lambda_3)(\lambda_3^2 + \sigma_{33}\lambda_{13} + S_{33}) \ . \end{aligned} \tag{11.20}$$

Substituting $\mathbf{n}^{(1)}$ and $\mathbf{n}^{(3)}$ into the expression (11.18), and after long transformations we obtain

$$m_{ij} = \frac{\lambda_2 I_0^2 \delta_{ij} / 2 - (\lambda_2^2 + 2\lambda_1\lambda_3)\sigma_{ij} + 3\lambda_2 S_{ij}}{2(\lambda_1 - \lambda_2)(\lambda_2 - \lambda_3)(\lambda_1 - \lambda_3)} . \tag{11.21}$$

The cumbersome form of the tensor (11.21) may suggest that the incompressibility condition has not been fulfilled as a result of the large number of transformations. We calculate the trace of the tensor m_{ij}:

$$\mathrm{Sp}(m_{ij}) = \frac{3\lambda_2}{4(\lambda_1 - \lambda_2)(\lambda_2 - \lambda_3)(\lambda_1 - \lambda_3)} \left[I_0^2 + 2(S_{11} + S_{22} + S_{33}) \right].$$

Substituting here I_0^2 from (11.15) and the values

$$S_{11} = \sigma_{22}\sigma_{33} - \sigma_{23}^2, \quad S_{22} = \sigma_{11}\sigma_{33} - \sigma_{13}^2, \quad S_{33} = \sigma_{11}\sigma_{22} - \sigma_{12}^2$$

we determine

$$\mathrm{Sp}(m_{ij}) = \frac{3\lambda_2}{4(\lambda_1 - \lambda_2)(\lambda_2 - \lambda_3)(\lambda_1 - \lambda_3)}\left[\sigma_{11} + \sigma_{22} + \sigma_{33}\right]^2 \equiv 0 .$$

Consequently, because of the relationship (11.18) the condition of zero divergence of the speed field is fulfilled.

We calculate the intensity of the tensor $\hat{m}$. Substituting into the determination of the intensity components (11.18)

$$\mathrm{Int}(m_{ik}) = \frac{1}{4}\sum_{i=1}^{3}\sum_{k=1}^{3}\left(n_i^{(1)}n_k^{(1)} - n_i^{(3)}n_k^{(3)}\right)^2 = \frac{1}{4}\sum_{i=1}^{3}\left[\left(n_i^{(1)}\right)^2 + \left(n_i^{(3)}\right)^2\right] = 1/2 .$$

This shows that the root of the intensity of the strain rate is indentical in accuracy with the strain rate in uniaxial loading.

Direct calculations show clearly that

$$\det(m_{ik}) = 0$$

i.e., the columns of the matrix, defining the tensor $\hat{m}$ are linearly dependent and determine the flat flow.

11.4. Degenerate cases and variants of coaxiality of the tensors

Equation (11.21) shows that the assumption on the coaxiality of the stress and strain rate tensors in superplastic deformation is justified only in a number of partial cases. For example, for $\lambda_2 = 0$ and, consequently, $\lambda_3 = -\lambda_1, \quad D = 0, \quad I_0^2 = 2\lambda_1^2$, we obtain

$$m_{ij} = \frac{\sigma_{ij}}{\sqrt{2I_0}} .$$

It is well known that in the degeneration case (i.e., the coincidence of two main stresses) the main feature is not the formation of two transversal sliding systems and it is the formation of a continuous set of planes passing through the generating lines of the cone with the apex angle $\pi/2$. The flow pattern has special features in comparison with the non-degenerate case. For the degenerate root, with the exception of (11.14), the following equality is also fulfilled

$$3\lambda^2 - I_0^2 / 2 = 0 .$$

Thus, the degenerate root is equal to either $I_0 / \sqrt{6}$ or $-I_0 / \sqrt{6}$. In the first case the ordered system of the rootes has the form $\lambda_1 = \lambda_2 = I_0 / \sqrt{6}$, $\lambda_3 = -2I_0 / \sqrt{6}$, and in the second case $\lambda_1 = 2I_0 / \sqrt{6}, \lambda_2 = \lambda_3 = -I_0 / \sqrt{6}$. Since the degenerate root also satisfies the equation (11.14), we obtain the condition for the matrix of the deviator leading to the coincidence of two main stresses. For the first system of the roots $3\sqrt{6}\det\hat{\sigma} = -I_0^3 < 0$, and for the second system $3\sqrt{6}\det\hat{\sigma} = I_0^3 > 0$. To construct the eigenvectors the conditions for the determinant and the modulus of the deviator are not sufficient and it is also necessary to fulfil the conditions for its components. Since the valuations of the appropriate columns of the adjoint matrix are identical zeros (see (11.20)) and, in the case of the real determined matrix all components of these vectors are also zeroes. These conditions define the relationships between the deviator components. The solutions of the equations determine the following variants.

1. All shear components – zeros; two main stresses coincide.

2. Any pair from the three shear components – zeros (it can be shown that the equality to zero of some shear component leads automatically to 'zeroing' of one of the remaining components). Thus, if $\sigma_{12}=\sigma_{13} = 0$, the following variants can exist:

$$\sigma_0 \begin{pmatrix} 1 & 0 & 0 \\ 0 & 0 & \pm\sqrt{2} \\ 0 & \pm\sqrt{2} & -1 \end{pmatrix}, \quad \sigma_0 \begin{pmatrix} 1 & 0 & 0 \\ 0 & -1 & \pm\sqrt{2} \\ 0 & \pm\sqrt{2} & 0 \end{pmatrix}.$$

The equality to zero of another pair results in identical matrices, differing from the previously mentioned ones by the permutation of the components.

3. All the shear components are not equal to zero. In this case, the diagonal components are determined unambiguously:

$$\sigma_{11} = \frac{2\sigma_{12}^2\sigma_{13}^2 - \sigma_{23}^2(\sigma_{12}^2 + \sigma_{13}^2)}{3\sigma_{12}\sigma_{13}\sigma_{23}}$$

$$\sigma_{22} = \frac{2\sigma_{12}^2\sigma_{23}^2 - \sigma_{13}^2(\sigma_{12}^2 + \sigma_{23}^2)}{3\sigma_{12}\sigma_{13}\sigma_{23}}$$

$$\sigma_{33} = \frac{2\sigma_{13}^2\sigma_{23}^2 - \sigma_{12}^2(\sigma_{13}^2 + \sigma_{23}^2)}{3\sigma_{12}\sigma_{13}\sigma_{23}} .$$

We construct the geometrical tensor for the third variant (the previous two variants are relatively elementary). The resultant matrix has eigenvalues

$$\lambda_1 = 2\frac{\sigma_{12}^2\sigma_{13}^2 + \sigma_{12}^2\sigma_{23}^2 + \sigma_{13}^2\sigma_{23}^2}{3\sigma_{12}\sigma_{13}\sigma_{23}}, \quad \lambda_2 = \lambda_3 = -\frac{\sigma_{12}^2\sigma_{13}^2 + \sigma_{12}^2\sigma_{23}^2 + \sigma_{13}^2\sigma_{23}^2}{3\sigma_{12}\sigma_{13}\sigma_{23}}$$

and eigenvectors

$$\mathbf{t}^{(1)} = \begin{pmatrix} \sigma_{12}\sigma_{13} \\ \sigma_{12}\sigma_{23} \\ \sigma_{13}\sigma_{23} \end{pmatrix}, \quad \mathbf{t}^{(2)} = \begin{pmatrix} -\sigma_{23} \\ \sigma_{13} \\ 0 \end{pmatrix}, \quad \mathbf{t}^{(3)} = \begin{pmatrix} -\sigma_{13}^2\sigma_{23} \\ -\sigma_{23}^2\sigma_{13} \\ \sigma_{12}(\sigma_{13}^2 + \sigma_{23}^2) \end{pmatrix} .$$

Equation (11.16) is maximum at

$$\alpha_1^2 = 1/2, \quad \alpha_2 = \cos(\varphi)/\sqrt{2}, \quad \alpha_3 = \sin(\varphi)/\sqrt{2} .$$

The arbitrary angle φ parametrises the sliding systems of the investigated set. The normal to the sliding plane and the sliding direction are determined by the vectors

$$\mathbf{n} = \alpha_1\boldsymbol{\tau}^{(1)} + \alpha_2\boldsymbol{\tau}^{(2)} + \alpha_3\boldsymbol{\tau}^{(3)}$$
$$\boldsymbol{\tau} = \alpha_1\boldsymbol{\tau}^{(1)} - \alpha_2\boldsymbol{\tau}^{(2)} + \alpha_3\boldsymbol{\tau}^{(3)}$$

where $\boldsymbol{\tau}^{(i)}$ is the vector $\mathbf{t}^{(i)}$ normalised with respect to unity. As a result, we obtain the following expression for the geometrical tensor:

$$m_{ik} = \frac{1}{2}\left[\tau_i^{(1)}\tau_k^{(1)} - \frac{\tau_i^{(2)}\tau_k^{(2)} + \tau_i^{(3)}\tau_k^{(3)}}{2} - \frac{\tau_i^{(2)}\tau_k^{(2)} - \tau_i^{(3)}\tau_k^{(3)}}{2}\cos(2\varphi) - \frac{\tau_i^{(2)}\tau_k^{(3)} + \tau_i^{(3)}\tau_k^{(2)}}{2}\sin(2\varphi)\right].$$

Assuming that the sliding systems, parameterised by the angle, act independently, after averaging with respect to the angle we have

$$m_{ik} = \frac{3}{8}\frac{\sigma_{ik}}{T},$$

where T is the maximum shear stress equal to $(\lambda_1 - \lambda_3)/2$. Thus, the assumption on the coaxiality of the stress and strain rate tensors is fulfilled in the degenerate case. We have formulated the general scheme of constructing the CE based on the deformation mechanism given by shear along the CGBS bands. Using relatively general assumptions it has been shown that the constitutive equations, based on the Taylor model, cannot be used for describing superplastic deformation. The geometrical properties of the large-scale flow pattern and its force characteristics were divided for the general uniform stress state. To examine its characteristics, it is sufficient to carry out uniaxial experiments. Investigations of the flow geometry for the given stress state are reduced to constructing the main axes of the deviator. It has been shown that the assumption on the coaxiality of the stress and strain rate tensors are usually not fulfilled. The general form of their relationship was determined. A number of partial cases in which the tensors are coaxial have been discussed. The direct application of these results requires conversion of the tensor relationships and reducing the CEs to the canonic form.

The model must be adapted to the requirements of mechanics. At present, the model includes a very large number of the parameters of the material and the structure and this creates difficulties in practical application. Some of these parameters are of the phenomenological origin (for example, the constant controlling the transition to the dislocation creep regime). It is necessary to minimise the number of parameters by defining their stable complexes whilst maintaining the physical content. In addition, it is necessary to evaluate the possibilities of the independent experimental determination of the complex parameters.

References

1. Vasin R.A. and Enikeev F.U., Introduction to mechanics of superplasticity, part 1, Gilem, Ufa, 1998.
2. Vasin R.A., Mater. Sci. Forum, 1997, vol. 243–245, 173–178.
3. Enikeev F.U., Rheological behaviour of microcrystalline materials in processes of pressure working of metals in the superplastic conditions, Disseration, Moscow, 2000.
4. Molinary A., et al., Acta Metall., 1987, vol. 35, No. 12, 2983–2994.
5. Pshenichnyuk A.I., et al., Mathematical modelling of systems and processes, No. 6, Perm' State Technical University, Perm', 1998, 92–98.
6. Pshenichnyuk A.I., et al., Mathematical modelling of systems and processes, No. 6, Perm' State Technical University, Perm', 1998, 99–109.

7. Kaibyshev O.A. and Pshenichnyuk A.I., Izv. RAN, Mekh. Tverdogo Tela, 1999, No. 5, 148–164.
8. Lifshits I.M., ZhETF, 1963, vol. 44, No. 4, 1349–1367.
9. Gantmakher F.R., Matrix theory, Nauka, Moscow, 1988.

Conclusion

We shall not present the results in the form of a brief description of the study, it is sufficient to refer to the introduction for this purpose. We shall better fantasise over the following subject: how would what we already know help us in analysis of new situations.

The highly promising directions of scientific research grouped under the term 'nanotechnology', are reduced within the framework of our subject to two main tasks: production of nanocrystalline materials with a homogeneous structure and analysis of the properties of these materials. If we discuss the mechanical properties of nanocrystalline materials, it is above all essential to describe and analyse the deformation mechanisms of these materials. The first task has been fulfilled by the application of methods developed for the production of ultrafine-grained materials. In other words, this is some variant of the method of severe plastic deformation. However, we do not yet know the minimum grain size, which can be produced by this prcessing. Of course, we cannot know this at the moment because we do not yet know the deformation mechanisms as these mechanisms determine the dynamics of formation of the structure of the material, including the grain structure. The two tasks (1 – processing and 2 – properties) represent a single inseparable problem. How the studies carried out so far could help us in analysing this problem or at least some aspects of this problem?

When analysing superplastic deformation we concluded that this phenomenon is characteristic of greatly differing materials: metals, ceramics, intermetallics. The point is that the deformation mechanism is based on shear on planes of cooperative grain boundary sliding, and the bands themselves form by the merger of the grain boundaries, i.e., the types of crystal lattices of the grains (intragranular properties) play some role which, however, is minimal. If we consider the grain sizes so small that the initiation of plastic flow in these grains requires the stress level close to the theoretical strength, we not only 'shift' all the deformation processes to the boundaries (and the appropriate analogues of the bands of cooperative grain boundary sliding) but on the level of the high acting stresses we no longer see the difference in the properties of the grain boundaries of different types. Saying the words 'the grain boundary' we no longer ask what the misorientation and the plane in which they are situated are. It is not surprising that the flow pattern of the material produced in this manner should resemble the flow pattern of the granulated materials. It is interesting to note that in the examination of superplasticity we have ignored to a large degree the lattice properties of the materials and obtained a sort of universal behaviour because in the case of nanocrystalline materials we can expect even greater unification. The network of the grain boundaries is an independent and only active phase. The properties of the materials in the nanocrystalline condition are determined by the sizes of the grains-granules (absolutely rigid in the first approximation) and by the properties of the grain boundary phase (at the moment, this phase has not been sufficiently defined and is relatively mysterious). It is highly likely that the rheology of these materials can be restored in the model of rheology of suspensions on the basis of 'soft' glasses.

We hope that thoughtful readers will also detect other useful properties of the proposed approach which can be used in solving new problems.

Index

O

P

R

S

T

Z